Lectures in Applied Mathematics

Proceedings of the Summer Seminar, Boulder, Colorado, 1960

VOLUME 1 LECTURES IN STATISTICAL MECHANICS
G. E. Uhlenbeck and G. W. Ford with E. W. Montroll

VOLUME 2 MATHEMATICAL PROBLEMS OF RELATIVISTIC PHYSICS
I. E. Segal with G. W. Mackey

VOLUME 3 PERTURBATION OF SPECTRA IN HILBERT SPACE
K. O. Friedrichs

VOLUME 4 QUANTUM MECHANICS
R. Jost

Proceedings of the Summer Seminar, Ithaca, New York, 1963

VOLUME 5 SPACE MATHEMATICS. PART 1
J. Barkley Rosser, Editor

VOLUME 6 SPACE MATHEMATICS. PART 2
J. Barkley Rosser, Editor

VOLUME 7 SPACE MATHEMATICS. PART 3
J. Barkley Rosser, Editor

Proceedings of the Summer Seminar, Ithaca, New York, 1965

VOLUME 8 RELATIVITY THEORY AND ASTROPHYSICS
1. RELATIVITY AND COSMOLOGY
Jürgen Ehlers, Editor

VOLUME 9 RELATIVITY THEORY AND ASTROPHYSICS
2. GALACTIC STRUCTURE
Jürgen Ehlers, Editor

VOLUME 10 RELATIVITY THEORY AND ASTROPHYSICS
3. STELLAR STRUCTURE
Jürgen Ehlers, Editor

Proceedings of the Summer Seminar, Stanford, California, 1967

VOLUME 11 MATHEMATICS OF THE DECISION SCIENCES, PART 1
George B. Dantzig and Arthur F. Veinott, Jr., Editors

VOLUME 12 MATHEMATICS OF THE DECISION SCIENCES, PART 2
George V. Dantzig and Arthur F. Veinott, Jr., Editors

Proceedings of the Summer Seminar, Troy, New York, 1970

VOLUME 13 MATHEMATICAL PROBLEMS IN THE GEOPHYSICAL SCIENCES
1. GEOPHYSICAL FLUID DYNAMICS
William H. Reid, Editor

VOLUME 14 MATHEMATICAL PROBLEMS IN THE GEOPHYSICAL SCIENCES
2. INVERSE PROBLEMS, DYNAMO THEORY, AND TIDES
William H. Reid, Editor

Proceedings of the Summer Seminar, Potsdam, New York, 1972

VOLUME 15 NONLINEAR WAVE MOTION
Alan C. Newell, Editor

Proceedings of the Summer Seminar, Troy, New York, 1975

VOLUME 16 MODERN MODELING OF CONTINUUM PHENOMENA
Richard C. DiPrima, Editor

Proceedings of the Summer Seminar, Salt Lake City, Utah, 1978

VOLUME 17 NONLINEAR OSCILLATIONS IN BIOLOGY
Frank C. Hoppenstead, Editor

Nonlinear Oscillations in Biology

Volume 17
Lectures in Applied Mathematics

Nonlinear Oscillations in Biology

Frank C. Hoppensteadt, Editor

1979
American Mathematical Society, Providence, Rhode Island

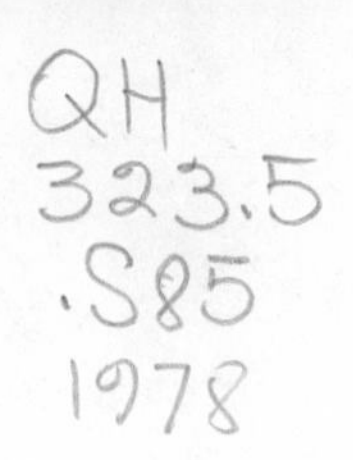

The proceedings of the Summer Seminar were prepared by the American Mathematical Society with partial support from National Science Foundation Grant MCS 78-03584.

Library of Congress Cataloging in Publication Data

Summer Seminar on Applied Mathematics, 10th, University of Utah, 1978.
Nonlinear oscillations in biology.
(Lectures in applied mathematics; v. 17)
"Sponsored jointly by the American Mathematical Society and the Society for Industrial and Applied Mathematics."
Includes bibliographies.
1. Biomathematics—Congresses. 2. Nonlinear oscillations—Congresses. I. Hoppensteadt, F. C. II. American Mathematical Society. III. Society for Industrial and Applied Mathematics. IV. Title. V. Series: Lectures in applied mathematics (Providence); v. 17.
QH323.5.S85 1978 574'.01'51 79–26469
ISBN 0-8218-1117-7

1980 Mathematics Subject Classification. Primary 34C15, 34C25, 34C27, 34C29, 34C45, 34E05, 34E20, 34E99, 34F05, 35G20, 65Lxx, 00A69, 92A05.

Printed in the United States of America

Contents

Foreword ix

Nonlinear oscillations 1
L. N. HOWARD

Studies of the ear 69
C. STEELE

24 hard problems about the mathematics of 24 hour rhythms 93
A. S. WINFREE

Stochastic modelling and nonlinear oscillations 127
D. LUDWIG

Computer studies of nonlinear oscillators 131
F. C. HOPPENSTEADT

Chaotic oscillations: an example of hyperchaos 141
O. E. RÖSSLER

Nonlinear oscillations in equations with delays 157
J. K. HALE

A brief introduction to dynamical systems 187
J. GUCKENHEIMER

Foreword

The Tenth Summer Seminar on Applied Mathematics, sponsored jointly by the American Mathematical Society and the Society for Industrial and Applied Mathematics, was held at the University of Utah from June 12 to June 23, 1978. The topic for the seminar was Nonlinear Oscillations in Biology.

The seminar was intended as an introduction to the theory and methods of nonlinear oscillations and how they are used to study oscillatory phenomena in the life sciences. A core series of lectures by L. Howard, in-depth case studies by A. Winfree and C. Steele and background lectures on mathematical topics by J. Guckenheimer, J. Hale, F. Hoppensteadt, D. Ludwig and O. Rössler are reproduced in these proceedings. Additional lectures on cell metabolism, population dynamics, perturbation theory, neural sciences, epidemiology and reaction-diffusion systems were given but without written record. The program for the seminar was organized by W. S. Childress (Courant Institute of Mathematical Sciences, New York University), D. S. Cohen (California Institute of Technology), F. C. Hoppensteadt, P. Waltman (University of Iowa), and A. Winfree (Purdue University).

The organizing committee wish to express their gratitude and appreciation to the Mathematics and Biology Divisions of the National Science Foundation for their fiscal support of the seminar, the University of Utah for its support in publication of these proceedings, the American Mathematical Society and its staff for administering the seminar, Mrs. Ann Reed and Ms. Carol Kohansky, who handled day to day operations of the seminar and entertainment, the Applied Mathematics Computing Laboratory and the Department of Mathematics for providing their excellent facilities for our use and for their kind hospitality, the speakers, who gave outstanding presentations, and the lively, active and interested group of students and scientists who participated in the seminar.

Finally, I thank the contributors to this volume for their hard work and cooperation in its preparation and production.

December 1978

Frank C. Hoppensteadt
Department of Mathematics
University of Utah

Salt Lake City, Utah

Lectures in Applied Mathematics
Volume 17, 1979

Nonlinear Oscillations

Louis N. Howard[1]

I. Examples of conservative and dissipative oscillators.

1. Oscillatory processes of a more or less regular kind are seen in very diverse forms in all parts of the biological world, as indeed elsewhere. Some of these, notably the circadian rhythms, are closely associated with external periodicities like day and night, the semidiurnal or fortnightly tides, or the seasons, but nevertheless seem often to be better described as essentially autonomous oscillations modified somewhat by the external periodicities, rather than simply as responses to the external forcing. Of course, the astronomical periodicities have probably played a more direct role in the *evolution* of biological systems possessing such rhythms, and many features of life on the earth might well be quite different now if the day and the year were of equal length, if we had no moon, if the earth's orbit were circular, and if its rotation axis were normal to the plane of the ecliptic. But in trying to understand better the workings of those biological oscillations which now exist, we need not be concerned with such long-term issues. Probably in many ways circadian and other externally influenced rhythmic processes are similar—at least from the rather abstract point of view of the structure of relevant mathematical models—to other biological oscillations, such as respiration or heartbeat, which do not have any obvious outside forcing.

2. In the first part of these lectures I propose to discuss, in a manner directed to the nonspecialist, some of the central mathematical ideas and techniques which have been found generally useful in studying nonlinear oscillations. Where convenient these will be illustrated by examples which have been proposed as models—even if very crude ones

AMS (MOS) subject classifications (1970). Primary 34C15, 34E15, 35K55; Secondary 92-02.

[1]Some of the research of the author reported herein was supported under National Science Foundation Grant #MCS 76 23 281.

—of biological phenomena. It is hoped that this will be helpful as background for some of the more specialized mathematical lectures at this seminar, and perhaps suggest certain approaches which might be useful in formulating and studying new mathematical models of some of the real oscillatory phenomena which will also be described here.

Later on, I shall also describe some of the work with which I have been associated concerned with the formation of spatially organized and temporally periodic processes, through the interaction of local oscillatory mechanisms with diffusion. Though probably on firmer ground as models for certain nonliving systems, these mathematical investigations are thought also to be of some interest in biological questions, for instance in the propagation of waves in interacting populations.

3. Among the most familiar mathematical oscillators are the *linear oscillator*, described by the differential equation

$$\ddot{x} + x = 0, \tag{1}$$

and the *pendulum*, described by

$$\ddot{x} + \sin x = 0. \tag{2}$$

The first may be thought of as an idealized model for a weight on a spring (as well as for millions of other things), the second for a real pendulum. These two equations have a great deal in common. They both are equivalent to autonomous systems

$$\dot{x} = y, \qquad \dot{y} = -x, \tag{1'}$$

$$\dot{x} = y, \qquad \dot{y} = -\sin x, \tag{2'}$$

which have the origin as critical points; the linearizations at the origin of the two systems are of course identical. All (nonconstant) solutions of (1) or (1′) are periodic, and if x is regarded as an angle the same is true (with one limiting case) of (2) or (2′). Both systems are conservative, and can be written in the form $\ddot{x} + V'(x) = 0$, with potential energy functions $V_1 = x^2/2$, $V_2 = 1 - \cos x$; they have the "energy integral" $\frac{1}{2}\dot{x}^2 + V(x) = E = \text{const.}$, which gives the equation of the orbits in the phase plane of (1′) or phase cylinder of (2′). Perhaps the most significant *differences* of the two systems are (a) (2′) also has a second critical point $x = \pi$, $y = 0$ (which is a saddle point) and (b) the periods of the periodic solutions of (2′) are *amplitude dependent* (ranging from about 2π for small E up toward ∞ as E approaches 2, and then coming down from ∞ toward 0 as E increases beyond 2 toward ∞) whereas all (nonconstant) solutions of (1′) have period 2π.

A curious aspect of this difference is that while the phase plane orbits of the linear oscillator are all concentric circles, those of the pendulum are of two different types; for $0 < E < 2$ they roughly resemble circles surrounding the critical point at $x = 0$, but for $E > 2$ the pendulum goes 'over the top', and the orbits encircle the phase cylinder. The two kinds are separated by two 'homoclinic' orbits, which go from the critical point at $x = \pi$ to itself, starting out as one branch of the unstable manifold of the saddle point, and ending up as one branch of its stable manifold. From an overall point of view, a particular oscillation of the pendulum is *characterized* by its energy E. Certain other quantities of some physical interest in such an oscillatory system are obtained by time averaging over the orbit, and may thus be regarded as functions of E. An example is the *mean potential energy* of the pendulum, which can be measured by the mean height

$$\bar{z} = \overline{1 - \cos x} = \frac{1}{T}\oint(1 - \cos x)\, dt,$$

this integral being over the particular orbit, of period T, corresponding to energy E. Some such quantities, in particular this one $\bar{z}(E)$, exhibit a rather peculiar nonanalytic behavior at the point ($E = 2$) corresponding to the boundary between the two different types of orbits. It is for instance clear that for small E, $\bar{z} \simeq \frac{1}{2}E$ (equipartition of energy for the linear oscillator), but as E increases toward 2, the pendulum spends a lot of time near $x = \pi$, so that $\bar{z}$ approaches 2. On the other hand, as E goes beyond 2, $\bar{z}$ will then decrease again, approaching $\bar{z} = 1$ when E is very large and the pendulum is whirling rapidly around its support. The precise behavior of $\bar{z}(E)$ can be readily computed in terms of elliptic integrals and its graph is found to have a cusp at $E = 2$ (see Figure 1). This is rather reminiscent of a phase change such as melting or the onset of ferromagnetism at which various quantities, e.g. specific heat, exhibit nonanalytic behavior. Thus, the pendulum might be regarded as one of the simplest dynamical systems with two phases. This comes about because of the two different types of orbits.

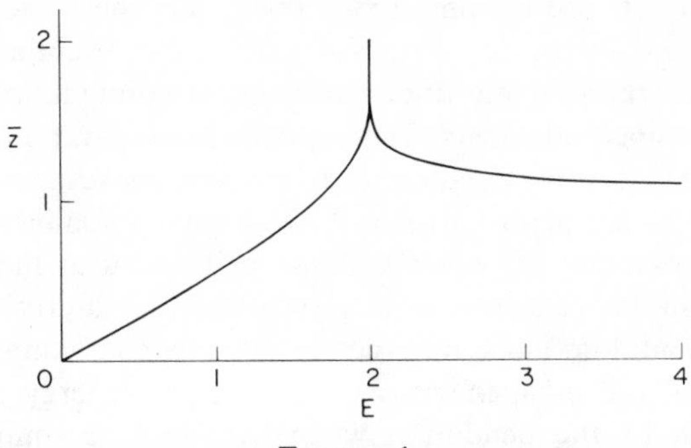

FIGURE 1.

Both the pendulum and the linear oscillator are *structurally unstable*: arbitrarily small perturbations of the vector fields defined by the right-hand sides of (1′) and (2′) can (and usually do) qualitatively alter the phase portrait. The addition of a bit of linear friction, for instance, to either system (giving $\ddot{x} + \varepsilon\dot{x} + \dot{x} = 0$ or $\ddot{x} + \varepsilon\dot{x} + \sin x = 0$) completely destroys all periodic solutions and makes every solution tend to $x = 0$ as $t \to \infty$ (except $x = \pi$ in the pendulum case). This is true of all two-dimensional systems whose phase planes contain a *center*: a critical point surrounded by a family of closed trajectories. It is sometimes asserted that structurally unstable systems of differential equations cannot be useful as models of real systems, because real systems are always subject to perturbations. This is probably true in the sense that such models do not tell the whole story, but in practice it is almost entirely false. We obtain some understanding of real systems always by relating them to idealizations which do not tell the whole story, and these idealizations are often most comprehensible when they are themselves built up from still further idealizations together with an (also idealized) description of the ways in which the further idealizations principally fail to tell the whole story. If we actually knew the whole story, we could not understand it. It appears to be almost a generic property of *interesting* models of real systems that they are not generic, and the linear oscillator, so central in almost all applied mathematics, is a prime example. Of course, the concept of 'generic' is dependent on the class of problems regarded as 'of interest', and most disagreements on these issues evaporate when this aspect of problem formulation is carefully considered.

4. Many aspects of a large class of real oscillatory systems are accurately described by mathematical models such as the linear oscillator. But such systems also frequently operate at a certain definite amplitude, and to understand this the incorporation of nonlinear (and frequently linear dissipative) effects into the model is necessary. When these other effects are in some sense *small*, the linear oscillator—suitably generalized—provides a useful and often essential element in obtaining a more complete understanding, in combination with other essentially nonlinear elements. The approximate separation of the linear oscillator element from the nonlinear element makes possible certain mathematical techniques of great value in studying nonlinear oscillators. When this separation is possible in the mathematical model it often, though not always, reflects a similar separation in the real system. For instance, a pendulum clock may be regarded as consisting primarily of the pendulum, the escapement, and the weight (energy supply). The actual motion of the pendulum when the clock is running is quite accurately described by one of the solutions of $\ddot{x} + \sin x = 0$ (or, for small oscillations, of $\ddot{x} + x = 0$). But the real clock does not run without the escapement and weight, and with them runs at a definite amplitude. The main effects of the escapement are that it communicates

small (and brief) impulses to the pendulum, transferring pulses of energy from the energy source, at particular phases of the cycle, and that it also introduces some friction; there are other causes of frictional dissipation as well. In the simplest mathematical models, the operation of the clock is still described by the angular displacement x of the pendulum; the effects of the escapement and energy source are modelled by small changes in the differential equation, i.e. by a small alteration of the vector field on the phase cylinder with coordinates $(x, \dot{x})$. The effect of these alterations is to change the phase portrait from one with a center to one which has a *stable limit cycle* (a periodic solution whose orbit is approached by all nearby solutions). Such a vector field is structurally stable, within the class of second-order autonomous systems, so this model is more satisfactory as a description of the clock's operation. Still, the motion of the pendulum on this limit cycle is very nearly the same as that described by *some* solution of the linear oscillator equation. From the problem-solving point of view the main role of the small nonlinear and frictional terms is to determine the *amplitude*, i.e. to specify which one of the solutions of the linear oscillator equation gives the 'correct' approximate description of the limit cycle. It is often possible to use relatively simple methods to obtain a good approximation to this amplitude; an example of this will be given shortly. An interesting discussion of several different models of a pendulum clock can be found in the book of Andronov, Chaiken and Witt [**1**]. A significant point brought out there is that different models may be equally successful in 'explaining' the *periodic* operation of the clock, but differ in other respects. Thus, for example, a real pendulum clock has two stable modes of operation (or nonoperation!) even when wound up; either with the pendulum oscillating at a certain definite amplitude, or with it motionless and hanging straight down. To start the clock the pendulum must be pushed to one side sufficiently far; otherwise it goes back to the motionless state. Thus, the phase portrait of a satisfactory model of a clock must qualitatively have the form indicated in Figure 2; there should be both a stable limit cycle and a stable critical point. Between these is an unstable periodic orbit, separating those initial conditions which lead back to the motionless state from those which lead to the limit cycle. Some models of the clock do not have this character, and thus are not satisfactory, however well they may agree on the limit cycle. This point is well illustrated by one of the simplest examples of a system of differential equations with a limit cycle:

$$\begin{aligned} \dot{x} &= -y + ax(1 - x^2 - y^2), \\ \dot{y} &= x + ay(1 - x^2 - y^2). \end{aligned} \tag{3}$$

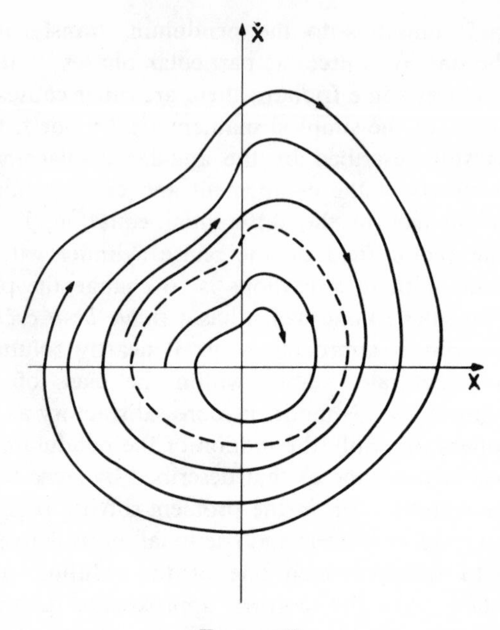

FIGURE 2.

The limit cycle solution is $x = \cos t$, $y = \sin t$; this nonlinear oscillator has perfectly sinusoidal oscillations indistinguishable from a solution of the linear oscillator equation (1). Unlike a pendulum clock, (3) is also 'self-starting': every solution except $x = y = 0$ eventually comes close to the limit cycle. An example with a phase portrait as in Figure 2 is

$$\dot{x} = -y + ax(1 - x^2 - y^2)\left(x^2 + y^2 - \tfrac{1}{4}\right),$$

$$\dot{y} = x + ay(1 - x^2 - y^2)\left(x^2 + y^2 - \tfrac{1}{4}\right). \tag{4}$$

In trying to assess the plausibility of a particular mathematical model for a biological oscillator it is thus important not only to compare with observations on the periodic oscillations but also to look at experiments designed to explore other parts of the phase plane. A good discussion of this in the context of models of the mitotic oscillator can be found in the article of Kauffman [2]. [Some of the subtleties and difficulties in interpreting experiments on biological oscillators whose detailed mechanism is quite obscure, such as circadian rhythms, are brought out by Winfree in his lecture at this seminar.]

Two other examples of oscillators which operate in a manner rather similar to a pendulum clock are the vacuum tube oscillator illustrated in Figure 3 and the toy woodpecker sketched in Figure 4. The circuit of Figure 3, familiar to anyone who experimented with electronics as a child (and who is sufficiently old to know what a vacuum tube is!)

contains the coil (L) and capacitor (C) connected between the grid and cathode of the tube. These provide an approximate linear oscillator analogous to the pendulum of the clock. It works roughly as follows: when the capacitor is charged so that the potential of the grid is positive with respect to the cathode, the tube acts like a rather low resistance, and current flows from the anode to the cathode, passing through the 'tickler' coil (L_1). As the capacitor discharges through L, the grid becomes negative with respect to the cathode and the tube rather suddenly becomes a very high resistance, thus switching off the current through L_1. This sudden change in the current through L_1 induces an impulsive voltage in L, in a sense tending to amplify the oscillation in the L–C circuit. A half cycle later, the grid once again becomes positive, the tube again conducts, and a second amplifying pulse is delivered. Thus the tube and L_1 act rather like the escapement, delivering small pulses of energy to the pendulum-like L–C circuit, and so maintaining its oscillations against resistive dissipation. The current in the L–C circuit is nearly sinusoidal, but at an amplitude determined by the ohmic losses and the nonlinear switching action of the vacuum tube.

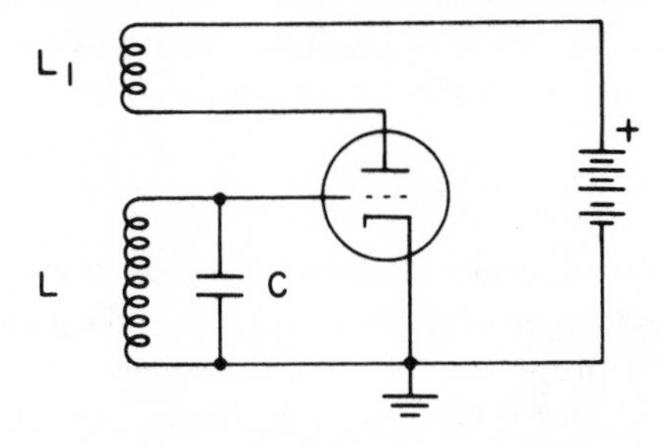

FIGURE 3.

The woodpecker toy consists of a small piece of wood with a hole in it through which passes a vertical rod. Attached to one side of the bit of wood is a coil spring, on the other end of which is the model woodpecker (actually a flicker, Colaptes auratus). The spring and bird are approximately a linear oscillator, like the pendulum. When the bird swings up, the lateral force exerted by the bit of wood on the rod is reduced, allowing it to start slipping downward, but on the next down swing it catches again (like the escapement) and the gravitational potential energy which had been converted to kinetic energy of the bird in its short fall is transferred to the spring, in such a phase as to tend to maintain the oscillation. Like the pendulum clock, this system has both a stable oscillatory state and a stable motionless state. The bird must be given a sufficiently large initial displacement to start the oscillation. This toy does not seem to be a very good model for a real woodpecker—a biomechanical oscillator which so far as I know has not been too fully investigated from a mathematical point of view.

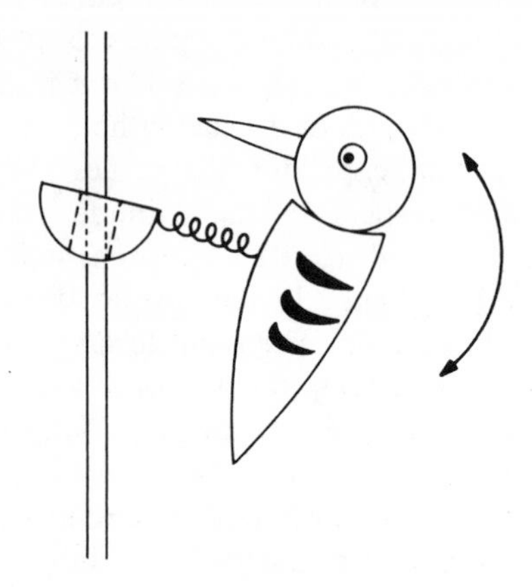

FIGURE 4.

Electronic oscillators closely related to that of Figure 3 were the original motivation for the investigations of van der Pol and others on nonlinear differential equations with limit cycles. The van der Pol equation

$$\ddot{x} + \dot{x}(x^2 - \varepsilon^2) + x = 0 \tag{5}$$

is one of the simplest and most carefully studied of such equations; it is obtained as a description of various vacuum tube circuits, after some idealization of the properties of vacuum tubes. (It is somewhat less satisfactory as a model for circuits using transistors, but is qualitatively applicable there, too.) It is easy to see in this differential equation a representation of the same basic elements of the 'pendulum' ($\ddot{x} + x$) and the 'escapement' ($\dot{x}(x^2 - \varepsilon^2)$), a nonlinear energy transferring mechanism which is a source of energy for the 'pendulum' at certain phases of its cycle, and a sink at others. Indeed, multiplication of (5) by $\dot{x}$ gives an equation which can be written as

$$\frac{d}{dt}\left(\frac{1}{2}\dot{x}^2 + \frac{1}{2}x^2\right) = \dot{x}^2(\varepsilon^2 - x^2). \tag{6}$$

This may be interpreted as saying that the pendulum-like part of the system acquires energy at the rate $\dot{x}^2(\varepsilon^2 - x^2)$, which is positive if $x^2 < \varepsilon^2$ and negative when $|x|$ is larger. Thus, the amplitude of the approximate oscillation may be expected to increase until it becomes rather greater than ε, if it starts small, but to decrease if it starts large enough, for then the energy transfer will be negative most of the time. When ε is small and the amplitude is also of this order, (6) shows that the energy transferred per cycle is of order ε^4, hence each cycle is changed by a small fraction of itself. This means that the solution of the equation is approximately given by $x = A \cos t$, $\dot{x} = -A \sin t$, where A changes only a little per cycle. Integrating (6) over a cycle thus gives a change of energy:

$$\Delta(A^2) \simeq \int_0^{2\pi} A^2 \sin^2 t(\varepsilon^2 - A^2 \cos^2 t)\, dt.$$

From this we can estimate the amplitude of the limit cycle by setting $\Delta(A^2) = 0$; this gives

$$A^2\left[\varepsilon^2 \cdot \pi - A^2 \cdot \tfrac{1}{4} \cdot \pi\right] = 0, \quad \text{hence } A = 2\varepsilon.$$

This little calculation is a simple example of the 'method of averaging' which can often be used to obtain, with rather little effort, a good approximation to the amplitude of an approximately sinusoidal nonlinear oscillation. Though possibly requiring more effort to carry through the details, similar ideas can be applied also to other systems, not especially sinusoidal, which can be regarded as regular perturbations of a system with a *family* of periodic orbits, e.g. a conservative oscillator with weak damping and driving.

II. Hopf bifurcations.

1. The van der Pol oscillator given by equation (5) does not have any (nonconstant) periodic solutions if the parameter denoted by ε^2 is $\leqslant 0$. This is immediately clear from (6), which then shows in fact that the critical point at $x = 0$ is *absolutely stable*, or is a 'global attractor': every solution tends to it as $t \to \infty$. This critical point however exhibits a change of stability as ε^2 crosses zero, and the linearization at $x = 0$ has a pair of conjugate complex eigenvalues which cross the imaginary axis as ε^2 crosses 0. The limit cycle then appears initially as a small amplitude (2ε) sinusoidal oscillation which grows in size and changes in shape as ε increases. It can be shown that this limit cycle exists for all positive ε^2, and is a global attractor: every solution except $x \equiv 0$ has an orbit which comes close to the limit cycle as $t \to \infty$ (the limit cycle is 'absolutely orbitally asymptotically stable'). This appearance of a periodic solution in the neighborhood of a critical point whose stability changes due to the crossing of a conjugate pair of eigenvalues over the imaginary axis is an example of a 'Hopf bifurcation'. As we shall see later, something like this happens whenever a pair of eigenvalues crosses the imaginary axis as a parameter varies, though sometimes the periodic solution occurs for parameter values where the critical point is stable. The form of the limit cycle as ε increases is somewhat more conveniently exhibited by rescaling (5) a bit so that it becomes (with a new ε)

$$\dot{x} = y,$$
$$\dot{y} = -x + \varepsilon y(a^2 - x^2). \tag{7}$$

In this form the limit cycle appears when $\varepsilon > 0$ but small as a circle of radius about $2a$. a may be chosen for convenient display on the screen of an analogue computer; pictures of the limit cycle obtained in this way are shown in Figure 5. When ε is large, the oscillation is far from sinusoidal, and is usually described as a 'relaxation oscillation', characterized by two (at least) quite disparate time-scales. Techniques for exploiting this relaxation oscillation character to obtain useful approximate descriptions will be discussed presently.

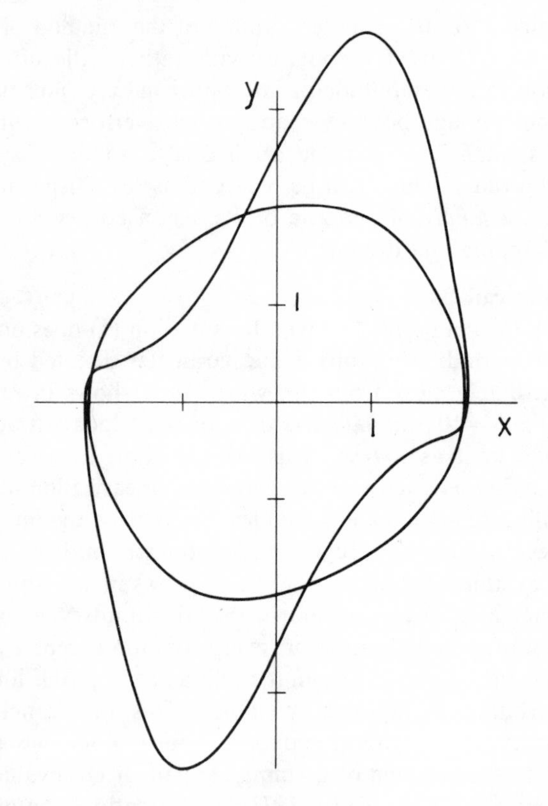

FIGURE 5.

2. A second example of a nonlinear oscillator is the 'Brusselator', a model chemical reaction suggested by Prigogine and Lefever [3] as perhaps the simplest oscillator obtainable on the basis of a 'chemical' model based on the law of mass action. The chemical equations are

$$
\begin{aligned}
A &\to X + B, \\
X &\to Y + C, \\
2X + Y &\to 3X + D, \\
X &\to E + F.
\end{aligned}
$$

Here the substances denoted by X and Y are both products and reactants; the other substances are supposed to be inactive when products and supplied at fixed concentration (e.g., by a buffer) when reactants. After suitable scaling, the differential equations of the chemical kinetics based on the law of mass action are

$$
\begin{aligned}
\dot{x} &= a - (b+1)x + x^2 y, \\
\dot{y} &= bx - x^2 y.
\end{aligned} \tag{8}
$$

This system has a critical point at $(a, b/a)$, and this is the only critical point. The linearization there has matrix

$$
\begin{bmatrix} -1+b & a^2 \\ -b & -a^2 \end{bmatrix}
$$

which has eigenvalues

$$
\tfrac{1}{2}(b - 1 - a^2) \pm \sqrt{\left(\tfrac{1}{2}(b - 1 - a^2)\right)^2 - a^2} \; .
$$

Thus the critical point is (linearly) *unstable* when $b > 1 + a^2$, and has complex eigenvalues when

$$
-a < b - (1 + a^2) < a,
$$

i.e. when

$$
\left(a - \tfrac{1}{2}\right)^2 + \tfrac{3}{4} < b < \left(a + \tfrac{1}{2}\right)^2 + \tfrac{3}{4}.
$$

For instance, when $a = 1$, the critical point is unstable for $b > 2$, and there are complex eigenvalues for $1 < b < 3$, so as b crosses 2 there is a conjugate pair crossing the imaginary axis (Hopf bifurcation). Here also (at this crossing) a stable limit cycle arises, and persists as b increases, gradually becoming more and more a relaxation oscillation. Graphs of the limit cycles for various values of b are shown for the case $a = 1$ in Figure 6. Some points are marked, corresponding to equal intervals in t, to make it apparent that for large b the portion of the trajectory which is approximately descending at 45° is traversed rapidly in comparison with the rest of it. Note that the smallest trajectory ($b = 2.15$) is fairly close to an ellipse, but that as b increases the shape is increasingly deformed.

3. For our next example consider the third order system:

$$
\begin{aligned}
\dot{x} &= -.1x + y, \\
\dot{y} &= -x - .1y + .8z - x^2 y, \\
\dot{z} &= \qquad .8y - \mu z.
\end{aligned} \tag{9}
$$

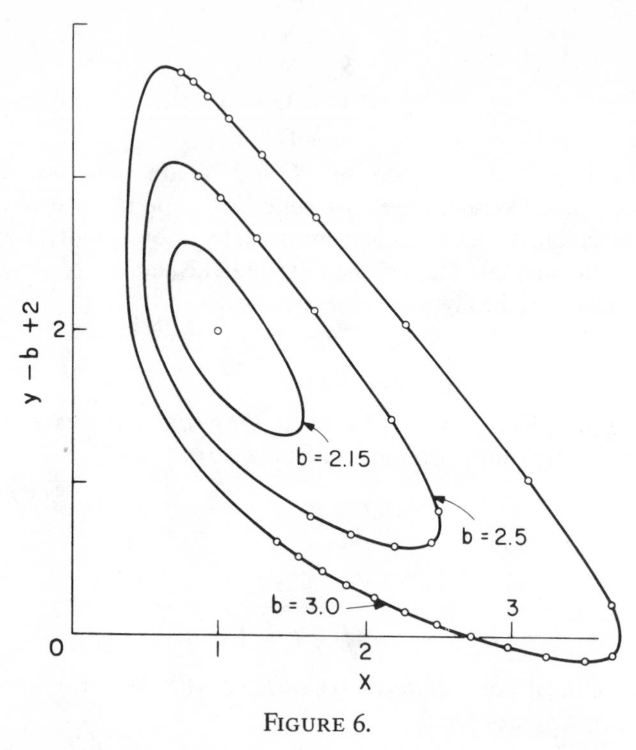

FIGURE 6.

(This example will be useful later to illustrate a kind of diffusive destabilization of a stationary state leading to a stable periodic solution.) It can be shown that the linearization at the origin of this system has a conjugate pair of pure imaginary eigenvalues when μ is a root of $\mu^2 - 3\mu + .69 = 0$, i.e. for $\mu = 0.251$ and 2.749, approximately. In fact, a conjugate pair crosses the imaginary axis from left to right as μ increases through 0.251, and later crosses back again when μ increases through 2.749. It turns out that between these two values of μ there is a stable limit cycle solution to (9), which appears to 'grow out' of the critical point at the origin as μ goes through 0.251, enlarges for a bit and then shrinks back onto the origin as μ approaches 2.749. Graphs of a few cases of this are shown in Figure 7, which shows the projection onto the (x, y) plane. For μ slightly above 0.251 the orbit in space is approximately an ellipse; this is also true for μ slightly below 2.749, but the two ellipses lie in different planes. The disappearance of the limit cycle at $\mu = 2.749$ may be described as an 'inverse Hopf bifurcation'.

4. Eberhard Hopf formulated and proved his theorem about the appearance of periodic solutions of the kind illustrated by the above examples in the paper [**4**]. This paper is now very difficult to find in this country, but a translation of it can be found in the book of Marsden

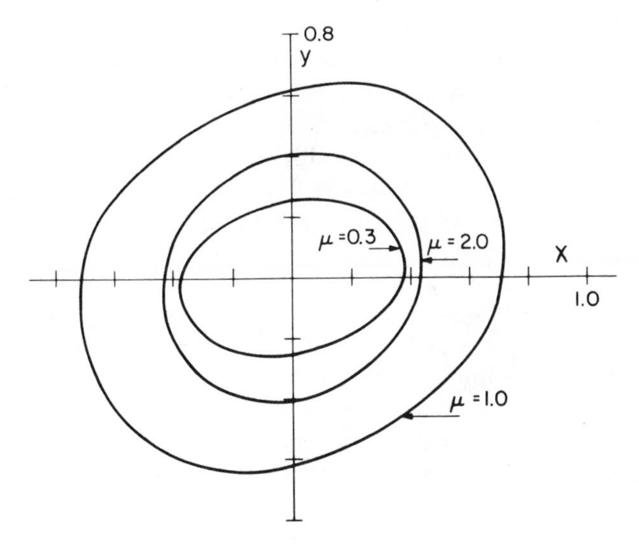

FIGURE 7.

and McCracken [5]. In his paper Hopf says,

> "Although I have not come across in the literature the consideration of the bifurcation problem on the basis of the hypothesis (1.2) [transversal crossing of the eigenvalues, see below], I scarcely think that there is something essentially new in the above theorem. The methods were developed by Poincaré perhaps 50 years ago, and belong today to the classical conceptual structure of the local theory of periodic solutions. Since, however, the theorem is of interest in nonconservative mechanics it seems to me that a thorough presentation is not without value. In order to facilitate the extension to systems with infinitely many degrees of freedom, for example the fundamental equations of motion of a viscous fluid, I have given preference to the more general methods of linear algebra rather than special techniques (e.g., choice of a special coordinate system)."

Thus Hopf himself might not entirely agree with the current usual designation of the result as the 'Hopf bifurcation theorem' or the description of the kind of oscillatory bifurcation to which it refers as a 'Hopf bifurcation'. Still, Hopf's clear formulation and presentation of the result was a significant contribution, and he was perhaps one of the first to understand clearly some features of it, particularly with regard to the stability properties of the periodic solution.

We are concerned here with an autonomous system of differential equations depending also on a parameter μ:

$$\dot{x} = F_\mu(x), \tag{10}$$

where x is a vector in an n-dimensional Euclidean space and μ is a real parameter. We shall assume that $x = 0$ is a critical point of this system for all values of μ in which we are interested, i.e., $F_\mu(0) = 0$. Perhaps the simplest 'bifurcation theorem' about such systems (also stated by Hopf in the above mentioned paper) is the following (it is about *stationary* rather than *periodic* bifurcations):

Suppose that the linearization of (10) at the origin has a matrix with a simple eigenvalue $\lambda(\mu)$ which crosses 0 transversely at a certain value of μ, say $\mu = 0$. (By 'transversely' we mean $\lambda(0) = 0$, $\lambda'(0) \neq 0$.) Then near the origin of the $(n + 1)$-dimensional (x, μ) space there always exists a locally unique curve of critical points $(x(\varepsilon), \mu(\varepsilon))$, distinct from the μ axis, and passing through $(0, 0)$. If F depends analytically on μ and x (this was assumed by Hopf, but with little change can be replaced by suitable differentiability hypotheses) then the parameter ε can be chosen so that $x(\varepsilon)$ and $\mu(\varepsilon)$ are analytic. If $\mu = \mu_1\varepsilon + \cdots$ and $\mu_1 \neq 0$ (the generic case) then there is a second critical point (besides 0) for all small $|\mu|$. This second critical point then also has a simple eigenvalue β which passes through 0 as μ does, and if $\beta = \beta_1\varepsilon + \cdots$ we have $\beta_1 = -\mu_1\lambda'(0)$.

This last result gives information about the *stability* of the second family of critical points. Thus if all the *other* eigenvalues of the linearization at $x = 0$, $\mu = 0$ are properly in the left half-plane and $\lambda'(0) > 0$, then $x = 0$ is stable for small $\mu < 0$ and unstable for small $\mu > 0$. Since β and μ then have opposite signs, and the other eigenvalues at the 'new' critical point are close to the other ones at $x = 0$, the new critical point is then stable for small $\mu > 0$ and unstable for small $\mu < 0$. We thus have an 'exchange of stabilities' at the bifurcation point, the 'new' critical point being stable when it is 'supercritical', i.e. when it occurs on the side of $\mu = 0$ for which $x = 0$ is unstable, and unstable when it is subcritical. The simplest example illustrating this is given by the one-dimensional differential equation $\dot{x} = \mu x - x^2$. Here the eigenvalue of the linearization at $x = 0$ is $\lambda(\mu) = \mu$, so $\lambda'(0) = 1$ and we have the required transversality. The new branch of critical points is evidently given in this case by $x = \mu$, or parametrically by $x = \varepsilon$, $\mu = \varepsilon$. Note that the linearization at the new critical point has the eigenvalue

$$\beta = \frac{\partial}{\partial x}\left[\mu x - x^2\right]_{x=\mu} = -\mu$$

which exhibits the exchange of stabilities.

It is not always the case that the new curve of critical points crosses the plane $\mu = 0$ in the (x, μ) space; the theorem only asserts that it passes through $x = 0$, $\mu = 0$, and exists somewhere in a neighborhood of this point in (x, μ) space. For instance, for the equation $\dot{x} = \mu x - x^3$ the new family of critical points $x = \varepsilon$, $\mu = \varepsilon^2$ lies entirely in $\mu \geqslant 0$; for the *linear* equation $\dot{x} = \mu x$ the two curves of critical points are the μ axis and the x axis. The basic similarity of all these cases is obscured if one looks only at the location of new critical points in x space for fixed μ, rather than using the formulation in (x, μ) space as above.

This general kind of stationary bifurcation occurs in many problems of physical interest. The example of the 'Taylor cells', formed in the motion of a viscous fluid between two concentric cylinders with the inner rotating (sufficiently) more rapidly than the outer, was cited already by Hopf; other important examples occur in the theory of thermal convection, and in buckling problems. Some models of biological pattern-formation are also of this nature. In many cases the mathematical formulation is in terms of partial differential equations and is really an infinite dimensional analogue of the above, but the analogy is close.

It should be noticed that this stationary bifurcation is not really about the differential equation $\dot{x} = F_\mu(x)$ but only about solutions of $F_\mu(x) = 0$. The Hopf bifurcation theorem, however, is genuinely about the differential equation. It may be stated as follows:

(a) We again have the system (10) with $F_\mu(0) = 0$. Suppose now that

$$F_\mu(x) = L_0 x + \mu L_1 x + Q(x, x) + C(x, x, x) + \cdots,$$

where the terms not written explicitly are of smaller order in μ and x than those which are; thus, Q and C (vector valued quadratic and cubic forms) are the second and third order (in x) terms at $\mu = 0$, $\mu L_1 x$ is the first order (in μ) term in the linearization at $x = 0$, etc. It is now assumed, and this is a central point, that L_0 has a conjugate pair $\pm i\omega_0$ of simple ($\omega_0 > 0$) pure imaginary eigenvalues, and no other eigenvalues on the imaginary axis. Let r be the right (column) and l the left (row) eigenvector with eigenvalue $i\omega_0$; their complex conjugates $\bar{r}$ and $\bar{l}$ are then eigenvectors for $-i\omega_0$, and we have $l \cdot \bar{r} = 0$ since the eigenvalues $\pm i\omega_0$ are different. Because the eigenvalues are simple we may assume $l \cdot r = 1$.

(b) The second main hypothesis was stated by Hopf in the form that the continuous extension to $\mu \neq 0$ of the eigenvalue $i\omega_0$, say $\lambda(\mu)$, should transversely cross the imaginary axis at $\mu = 0$, i.e., $\operatorname{Re}(d\lambda/d\mu)|_{\mu=0} \neq 0$. An equivalent form of this is the inequality $\operatorname{Re}(l \cdot L_1 r) \neq 0$. Indeed, it is easy to see that $(d\lambda/d\mu)_0 = l \cdot L_1 r$. [Proof: Let $L(\mu)$ be the matrix of the linearization at $x = 0$, and suppose $L(\mu)e(\mu) = \lambda(\mu)e(\mu)$. Differentiating with respect to μ and setting $\mu = 0$ gives

$$L_0 e'(0) + L_1 r = \lambda'(0) r + i\omega_0 e'(0),$$

and multiplying this on the left by l, using $lL_0 = i\omega_0 l$ and $l \cdot r = 1$, gives the result.]

The conclusion of the theorem then is that in a neighborhood of the origin in (x, μ) space there always exists a one-parameter family $x(t, \varepsilon)$,

$\mu(\varepsilon)$ of periodic solutions of (10) (with $x(t, \varepsilon) \to 0$, $\mu \to 0$ as $\varepsilon \to 0$), having periods $T(\varepsilon)$ (with $T \to 2\pi/\omega_0$ as $\varepsilon \to 0$). When $F_\mu(x)$ is analytic, ε can be chosen so that x, μ and T are analytic, and

$$x = \varepsilon x_1(t) + \cdots,$$
$$\mu = \mu_2 \varepsilon^2 + \cdots,$$
$$T = (2\pi/\omega_0)(1 + T_2\varepsilon^2 + \cdots).$$

If it turns out that $\mu_2 \neq 0$, which is the generic situation, then for small ε the periodic solutions exist either *supercritically*, when $\mu_2 \cdot \text{Re}\, \lambda'(0) > 0$, or else subcritically, when $\mu_2 \cdot \text{Re}\, \lambda'(0) < 0$. In this case the periodic solutions have a Floquet exponent $\beta(\varepsilon) = \beta_2\varepsilon^2 + \cdots$ with $\beta_2 = -2\mu_2 \text{Re}\, \lambda'(0)$. [If the other eigenvalues of L_0 are all in the left half-plane, then, for small ε, $n - 2$ of the Floquet exponents of the periodic solution are near them, hence also in the left half-plane, one is zero, of course, and the last one (β) is thus negative when the periodic solutions exist supercritically and positive when they are subcritical. Consequently, here also there is a kind of 'exchange of stabilities' on passing through the bifurcation: if the periodic solution appears as the critical point loses stability, it 'acquires' the stability, while if it was already there, it was unstable and 'transfers' its instability to the critical point on going out of existence.]

5. Because of the last mentioned facts the determination of μ_2 is of particular interest (when it is not zero) in that it indicates where the periodic solutions occur, whether for positive or negative μ, and gives some information about their stability. The part of the theorem which asserts the *existence* of the one-parameter family of periodic solutions near $x = 0$, $\mu = 0$ is a statement about solutions to the nonlinear problem which can be verified by examining only its linearization at $x = 0$. But the more quantitative part, referring to μ_2, β_2 etc., depends on the higher order terms, and as far as the first significant terms are concerned only on the quadratic and cubic terms. The actual calculation of μ_2 and T_2 will be discussed in the next section. A convenient way to organize the calculation will be presented formally, assuming analyticity, but proofs of the theorem usually depend on related considerations. The power series is typically replaced by an iteration or appeal to some implicit function theorem whose own proof might be made by iteration or contraction-mapping techniques.

III. Construction of weakly nonlinear periodic solutions.

1. Let $t = Ts/2\pi$ so that the periodic solution to be constructed has period 2π in s. Let

$$x = \varepsilon y = \varepsilon y_0 + \varepsilon^2 y_1 + \cdots,$$

$$T = (2\pi/\omega_0)(1 + \varepsilon^2 T_2 + \cdots),$$
$$\mu = \varepsilon^2 \mu_2 + \cdots.$$

The parameter ε is essentially defined by the normalization: $l \cdot y(0) = \frac{1}{2}$, i.e., $l \cdot y_0(0) = \frac{1}{2}$, $l \cdot y_i(0) = 0$, $i > 0$. Inserting these into the differential equations (10), one obtains from the first three orders appearing:

$$\omega_0 \frac{dy_0}{ds} = L_0 y_0, \tag{11}$$

$$\omega_0 \frac{dy_1}{ds} = L_0 y_1 + Q(y_0, y_0), \tag{12}$$

$$\omega_0 \frac{dy_2}{ds} = L_0 y_2 + T_2 L_0 y_0 + \mu_2 L_1 y_0 + 2Q(y_0, y_1) + C(y_0, y_0, y_0). \tag{13}$$

(11) has a unique real 2π-periodic solution satisfying the normalization condition $l \cdot y_0(0) = \frac{1}{2}$, namely

$$y_0 = \operatorname{Re}(re^{is}). \tag{14}$$

With this,

$$Q(y_0, y_0) = \tfrac{1}{2} Q(r, \bar{r}) + \tfrac{1}{2}\operatorname{Re}\left[e^{2is} Q(r, r) \right]$$

so the periodic solutions of (12) have the form

$$y_1 = a + \operatorname{Re}(ce^{2is}) + \operatorname{Re}(C_1 re^{is})$$

where a and c satisfy

$$- L_0 a = \tfrac{1}{2} Q(r, \bar{r}), \tag{15}$$

$$(2i\omega_0 - L_0)c = \tfrac{1}{2} Q(r, r). \tag{16}$$

With the assumption that $\pm i\omega_0$ are the only eigenvalues of L_0 on the imaginary axis, we see that (15) and (16) can be solved uniquely for a and c. With these known, the number C_1 is then obtained (uniquely) so as to satisfy $l \cdot y_1(0) = 0$. Actually, using $lL_0 = i\omega_0 l$, one can find an explicit formula for C_1 without explicitly calculating a and c; a short calculation gives:

$$C_1 = (i\omega_0)^{-1} l \cdot \left[Q(r, \bar{r}) - \tfrac{1}{2} Q(r, r) + \tfrac{1}{6} Q(\bar{r}, \bar{r}) \right].$$

Equation (13) now becomes, using the above:

$$\begin{aligned} \omega_0 \frac{dy_2}{ds} = L_0 y_2 + \operatorname{Re}\Big\{ & C_1 Q(r, \bar{r}) \\ & + e^{is}\left[(i\omega_0 T_2 + \mu_2 L_1) r + 2Q(r, a) + Q(\bar{r}, c) + \tfrac{3}{4} C(r, r, \bar{r}) \right] \\ & + e^{2is}\left[C_1 Q(r, r) \right] + e^{3is}\left[Q(r, c) + \tfrac{1}{4} C(r, r, r) \right] \Big\}. \end{aligned} \tag{17}$$

Now this equation has no periodic solutions at all unless the terms in e^{is} and e^{-is} are orthogonal to l and $\bar{l}$, respectively; if the latter is true it has a unique one satisfying the normalization condition. Thus, μ_2 and T_2 are to be determined from the equation

$$\mu_2(l \cdot L_1 r) + i\omega_0 T_2 = -2l \cdot Q(r, a) - l \cdot Q(\bar{r}, c) - \tfrac{3}{4} l \cdot C(r, r, \bar{r}) \equiv B, \quad \text{say.} \tag{18}$$

Since the transversality condition gives $\mathrm{Re}(l \cdot L_1 r) \neq 0$, (18) evidently uniquely determines μ_2 and T_2, and y_2 can then be found as well. Thus we have the following algorithm for calculating μ_2 (and so β_2) and T_2:

(1) Find ω_0, r and l from L_0.
(2) Find a and c from (15) and (16).
(3) Calculate B, and solve (18) for μ_2 and T_2.

2. EXAMPLE.

$$\dot{x} = 2\mu x + y + 2x^2,$$
$$\dot{y} = -x - 2y^2.$$

Here we have

$$L_0 = \begin{bmatrix} 0 & 1 \\ -1 & 0 \end{bmatrix}, \quad L_1 = \begin{bmatrix} 2 & 0 \\ 0 & 0 \end{bmatrix}, \quad Q = \begin{bmatrix} 2x^2 \\ -2x^2 \end{bmatrix}, \quad C = \begin{bmatrix} 0 \\ 0 \end{bmatrix},$$

so step (1) gives us

$$\omega_0 = 1, \qquad r = \begin{bmatrix} 1 \\ i \end{bmatrix}, \qquad l = \tfrac{1}{2}[1, -i].$$

Thus $l \cdot L_1 r = 1$, verifying the transversality condition. We then calculate

$$\tfrac{1}{2} Q(r, \bar{r}) = \frac{1}{2}\begin{bmatrix} 2 \cdot 1 \cdot 1 \\ -2 \cdot 1 \cdot 1 \end{bmatrix} = \begin{bmatrix} 1 \\ -1 \end{bmatrix},$$

and similarly

$$\tfrac{1}{2} Q(r, r) = \begin{bmatrix} 1 \\ -1 \end{bmatrix}.$$

Equations (15) and (16) are then:

$$\begin{aligned} -a_2 &= 1, \\ a_1 &= -1, \end{aligned} \qquad \text{and} \qquad \begin{aligned} 2ic_1 - c_2 &= 1, \\ c_1 + 2ic_2 &= -1, \end{aligned}$$

so

$$a = \begin{bmatrix} -1 \\ -1 \end{bmatrix} \quad \text{and} \quad c = \frac{1}{3}\begin{bmatrix} 1 - 2i \\ 1 + 2i \end{bmatrix},$$

completing step (2). From these we find

$$Q(r, a) = 2 \cdot 1 \cdot (-1)\begin{bmatrix} 1 \\ -1 \end{bmatrix} = \begin{bmatrix} -2 \\ 2 \end{bmatrix}, \qquad Q(\bar{r}, c) = \tfrac{2}{3}(1 - 2i)\begin{bmatrix} 1 \\ -1 \end{bmatrix}$$

so equation (18) becomes:

$$\mu_2 + iT_2 = (-1)\cdot(-2) + (i)\cdot(2) - \left(\tfrac{1}{2}\right)\cdot\tfrac{2}{3}(1-2i) - \left(\tfrac{1}{2}i\right)\cdot\tfrac{2}{3}(1-2i)$$
$$= 1 + \tfrac{7}{3}i.$$

Thus $\mu_2 = 1 > 0$, the bifurcation is supercritical, and there must be a stable limit cycle for sufficiently small positive μ.

The results of direct numerical calculations of this limit cycle are shown for $\mu = (.05)^2$, $(.1)^2$ and $(.15)^2$ in Figure 8. To compare these with the truncated series $\varepsilon y_0 + \varepsilon^2 y_1 + \varepsilon^3 y_2$ some points are shown which have been calculated using this series with $\varepsilon = .05$ and $\varepsilon = .13$. Since the meaning of ε (given by the normalization) is that it is the (positive) value of x when $y = 0$ the value .13 has been used rather than .15 because this is the actual value of x for $y = 0$ in the case of the limit cycle with $\mu = (.15)^2$. Since the series for μ *starts* with the ε^2 term and the higher terms were not calculated, the value of μ given by this truncation to a single term is rather less accurate than are the values given by the three-term y series. It is evident however that the deformation from a circle which occurs as μ increases is fairly well represented by the three terms, in the case shown.

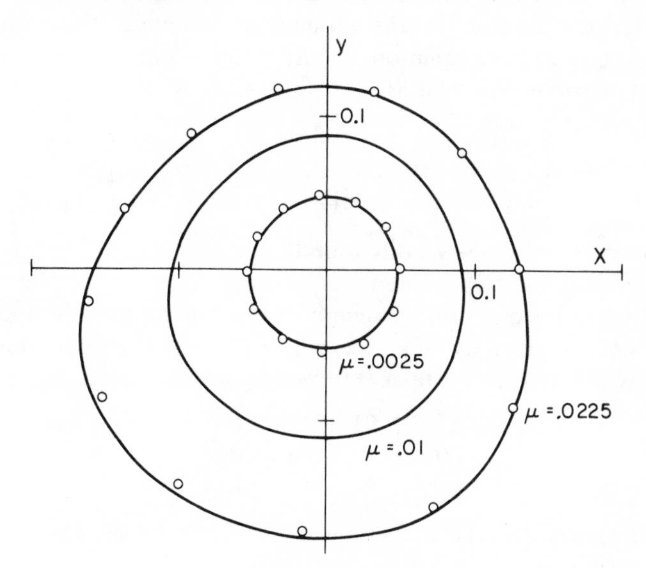

FIGURE 8.

It will be noticed that if no quadratic terms are present the calculation of μ_2 and T_2 is rather easier, for then the solution of equations (15) and (16) becomes trivial: $a = c = 0$. Also B is easier to determine. This is probably the reason that many examples cited in illustration of the

Hopf theorem contain only cubic terms. On the other hand, if quadratic terms *are* present, it makes rather little difference to the difficulty of calculation whether or not cubic terms occur, also. For this reason the example just cited, in which we have not bothered to include cubic terms, is still fairly representative of the general case.

IV. Diffusive instability and relaxation oscillations.

1. As an example with some vaguely biological interest which may be illuminated some by using bifurcation methods, we give a slightly simpler version of an example constructed by Smale of two identical 'cells' containing some reacting chemicals. Under certain conditions, these systems when isolated from each other always tend to a stable stationary state. But when they are coupled together by diffusion across a membrane this stationary state becomes unstable, and a stable limit cycle occurs instead. Smale's example, which is presented in [**5**], has four chemicals in each cell. The example below has only three, but is not as well adapted as is Smale's to a global treatment. [It seems to be impossible to construct an example like this with only two chemicals in each of two cells, although it can probably be done with three cells.] Let x be the vector of concentrations of the three chemicals in the first cell, y that for the second. In the absence of coupling, these are both supposed to satisfy the equation $\dot{z} = R(z)$, describing the kinetics of the chemical reactions. We take $R(z) = Az + C(z)$, where

$$A = \begin{bmatrix} -.1 & -1 & .8 \\ 1 & -.1 & 0 \\ .8 & 0 & -.1 \end{bmatrix}, \qquad C(z) = \begin{bmatrix} -z_1 z_2^2 \\ 0 \\ 0 \end{bmatrix}.$$

The eigenvalues of A are readily found to be $-.1$ and $-.1 \pm .6i$, so this system has a stable critical point at the origin. Now let the two cells be coupled by diffusion across a membrane, assuming that the flux from cell 2 to cell 1 is given by $D(y - x)$ where D is a positive definite matrix of diffusivities. In fact, we take D to be diagonal, and given by:

$$D = \begin{bmatrix} .01 & 0 & 0 \\ 0 & .01 & 0 \\ 0 & 0 & .5\mu \end{bmatrix}$$

for some constant μ. Then the equations of the six dimensional combined system are:

$$\dot{x} = Ax + C(x) + D(y - x),$$
$$\dot{y} = Ay + C(y) + D(x - y).$$

This system has $x - y = 0$ and $x + y = 0$ as invariant manifolds, and on $x - y = 0$ we get the uncoupled system with the origin stable. On $x + y = 0$ (which, at least close to the origin, is approached as $t \to \infty$) x

satisfies the equation

$$\dot{x} = Ax + C(x) - 2Dx = \begin{bmatrix} -.12 & -1 & .8 \\ 1 & -.12 & 0 \\ .8 & 0 & -.1-\mu \end{bmatrix} x - \begin{bmatrix} x_1 x_2^2 \\ 0 \\ 0 \end{bmatrix}. \tag{19}$$

This system is a minor variant of the one described above in §II.3. The characteristic polynomial of its linearization is found to have a conjugate pair of pure imaginary eigenvalues when μ is one of the roots of $\mu^2 - 2.2267\mu + .461733 = 0$, i.e., $\mu \simeq 1.99525$ or 0.231416. At each of these values the eigenvalue pair crosses the imaginary axis transversely, and between them it is in the right half-plane. Both of these Hopf bifurcations are found to be supercritical, with, for instance, at $\mu_0 = 0.231416$:

$$\omega_0 = 0.67375,$$
$$l \cdot L_1 r = .2712 + .4081i,$$
$$\mu_2 = 2.5426,$$
$$T_2 = -1.4684.$$

Thus we can be sure of the existence of a stable limit cycle for some μ above .231416 and some less than 1.999525, and, of course, can expect it for all μ between these values, as is indeed confirmed by numerical calculation. For some interesting related examples and references see Othmer [**6**].

The diffusive destabilization of this example is an oscillatory version of the mechanism first suggested by Turing [7] as a possible basis for the development of spatial structure in morphogenesis. It depends in an essential way on the difference between the diffusivities of the third chemical and the other two, and is closely related to similar phenomena known in convection driven by gradients of both temperature and salinity. An intuitive understanding of the reason for the instability can be obtained as follows: if x_3 were zero, x_1 and x_2 would evidently form a weakly damped oscillator. The third of equations (19) however shows that if μ is large enough x_3 tends to 'follow' x_1, being roughly equal to $.8x_1/(.1 + \mu)$. Since x_3 occurs in the first equation with the positive coefficient .8, this term tends to compensate the damping term $-.12x_1$, and thus may be able to destabilize the oscillator. If μ is too large, however, x_3 is too small a multiple of x_1 to do this. (If μ is too small, x_3 does not follow x_1 closely enough to be a destabilizing influence.)

2. The bifurcation methods discussed so far provide the most important technique for showing the existence–and calculating approximately–oscillations which are *weakly* nonlinear. Systems for which

these methods are suitable often exhibit in their physical structure an element which is essentially a linear oscillator, and even when this is not apparent on the surface, the occurrence of a Hopf bifurcation indicates that something like this is always present: the space spanned by the real and imaginary parts of r constitutes (mathematically) a linear oscillator (near the stationary solution), at the bifurcation point. The method of averaging, alluded to above, is often useful for obtaining information about more strongly nonlinear oscillators when they can be regarded as regular perturbations of a system with a (nonlinear) *center*, provided this 'unperturbed' system is sufficiently tractable. Such systems are qualitatively rather similar to weakly nonlinear ones. However, strongly nonlinear oscillations frequently occur in other circumstances, often without any physically obvious oscillatory component such as a pendulum. Available techniques for studying such systems appear to be principally either numerical or based on singular perturbation methods. We turn now to some examples of the latter.

The circuit of Figure 9 is an example of a *relaxation oscillator*.

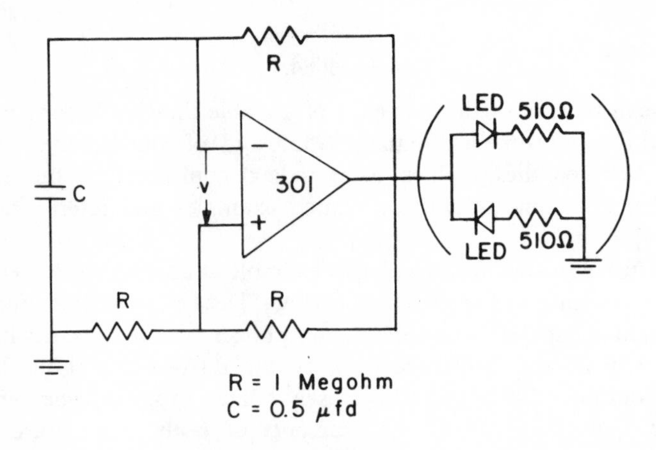

FIGURE 9.

An idealization of the behavior of the operational amplifier (301) in this circuit is that the output voltage $V_0 = V_s \cdot \operatorname{sgn}(v)$, where V_s (= 15 volts, say) is the supply voltage (the amplifier is powered by two 15-volt batteries, not shown on the diagram) and v is the voltage difference between the positive and negative inputs. Also, the current in the input circuit is taken to be negligible. Thus the voltage at the + input is $V_0/2$, and that at the − input is q/C, where q is the charge on the capacitor. Consequently we have the following two equations:

$$v = \frac{V_0}{2} - \frac{q}{C}, \tag{20}$$

$$R\frac{dq}{dt} = V_0 - \frac{q}{C}, \tag{21}$$

the latter coming from Ohm's law applied to the resistor at the top, neglecting input current to the amplifier. Now if we start say, with $v > 0$ and $q = 0$, we have $V_0 = V_s$, and so $q/C = V_s \cdot (1 - e^{-t/RC})$, so long as $q/C < V_s/2$. But when q/C reaches $V_s/2$, i.e. at $t = t_1 \equiv RC \ln 2$, v reaches 0 and then drops suddenly to $-V_s$ as does V_0 also. Thereafter, so long as $q/C > -V_s/2$, we have $q/C = -V_s + \frac{3}{2}V_s e^{-(t-t_1)/RC}$. Thus, V_0 remains at $-V_s$ until $3e^{-(t-t_1)/RC} = 1$, i.e. until $t = t_2 \equiv t_1 + RC \ln 3$; then v becomes positive, and jumps up to V_s as does V_0. We then have $q/C = V_s - \frac{3}{2}V_s e^{-(t-t_2)/RC}$, until $t = t_3 \equiv t_2 + RC \ln 3$, etc. We thus get discontinuous oscillations with period $2RC \ln 3$—about one second for the values of R and C shown on the diagram. The part of the circuit in parentheses serves only to make the oscillations visible: the two light emitting diodes turn on alternately depending on the sign of V_0. The waveforms of q/C and V_0 are sketched in Figure 10. The oscillation is made up of an alternation of slow 'relaxation' phases alternating with fast, almost discontinuous, jumps. These jumps are, of course, not really discontinuous, but their more exact description requires the inclusion of aspects of the system not present in the above idealized mathematical model. As an example of the kind of thing to be expected in a more complete model, we might suppose the function $\operatorname{sgn}(v)$ to be rounded off a bit, and modify equation (20) to something like

$$\varepsilon\frac{dv}{dt} = -v + \frac{V_0}{2} - \frac{q}{C}. \tag{22}$$

Here ε is to be thought of as another characteristic time-scale of the system, short compared with RC. If the right-hand side of (22) is not nearly zero, v must change rapidly. The direction of this change is indicated in Figure 11. If one starts off the curve labeled SM in Figure 11, the "slow manifold" where the right-hand side of (22) vanishes, v changes rapidly so as to move the phase point close to either the left or the right downward sloping part of it. Once SM is approached, the rapid motion ceases but the phase point continues to move slowly, close to SM, in accordance with (21). This motion inevitably drives the phase point above the peak of the curve, or below the valley, leading to another fast phase. Many other relaxation oscillators can be described similarly, i.e. in terms of a 'folded over' slow manifold. For another example consider the van der Pol equation (5). If we let $x = \varepsilon\dot{y}$ and integrate the equation once it takes the form (Rayleigh equation):

$$\ddot{y} + \varepsilon^2\dot{y}\left(\tfrac{1}{3}\dot{y}^2 - 1\right) + y = 0.$$

Setting $\dot{y} = v$, $y = \varepsilon^2 z$ and $t = \varepsilon^2\tau$ this equation can be written as the

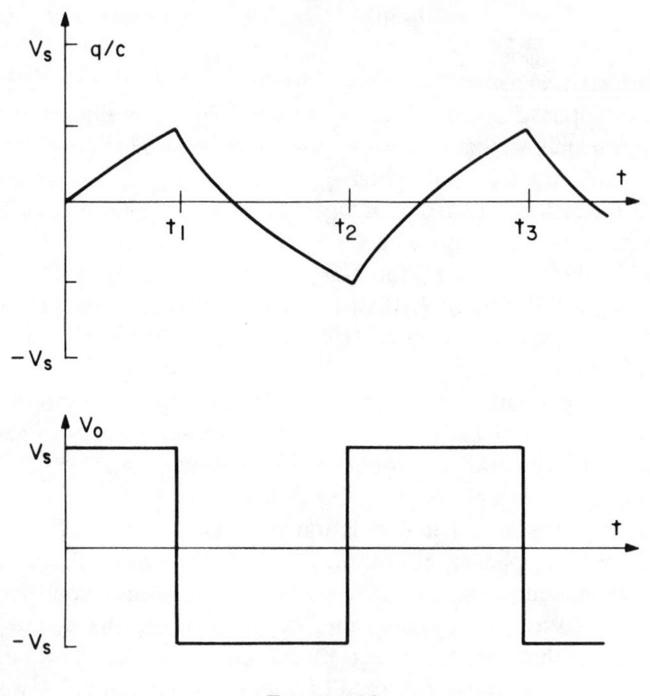

FIGURE 10.

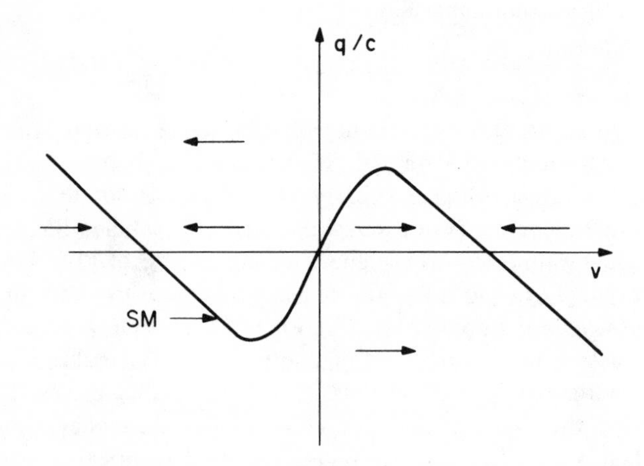

FIGURE 11.

system

$$\frac{dz}{d\tau} = v, \qquad \frac{dv}{d\tau} = \varepsilon^4\left[-\tfrac{1}{3}v^3 + v - z\right]. \tag{23}$$

Consider now the case of *large* ε; then v will change rapidly except on the slow manifold $z = v - v^3/3$, whose graph is qualitatively similar to the curve SM in Figure 11. Also, the direction of this rapid change is as shown in that picture, and in fact, the system (23) has, for large ε, a limit cycle of relaxation oscillator type. Its orbit consists approximately of two 'slow' segments given by $z = v - v^3/3$ for $1 \leqslant v \leqslant 2$ and $-2 \leqslant v \leqslant -1$, joined by 'fast' segments on $z = \frac{2}{3}$, $-2 \leqslant v \leqslant 1$, and $z = -\frac{2}{3}$, $-1 \leqslant v \leqslant 2$. Along the right-hand slow segment we have

$$\frac{dz}{d\tau} = \frac{d}{d\tau}\left(v - \frac{v^3}{3}\right) = (1 - v^2)\frac{dv}{d\tau} = v.$$

Thus the time occupied in traversing it is

$$\tau_1 = \int_2^1 \left(\frac{1 - v^2}{v}\right) dv = \tfrac{3}{2} - \ln 2.$$

Since very little time is involved in traversing the fast segments, the period of the oscillation (in τ) is given approximately, for large ε, by

$$T = 3 - 2\ln 2 \simeq 1.6137.$$

While the above calculation is very simple, obtaining higher terms in the asymptotic behavior for $\varepsilon \to \infty$ turns out to be quite difficult. This is because it is then necessary to investigate carefully the transition between the slow and fast segments, and this is in fact a rather complicated process. A discussion of this and references can be found in the book of Cole [**8**]. Other interesting examples of relaxation oscillators can be found in Andronov, Chaiken and Witt [**1**].

V. Unstable periodic solutions and wave trains. FitzHugh-Nagumo model. Flatto-Levinson theorem.

1. The examples of relaxation oscillators just discussed both have a limit cycle. However one is sometimes interested in periodic solutions which are not limit cycles, i.e., unstable periodic solutions, and when these arise in a singular perturbation context a closely related approach, again involving the idea of a slow manifold, can be used to advantage. This may be illustrated with the FitzHugh-Nagumo model [**9**] for the propagation of pulses along an idealized nerve. This is based on the following equations:

$$\frac{\partial \nu}{\partial t} + f(\nu) + z = \frac{\partial^2 \nu}{\partial x^2}, \qquad \frac{\partial z}{\partial t} = \varepsilon \nu, \tag{24}$$

where $f(\nu) = \nu(1 - \nu)(a - \nu)$, or at any rate is a function whose graph is qualitatively similar to that of this cubic, having zeros at 0, 1 and an intermediate point a (for definiteness take $a < \frac{1}{2}$, or $\int_0^1 f d\nu < 0$). We are here interested in solutions of these partial differential equations which have a 'permanent form', i.e. which appear steady in a suitably moving coordinate system. In this subject it is customary to think of the permanent wave as propagating to the left with speed θ, so the solutions of interest are those which are functions of the single variable $\xi = x + \theta t$. Such ν and z are solutions of the ordinary differential equations

$$\nu_{\xi\xi} = \theta \nu_\xi + f(\nu) + z,$$

$$\theta z_\xi = \varepsilon \nu. \tag{25}$$

In studying this topic, most attention has been directed to solutions representing a single propagating pulse, with ν and z tending to 0 at $\pm\infty$. However these equations also have solutions corresponding to infinite trains of waves–periodic solutions of (25)–and the latter are more relevant for our present purpose. As solutions of (25) these wave trains are not stable periodic solutions, but one must remember that stability with respect to ξ is not the same thing as stability with respect to t in the original partial differential equations. Thus at least some of the wave trains may well be stable in the latter sense and so of real physical interest. Now if we set $\nu_\xi = \omega$ and introduce $\zeta = \varepsilon\xi$ as a new independent variable (25) can be written

$$\varepsilon \nu_\zeta = \omega,$$

$$\varepsilon \omega_\zeta = z + f(\nu) + \theta\omega,$$

$$z_\zeta = \frac{1}{\theta}\nu. \tag{26}$$

In this form we recognize that for small ε we have a 'slow manifold' $\omega = 0$, $z = -f(\nu)$; unless the phase point is close to this curve, ν and/or ω must change rapidly. The equations (26) have a single critical point, the origin, and, assuming θ is positive and of order unity in ε, this is found to have two eigenvalues of moderate size and opposite signs and one small negative eigenvalue. There is thus a single trajectory leaving this critical point, i.e., just one solution (up to a translation in ζ) which tends to 0 as $\zeta \to -\infty$. If this happens to be a 'homoclinic trajectory', i.e. one that returns to the origin as $\zeta \to +\infty$, then the corresponding solution gives the single pulse referred to above. In general, this does not happen: there is no homoclinic trajectory and no pulse solution. But this does actually happen for a certain special value θ_0 of θ (actually two of them, but one is not $O(1)$ in ε), and a central aspect of the single pulse problem is to find this propagation speed. For other values of θ however there may be periodic solutions, corresponding to wave trains, and it is in these that we are here interested.

The graph of the slow manifold is sketched in Figure 12. If one is off this slow manifold, the motion of the phase point is more conveniently described in terms of the fast variable ξ for which the equations (26)

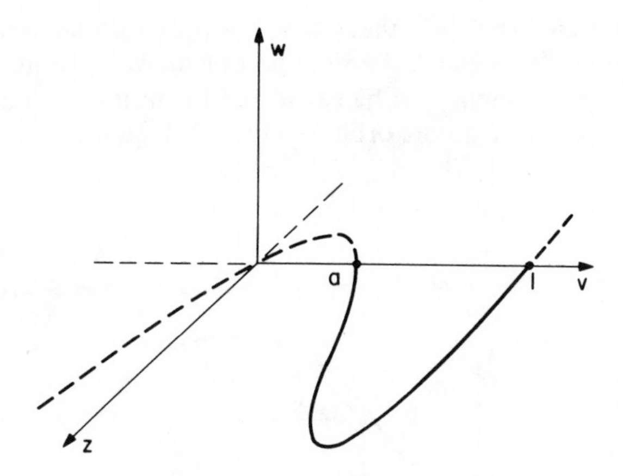

FIGURE 12.

become

$$v_\xi = \omega,$$
$$\omega_\xi = z + f(v) + \theta\omega,$$
$$z_\xi = \frac{\varepsilon}{\theta} v. \tag{27}$$

Thus for small ε, a fast portion of a trajectory lies approximately in a plane z = const., and there is an integral curve of the first two equations of (27) with z regarded as constant, say $z = z_1$. Assuming that z_1 lies between the local extrema of the cubic $-f(v)$, the three points at which the slow manifold intersects the plane $z = z_1$ are critical points of this two-dimensional system. The linearization at such a point, $v = v_*$, $\omega = 0$, say, has matrix:

$$\begin{bmatrix} 0 & 1 \\ f'(v_*) & \theta \end{bmatrix}.$$

Thus for the smallest and largest values of v_*, where $f' > 0$, we have saddle points, while for the one in between, where $f' < 0$, we have an unstable spiral (if $\theta^2 < -4f'(v_*)$) or an unstable node (if $\theta^2 > -4f'(v_*)$). Now it turns out that for a given θ, smaller than the homoclinic value θ_0, there is a certain value $z_1 > 0$ of z for which the trajectory leaving the left-hand saddle point (upwards and to the right) goes to the one on the right. There is also a (larger) value z_2 at which the trajectory leaving the right-hand saddle point (downward and to the left) goes to the one on the left. This suggests the existence, for small ε, of a periodic solution whose orbit tends (as $\varepsilon \to 0$) to the above two fast segments in the planes $z = z_1$ and $z = z_2$, joined by slow segments from z_1 to z_2 and z_2 to z_1 along the slow manifold $\omega = 0$, $z + f(v) = 0$. (On one of the slow segments, the third of equations (26) thus gives us $-f'(v)v_\zeta = v/\theta$, from which the slow part of the wave form, and so the approximate wavelength, can be calculated by an integration.) A qualitative sketch of this periodic orbit is shown in Figure 13.

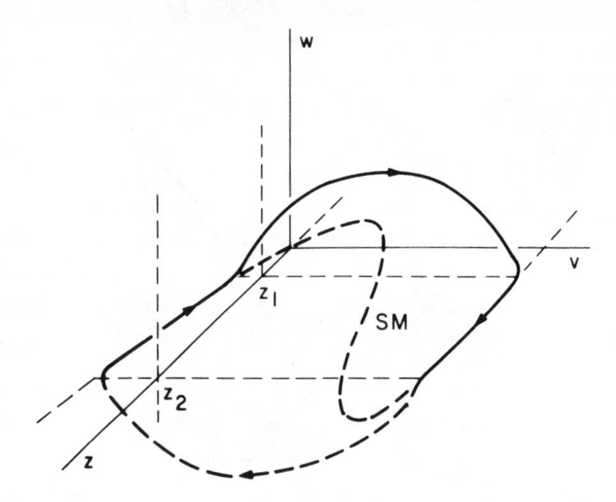

FIGURE 13.

In the case $f(v) = v(1 - v)(a - v)$ the connecting trajectories of the fast portions of the orbits can fairly easily be determined analytically, together with the relationship between z_1, z_2 and the propagation speed θ. Dividing the second of equations (27) by the first leads to:

$$\omega \frac{d\omega}{dv} = z + f(v) + \theta\omega. \tag{28}$$

Let the three roots of $z + f(v) = 0$ be $v_1 < v_2 < v_3$, so that (28) can be written as

$$\omega\left(\frac{d\omega}{dv} - \theta\right) = (v - v_1)(v - v_2)(v - v_3).$$

This has a solution, with $\omega = 0$ when $v = v_1$ and v_3, of the form $\omega = \lambda(v - v_1)(v - v_3)$, provided

$$\lambda[\lambda(2v - v_1 - v_3) - \theta] = v - v_2,$$

i.e. provided $\lambda^2 = \frac{1}{2}$ and $\theta/\lambda = v_1 + v_3 - 2v_2$. If $\lambda > 0$, $\omega < 0$ for v between v_1 and v_3 so this case corresponds to the "down-jump" at z_2. If $\lambda < 0$ we get the "up-jump" at z_1, i.e. we have as a condition for an up-jump:

$$\text{Up:} \quad \theta_u = -\sqrt{2}\left(\tfrac{1}{2}(v_1 + v_3) - v_2\right)$$

and for a down-jump:

$$\text{Down:}\quad \theta_d = \sqrt{2}\left(\tfrac{1}{2}(\nu_1 + \nu_3) - \nu_2\right).$$

By solving the cubic $z + f(\nu) = 0$ for $\nu_i(z)$ we can obtain $\theta_u(z)$ and $\theta_d(z)$; a typical graph of these functions (for $a = \frac{1}{4}$) is shown in Figure 14. Evidently for a particular $\theta = \theta_1$, say, on $(0, \theta_0)$, we get values z_1 and z_2 for the two jumps. It can be shown that these are the only solutions whose trajectories join ν_1 and ν_3. The limiting case of θ_0 which corresponds to $z_1 = 0$ (hence $\nu_1 = 0$, $\nu_2 = a$, $\nu_3 = 1$ and $\theta_0 = \frac{1}{\sqrt{2}}(1 - 2a)$; also then $z_2 = \frac{2}{27}(1 - 2a)(2 - a)(1 + a)$) gives the homoclinic orbit going from the origin back to itself. It may be regarded as the limit of wave trains when the wavelength goes to infinity. Although up- and down-jumps exist for some $\theta > \theta_0$, these have $z_1 < 0$ and the path along the slow manifold is interrupted by the critical point, so these do not seem to correspond to periodic wave trains.

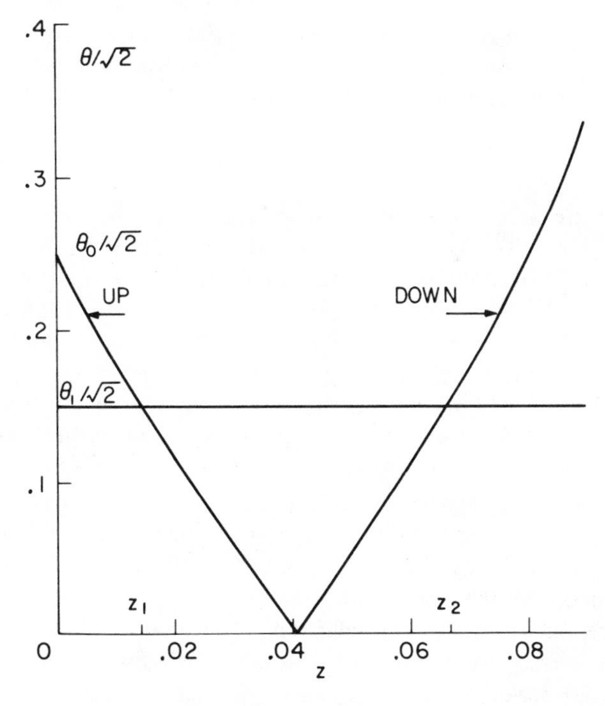

FIGURE 14.

It should be noticed that although a cubic slow manifold is involved in both cases, these periodic solutions are rather different from the relaxation oscillations of the van der Pol equation. In that case the up- and down-jumps always leave the slow manifold at the local extrema of the cubic, and indeed no trajectories leave it except on the middle branch where they all do. In the present case some trajectories everywhere are going away from the slow manifold: the periodic solution is unstable as a solution of (25), although possibly not as a solution, bounded in space, of (24), but only at z_1 and z_2 do they subsequently return to it.

The singular perturbation calculations for this problem are given by Casten, Cohen and Lagerstrom in **[10]**. Proofs of the existence of genuine solutions close to the singular solution when ε is small are given, in considerably greater generality, by Carpenter **[11]**. She uses topological methods.

2. There is a third kind of nonlinear oscillation in which singular perturbations and a slow manifold are involved. This has to do with systems of the form

$$\dot{x} = f(x, y, \varepsilon),$$
$$\varepsilon\dot{y} = g(x, y, \varepsilon), \tag{29}$$

where x is, say, n dimensional and y is m dimensional. In this case the 'slow manifold', for $\varepsilon \to 0$, is given by the m equations $g(x, y, 0) = 0$, and here we suppose that these equations can be solved for y in the form

$$y = h_0(x). \tag{30}$$

(Thus we are here *not* dealing with a case where the slow manifold is 'folded over'.) To assure this we may assume that the Jacobian $\det|g_y(x, y, 0)|$ is not zero, at least in the range of interest. We then assume also that the equation $\dot{x} = f(x, h_0(x), 0)$ has a solution $x_0(\sigma_0 t)$, 2π-periodic in its argument $\tau \equiv \sigma_0 t$, i.e. periodic with period $2\pi\sigma_0^{-1}$ in t. This solution is also supposed to be a stable limit cycle, or at any rate to be such that all but one of the Floquet multipliers of the linearization about it are off the unit circle. Actually we assume a little more than that g_y is nonsingular, namely that there is a positive number δ such that for $0 \leqslant \theta \leqslant 2\pi$ all eigenvalues of $g_y(x_0(\theta), h_0(x_0(\theta)), 0)$ have real parts exceeding δ in absolute value. Assuming suitable continuity this is then true of the eigenvalues of $g_y(x, y, \varepsilon)$ for all small ε and all (x, y) in some neighborhood of the set $\{(x_0(\theta), h_0(x_0(\theta))): 0 \leqslant \theta \leqslant 2\pi\}$. The basic result here, due to Flatto and Levinson **[12]**, is that then for small ε there is also a locally unique (up to a time translation) solution $x(\sigma t, \varepsilon)$, $y(\sigma t, \varepsilon)$, 2π-periodic in the first argument, which is close to $x_0(\sigma_0 t)$, $h_0(x_0(\sigma_0 t))$ and whose angular frequency $\sigma(\varepsilon)$ is close to σ_0.

A remarkable feature of this problem is that the periodic solution, even though it comes from a singular perturbation problem, does not have any 'boundary layer' character; the fact that here there is a periodic solution without *discontinuous* jumps at $\varepsilon = 0$ is paralleled by the existence of one without *fast* jumps for small ε. Of course it still is a singular perturbation problem, and *most* solutions of (29) have boundary layers, typically a rapid approach to (or flight from) something near the manifold $y = h_0(x)$. But some solutions, and in particular the periodic solution which is here of main interest, are so to speak already there, and do not need a boundary layer. The following geometric description of this kind of problem (shown to me by N. Kopell) clarifies this situation and 'explains' why the theorem is true.

If we rewrite (29) in terms of the fast time variable $\tau = t/\varepsilon$ we get

$$\frac{dx}{d\tau} = \varepsilon f(x, y, \varepsilon), \qquad \frac{dy}{d\tau} = g(x, y, \varepsilon),$$

and at $\varepsilon = 0$ the phase portrait of this system looks (schematically) like Figure 15.

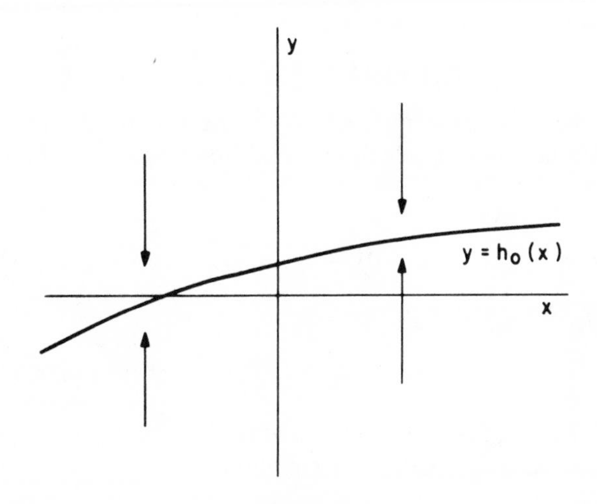

FIGURE 15.

The vector field is 'vertical' (in $x =$ const. surfaces), and there is an invariant slow manifold $y = h_0(x)$ (consisting entirely of critical points) which is hyperbolic in the sense that any center manifold of any of these critical points is n dimensional and tangent to $y = h_0(x)$. (This is because g_y is nonsingular.) It is then 'geometrically obvious' that if the vector field is perturbed by taking $\varepsilon > 0$ and small, there will still be an invariant manifold of dimension n, $y = h(x, \varepsilon)$, close to $y = h_0(x)$. [Actually, this is not unique, but all of them have $\lim_{\varepsilon \to 0} h(x, \varepsilon) = h_0(x)$, and, in fact, if we assume that f and g possess asymptotic expansions in powers of ε, uniform in x and y on some compact set large enough to contain a neighborhood of the range of $[x_0(\cdot), y_0(\cdot)]$, $y_0(\cdot) \equiv h_0(x_0(\cdot))$, then all $h(x, \varepsilon)$ have a common uniform asymptotic expansion so we have an 'asymptotically unique' invariant manifold.] If now we restrict the vector field of (29) to the invariant manifold $y = h(x, \varepsilon)$, we get, on this manifold, the equation $\dot{x} = f(x, h(x, \varepsilon), \varepsilon)$ which is a *regular* perturbation of $\dot{x} = f(x, h_0(x), 0)$. Then by a standard theorem on the perturbation of periodic solutions we obtain the existence of a periodic solution for small positive ε. These ideas provide the basis for a proof of the Flatto-Levinson theorem, if we had a proof of the 'geometrically obvious' facts stated above. Analytic demonstration of the existence of invariant manifolds of outer solutions was given by Vasil'eva and a geometric approach has been provided recently by Fenichel (see **[13]**).

For the actual construction of the periodic solution it may be preferable not to try to construct the whole invariant manifold first, but rather to go directly after the periodic solution itself. A sketch of a way to do this, which also follows the lines of the original Flatto-Levinson proof, is as follows. Let $\sigma t = \theta$ (so that we are looking for a solution of period 2π in θ) and set $\sigma = \sigma_0 + \varepsilon s$, $x = x_0(\theta) + \varepsilon\xi$,

$$y = h_0(x_0(\theta)) + \varepsilon h_{0x}(x_0)\xi + \varepsilon\eta.$$

Recalling that $\sigma_0 x_0'(\theta) = f(x_0, h_0(x_0), 0)$, the system (29) then takes the form (assuming a reasonable degree of differentiability of f and g):

$$\varepsilon\eta' - \sigma_0^{-1} g_y(x_0, h_0(x_0), 0)\eta = \sigma_0^{-1} g_\varepsilon(x_0, h_0(x_0), 0) - h_{0x}(x_0)x_0' + \varepsilon g_1(\theta, \xi, \eta, s, \varepsilon), \tag{31}$$

$$\xi' - \sigma_0^{-1}(f_x + f_y h_{0x})\xi - s\sigma_0^{-1}x_0' - \sigma_0^{-1}f_y\eta = \sigma_0^{-1} f_\varepsilon(x_0, h_0(x_0), 0) + \varepsilon f_1(\theta, \xi, \eta, s, \varepsilon). \tag{32}$$

Now the following facts can be verified fairly easily:

(A) For sufficiently small ε, the equation

$$\varepsilon\eta' - \sigma_0^{-1} g_y(x_0, h_0, 0)\eta = p(\theta),$$

where p is a 2π-periodic vector, has a unique 2π-periodic solution η, and $\sup_\theta|\eta| \leqslant K_1 \sup_\theta|p|$ for some constant K_1.

(B) Let $\Phi(\theta)$ be the matrix solution with $\Phi(0) = I$ of $\Phi' - \sigma_0^{-1}(f_x + f_y h_{0x})\Phi = 0$, i.e. a fundamental matrix for the operator on the left of (32). With our assumptions $\Phi(2\pi)$ has 1 as a simple eigenvalue, with $x_0'(0)$ as (right) eigenvector. Let l be the corresponding left eigenvector. Then there is a unique value of C such that the equation

$$\xi' - \sigma_0^{-1}(f_x + f_y h_{0x})\xi - Cx_0' = q(\theta)$$

(where $q(\theta)$ is a 2π-periodic vector) has a 2π-periodic solution ξ satisfying $l \cdot \xi(0) = 0$, and this ξ is also unique. Also there are constants K_2 and K_3 such that $|C| \leqslant K_2 \sup_\theta|q|$ and $\sup_\theta|\xi| \leqslant K_3 \sup_\theta|q|$. The periodic solution of (31) and (32) is then constructed using the iteration:

$$\varepsilon\eta'_{n+1} - \sigma_0^{-1}g_y\eta_{n+1} = \sigma_0^{-1}g_\varepsilon - h_{0x}x'_0 + \varepsilon g_1(\theta, \xi_n, s_n, \varepsilon), \tag{33}$$

$$\xi'_{n+1} - \sigma_0^{-1}(f_x + f_y h_{0x})\xi_{n+1} - s_{n+1}\sigma_0^{-1}x'_0$$
$$= \sigma_0^{-1}f_y\eta_{n+1} + \sigma_0^{-1}f_\varepsilon + \varepsilon f_1(\theta, \xi_n, \eta_n, s_n, \varepsilon). \tag{34}$$

Starting with, say, $\xi_0 = 0$, $\eta_0 = 0$, $s_0 = 0$ on the right in (33) we find η_1 as the unique periodic solution of (33) according to fact (A) above. Using this in the first term on the right in (34) and ξ_0, η_0, s_0 in f_1, we then find s_1 and ξ_1 in accordance with fact (B). This completes the first step of the iteration, and we then continue in the same way. Using the estimates cited in (A) and (B) one can show that for small enough ε this iteration always converges to a unique periodic solution of (31) and (32), normalized by $l \cdot \xi(0) = 0$. (Some such normalization is required for uniqueness, since the original problem was an autonomous system.) This determines the angular frequency σ and the periodic solution (x, y) of (29).

3. In carrying out this procedure in a particular case one has to solve–usually numerically–the linear problems mentioned in (A) and (B). Although straightforward in principle, there are certain complications in practice which should be mentioned. First of all, some of the solutions of the homogeneous form of the problem in (A) may grow, rapidly for small ε, as θ increases; others may decay, also rapidly. If it happens, say, that all decay, one can in principle calculate the periodic solution simply by integrating forward "until the transients die out"; this will happen very soon when ε is small, so that by integrating for a little more than 2π the periodic solution should be accurately obtained, no matter what initial conditions (within reason) are used. (Similarly if all solutions of the homogeneous problem grow, one can integrate backwards.) *However*, when ε is small we have a "stiff" system, and a serious numerical problem will arise if this is not recognized. The point is that the forcing term $p(\theta)$, and the periodic solution we are trying to compute, have a characteristic time scale of order unity in θ, but the "response times" of the system are very fast, of order ε. In order for the more familiar numerical methods to work satisfactorily it is necessary to use a step size small enough to deal adequately with variations which occur on this short time scale, even though no such are present in the actual solution of the differential equation which the numerical method is trying to approximate. The reason for this can be seen in the simplest example: suppose we wish to solve $\varepsilon\dot{y} + y = 0$ on $0 \leqslant t \leqslant 1$ with initial condition $y(0) = 0$, using the simple Euler method. The exact solution is of course $y \equiv 0$; nothing could be smoother. The difference equation of the Euler method for this example is $\varepsilon(y_{n+1} - y_n)/h = -y_n$, or $y_{n+1} = (1 - h/\varepsilon)y_n$. If, because of a small numerical error, for instance due to round-off, some y_n is slightly different from zero, then this error will

propagate but decrease in magnitude if h is small enough, and no great difficulty will arise. But if $h/\varepsilon > 2$ the error will be amplified exponentially, and soon destroy the validity of the approximation. Evidently even though the exact solution does not vary at all, a small step size is needed with this method when ε is small. Although the numerical problem could be eliminated by using a small enough h, a financial one might be created, and this is obviously not the right approach. Numerical methods suitable for such stiff systems exist. The simplest is the "backward Euler method", which for the above example gives the difference equation $\varepsilon(y_{n+1} - y_n)/h = -y_{n+1}$, or $y_{n+1} = y_n/(1 + h/\varepsilon)$. Evidently this can be used with any step size without danger of numerical instability, though of course it will not give accurately the *transient* part of some *other* solution, such as the one with $y(0) = 1$, unless h is appreciably smaller than ε over the interval where the transient is important. But in finding the periodic solution to the problem in (A), when all solutions to the homogeneous equation decay as θ increases, we are not concerned with an accurate representation of the transient and a step size suitable to an adequate representation of $p(\theta)$ could be used from the start. In practice it would no doubt be much better to use a method with a higher order of accuracy than the backward Euler method; such methods, adapted to stiff systems, also exist, but it does not seem appropriate to go further into these numerical issues here. It is perhaps enough to emphasize that this is a genuine problem that must be dealt with if a system of this sort is to be actually solved on a computer.

A second complication arises for the problem in (A) when some solutions of the homogeneous problem grow rapidly in the forward θ direction and others do so in the backward θ direction. We still have a stiff system, but the periodic solution we want to compute is here also unstable as a solution of the differential equation, both forward and backward. This is not a numerical instability which might be prevented by a different integration algorithm; in such cases the procedure of integrating until the transients decay cannot be used directly. One way to deal with this difficulty is the following. Consider the system

$$\varepsilon\eta' = A(\theta)\eta + p(\theta)$$

(for our problem $A = g_y(c_0, h_0, 0)/\sigma_0$) and suppose for each θ we find a matrix $P(\theta)$ such that

$$P^{-1}AP = \operatorname{diag}(A_1(\theta), A_2(\theta))$$

where the block A_1 has all its eigenvalues strictly in the left half plane and A_2 has all its strictly in the right half-plane. (Under our assumptions the dimensions of A_1 and A_2 do not vary with θ.) Then setting $\eta = Pu$

we get

$$\varepsilon P u' + \varepsilon P' u = APu + p(\theta),$$

and decomposing u into two parts corresponding to the blocks A_1, A_2 this can be written in the form

$$\varepsilon u_1' = A_1 u_1 + \varepsilon(B_{11}u_1 + B_{12}u_2) + p_1, \tag{35}$$

$$\varepsilon u_2' = A_2 u_2 + \varepsilon(B_{21}u_1 + B_{22}u_2) + p_2. \tag{36}$$

Now although it is not always true that all solutions of the equation $\varepsilon u_1' = A_1(\theta)u_1$ decay as θ increases merely because all eigenvalues of $A_1(\theta)$ have real parts less than $-\delta$, this *is* true if ε is small enough. Thus the periodic solution of (35) and (36) can be computed iteratively by solving

$$\varepsilon u_1^{(n+1)'} = A_1 u_1^{(n+1)} + \varepsilon\big(B_{11}u_1^{(n)} + B_{12}u_2^{(n)}\big) + p_1, \tag{37}$$

$$\varepsilon u_2^{(n+1)'} = A_2 u_2^{(n+1)} + \varepsilon\big(B_{21}u_1^{(n)} + B_{22}u_2^{(n)}\big) + p_2, \tag{38}$$

starting with, say $u_1^{(0)} = 0$, $u_2^{(0)} = 0$ and at each stage integrating (37) forward and (38) backward until the transients die out. This iteration can be shown to converge when ε is small enough. Naturally a suitable stiff-adapted numerical method should be used for these integrations. In the context of our problem, this iteration could well be incorporated into the iteration discussed in connection with equations (33) and (34) writing (33) as a pair of equations corresponding to the blocks A_1 and A_2 and putting the terms involving the blocks B_{ij} in (31) and (36) together with the terms from g_1 in (33). Apart from the numerical issues, this procedure is essentially that used by Flatto and Levinson in their proof.

Finally we describe how the solution of the problem in (B) can be obtained. First compute the fundamental matrix $\Phi(\theta)$ and the left eigenvector l. Then compute the particular solution ξ_1 of

$$\xi' - \sigma_0^{-1}(f_x + f_y h_{0x})\xi = q(\theta)$$

determined by $\xi_1(0) = 0$. Now it is readily verified that $\xi_2 = \theta x_0'(\theta)$ satisfies

$$\xi_2' - \sigma_0^{-1}(f_x + f_y h_{0x})\xi_2 = x_0'.$$

Thus the general solution of

$$\xi' - \sigma_0^{-1}(f_x + f_y h_{0x})\xi - Cx_0' = q(\theta)$$

is $\xi = \Phi(\theta)\nu + \xi_1 + C\xi_2$, where ν is an arbitrary constant vector. Such a solution is periodic when and only when $\xi(2\pi) = \xi(0)$, i.e. when

$$\begin{aligned}(\Phi(2\pi) - I)\nu &= -\xi_1(2\pi) - C\cdot 2\pi x_0'(2\pi)\\ &= -\xi_1(2\pi) - 2\pi C x_0'(0).\end{aligned}$$

This system of linear equations has a solution for ν when and only when the right-hand side is orthogonal to l; since $l \cdot x_0'(0) \neq 0$ (because of the simplicity of the eigenvalue 1 of $\Phi(2\pi)$), C is determined by $C = -l \cdot \xi_1(2\pi)/(2\pi l \cdot x_0'(0))$. With this value of C, the equations for ν are then to be solved with the restriction $l \cdot \nu = 0$, which uniquely determines ν and so the periodic solution ξ.

VI. A model of a chemical reaction. One of the most realistic mathematical models for a chemical oscillator is the "Oregonator", which was suggested by Field and Noyes **[14]** as giving a reasonable qualitative description of the major features of the Belousov reaction. In chemical terms it is described by the following five reactions:

$$\begin{aligned} A + Y &\to X, \\ X + Y &\to B, \\ C + X &\to 2X + Z, \\ 2X &\to D, \\ E + Z &\to fY + F. \end{aligned}$$

Here the substances denoted by X, Y, and Z which are both products and reactants are the ones whose concentrations oscillate periodically. In comparing this model with the Belousov reaction they correspond to Bromous Acid ($HBrO_2$), Bromide ion (Br^-) and Ceric ion (E^{++++}), respectively. The substances denoted by the letters A through F are supposed to be supplied at constant concentrations when reactants, and to be inert when products, so they play no important role in the kinetics of this reaction scheme. The fifth reaction is supposed to represent a process which is effectively of first order in Z; to best model the Belousov reaction the stoichiometric factor f (possibly "fudge factor" is a better name) is a number about $\frac{1}{2}$ or 1; in the original numerical study of the kinetics of this system, Field and Noyes took $f = 1$, and we shall do so here as well. These model reactions resemble closely some of the actual reactions which are thought to be of importance in the real Belousov reaction, though the more refined models also involve about 15 other reactions.

After suitable scaling, the differential equations of the chemical kinetics (based on the law of mass action) of this set of reactions can be written as (with $f = 1$):

$$\dot{x} = s^2(x + y - xy - qx^2), \tag{39}$$

$$\dot{y} = (-y - xy + z), \tag{40}$$

$$\dot{z} = sw(x - z). \tag{41}$$

The range of interest in the parameters (for comparison with the Belousov reaction) has $s \sim 100$, $q \sim 10^{-5}$, $sw \sim 10$. It was found numerically by Field and Noyes that in this range these equations have a limit cycle solution. We shall illustrate the application of the Flatto-Levinson theorem by considering this problem for fixed q and fixed ws, taking $\varepsilon \equiv s^{-2}$ as our small parameter for the singular perturbation. It is apparent from (39) that we have here as slow manifold the portion of the hyperbolic cylinder $x + y - xy - qx^2 = 0$ which lies in the first octant, i.e., $y = (x - qx^2)/(x - 1)$ for $1 < x \leqslant q^{-1}$ and $z > 0$; we describe this by

$$x = X(y) \equiv (2q)^{-1}\left[1 - y + \sqrt{(1 - y)^2 + 4qy}\,\right].$$

On this manifold equations (40) and (41) become

$$\dot{y} = x - y(1 + X(y)), \tag{42}$$

$$\dot{z} = ws(X(y) - z). \tag{43}$$

We shall see that this system has a limit cycle solution; actually this limit cycle itself is a fairly strong relaxation oscillation. Its relaxation character depends primarily on the smallness of the parameter q; it is not of course associated with small ε. The phase plane of (42) and (43) is also somewhat difficult to draw when q is small because of the large range of values of y and z which are of interest—this is a reflection of one of the striking features of the real Belousov oscillation, in which concentration changes of the order of a factor of 10^3 really do occur each cycle. A qualitative sketch of the $\log y$ vs. $\log z$ plane is shown in Figure 16. There is a critical point P at

$$z = \sqrt{\tfrac{1}{4} + q^{-1}} - \tfrac{1}{2}, \qquad y = \left(\sqrt{\tfrac{1}{4} + q^{-1}} - \tfrac{1}{2}\right)\Big/\left(\sqrt{\tfrac{1}{4} + q^{-1}} + \tfrac{1}{2}\right),$$

and the loci of $\dot{y} = 0$, $\dot{z} = 0$ are also shown, together with an indication of the vector field.

The line BF passes through the critical point P, but is almost the same as the $\log z$ axis. The line ABC is the $\log y$ axis ($z = 1$) and A is the point on it where $y(1 + X(y)) = 1$, approximately $y = q$. The line GFE is at $z = q^{-1}$, and E is the point on it at which $y(1 + X(y)) = q^{-1}$, approximately $y = 1/2q$. It is easy to verify that the rectangle $ACEG$ is (as indicated in the figure) positively invariant under the flow described by the differential equations and one can also readily check that the critical point is an unstable spiral. Thus the existence of a periodic solution can be deduced from the Poincaré-Bendixson theory; with some more effort it can be shown to be a stable limit cycle. As an alternative to the Poincaré-Bendixson approach one could use the following argument: A trajectory starting on the segment PD (not at P) must enter the rectangle $PDCB$, but since it cannot approach the critical point it must, in a finite time, leave this rectangle and can only do so through the edge BP, where it enters $BPHA$. Similarly, it must leave this rectangle in a finite time by crossing HP, then leave $HPFG$ across PF,

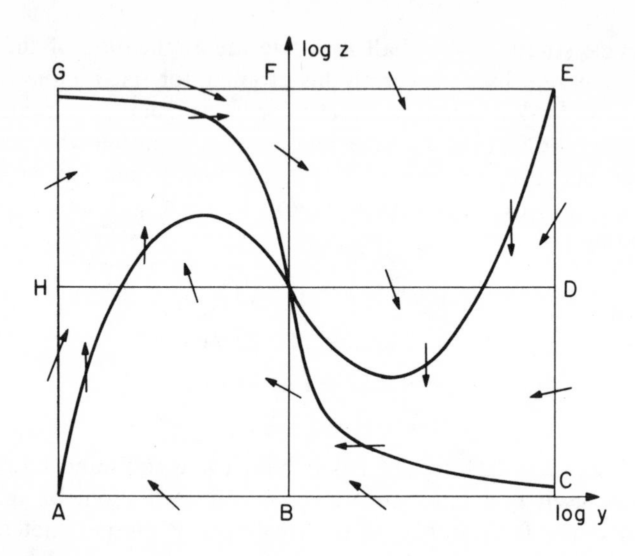

FIGURE 16.

and finally once again reach PD. Thus we get a mapping of PD into itself, and in view of the unstable spiral at P, of a closed subsegment (omitting a bit near P) into itself. This map must have a fixed point according to the simplest form of the Brouwer fixed point theorem, which again gives us a periodic orbit. In the present case there is no particular advantage to this topological argument over the Poincaré-Bendixson one–like the latter it does not in itself give us information about uniqueness or stability. However, this example gives a simple way to illustrate this 'box' method, which has been used to good advantage, particularly by Hastings, in considerably more complicated higher dimensional cases.

After verifying somehow that the periodic solution of (42) and (43) is indeed a stable limit cycle, and checking that the slow manifold is hyperbolic, we can appeal to the Flatto-Levinson theorem to show that for small $\varepsilon = s^{-2}$ there is also a periodic solution of (39)–(41) nearby.

Hastings and Murray **[15]** used the box method to show directly the existence of a periodic solution of (39)–(41). They did this by showing how to construct a large box in the x, y, z space made up of eight smaller boxes meeting at the critical point as in Figure 17. They showed (a) that the large box was positively invariant, (b) that no trajectory could stay in sub-boxes (4) or (5) unless it was on the (one dimensional) stable manifold of the critical point, (c) that no trajectory could stay forever in any of the other sub-boxes, and (d) that such trajectories had to cycle through those 6 boxes in the order $3 \to 1 \to 2 \to 6 \to 8 \to 7 \to 3$. From this they concluded then (in the parameter range where the critical point *has* a one-dimensional stable manifold) that a periodic solution must exist.

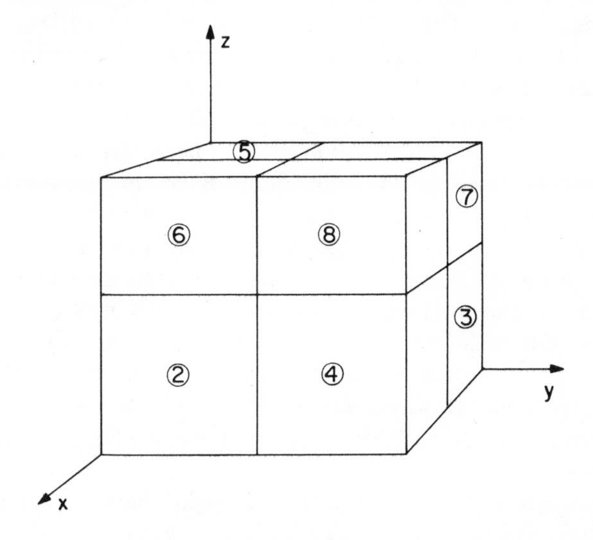

FIGURE 17.

Finally it should be noted that, in his recent very interesting thesis, Wolfe [**16**] has shown how the considerably more complicated kinetic equations of the most realistic chemical models of the Belousov reaction can be regarded as perturbations, in part singular and in part regular, of the Oregonator system. This provides a kind of mathematical basis for the chemical intuition which led Noyes and Field to propose this model as relevant to understanding the real Belousov reaction. Wolfe's work is also of direct mathematical interest, extending the general ideas of Flatto and Levinson in new directions.

VII. Reaction-diffusion wave trains and slowly varying waves.

1. I turn now to a description of some of the work of Kopell and myself on temporally periodic and spatially heterogeneous solutions of certain reaction-diffusion equations. We became interested in these questions in the course of attempting to understand better the spatial patterns sometimes formed in the Belousov reaction. These patterns are best seen when the reaction mixture is placed in a thin (1 or 2 mm) layer in a flat dish, and covered with a sheet of glass, with some of the oxidation-reduction indicator ferroin added to the solution. After a little time the layer is found to be essentially filled by a number of roughly polygonal regions, within each of which is an organized pattern of propagating waves of color. Typically these take the form of concentric circles, radiating outward from the center; sometimes they are spirals. Within each 'polygon' the period, wavelength, and propagation speed of

the waves are very uniform spatially, and the 'circles' are usually very round. A typical period is 15 sec and wavelength 1 mm; these characteristics change slowly with time, probably due to the variation in concentration of the major ingredients as the overall reaction runs down, but are fairly constant for a number of periods. However, there is usually a considerable difference from one 'polygon' to another in period, wavelength and propagation speed. The centers of the individual 'target patterns' seem to remain quite fixed in position, but the boundaries of the polygons slowly move: the higher frequency ones encroach upon, and may completely devour, their lower frequency neighbors. Although there is a superficial resemblance to patterns of waves on a pond in the rain, actually there is no sign of overlapping of wave trains from different centers. Wave fronts which meet at a polygon boundary seem simply to disappear. This indicates that, unlike water waves, these chemical waves are highly nonlinear. Also there is no decrease in amplitude with distance from the center as there is with water waves.

Without the ferroin indicator there is a slight color change associated with the pale yellow color of Ce^{++++}, but in thin layers this is not visible. However the patterns also are present then: they can be seen by setting the (glass bottomed) dish on a fluorescent screen, and illuminating it from above with a Halloween black light. Ce^{++++} absorbs strongly in the near-visible ultraviolet, and the wave patterns cast ultraviolet shadows on the fluorescent screen, which then makes them visible to the eye. Ultraviolet light has some effect on the reaction, so it is probably better (as well as safer) to use a not too bright Halloween light rather than a quartz lamp.

Wave patterns also occur in deep layers, but are quite difficult to see, probably largely because it is then almost impossible to avoid small motions (driven no doubt by thermal convection) which distort or destroy the patterns. An ingenious technique for studying some three-dimensional wave patterns was developed by Winfree; it, as well as other aspects of these patterns, will no doubt be described by him at this seminar. Photographs of the patterns can be found, for instance, in **[17]**, **[18]**, **[19]**. [At the seminar a movie illustrating the formation and evolution of these patterns was shown.]

Perhaps the most striking features of these patterns and some questions they raise are the following:

(a) The occurrence of trains of waves. How is it that such phenomena can occur at all in a system that seems to be a combination of oscillatory chemistry with diffusion, usually thought of as a smoothing and homogenizing influence?

(b) The visually striking *center* of the targets, and the uniformity of the wave characteristics throughout a particular target region. Is there something special like a foreign 'catalytic particle' associated with the center, or is it autonomous and somehow self-sustaining? The uniformity, and indeed also the initial development of a target region, suggest that the centers somehow act as 'pacemakers', controlling at least the period throughout their 'region of influence'. Why is this?

(c) The quite sharp boundaries between the different target regions. Why do the wave fronts seem to disappear there, and why do these boundaries propagate in the way they do when separating targets of different frequencies?

2. We have attempted to illuminate these questions by somewhat idealizing the above features, and by seeking to construct mathematical models based on reaction-diffusion equations which are appropriate to describe these idealizations. The underlying mathematical model then is a system of equations of the form

$$c_t = F(c) + K\nabla^2 c. \tag{44}$$

Here c is a vector of concentrations of the various substances participating as reactants and products, and $F(c)$ a vector function describing the chemical kinetics. We generally assume that we are dealing with an oscillating chemical reaction, and thus suppose that the ordinary differential equations $c_t = F(c)$ have a stable limit cycle solution, say $c = y_0(\sigma_0 t)$ where y_0 is a 2π-periodic function of its argument, so σ_0 is the angular frequency of this oscillator. There is already some idealization here; the real reaction runs down as the major ingredients are used up, and the real chemical kinetics does not have a truly periodic solution, anymore than an electronic oscillator run by a battery does. But as in the latter case, if there is a sufficiently great difference between the time scale of running-down and the period of oscillation it is a reasonable approximation to assume that concentrations of some 'fuel' ingredients are constant, and include only the others in c. [In an actual particular case, other approximations in the chemical kinetics, such as regarding certain fast reactions as being always in equilibrium, might also be appropriate and permit the elimination of certain variables. In this case $F(c)$ need not have the polynomial form obtained from the law of mass action.] For much of the following it is not necessary to take a specific form for $F(c)$; the assumption that $c_t = F(c)$ has a limit cycle is the main one. Likewise the (positive definite) diffusivity matrix K need not be scalar, though in certain examples it is taken to be so.

To investigate the existence of wave-like solutions in systems of this kind we take as our idealization the case of an *infinite train* of *plane waves*. Thus we look for solutions of (44) of the form $c = y(\sigma t - \alpha x)$ where $y(\theta)$ is a 2π-periodic function of its argument. Such exist if, for suitable σ and α, the system

$$\sigma y' = F(y) + \alpha^2 K y'' \tag{45}$$

has a 2π-periodic solution. In **[20]** we showed that under the assumptions mentioned above this was always the case for sufficiently small α^2, and for each such α^2 there is just one such solution, $y_{\alpha^2}(\theta)$ (up to a phase shift), and a corresponding angular frequency

$$\sigma = H(\alpha^2). \tag{46}$$

Note that here, unlike nonlinear water waves, there is just a *one-parameter* family of wave trains. With water waves, for each wave number *and* amplitude, in certain ranges, there is a wave train, and the nonlinear 'dispersion relation' has the form $\sigma = H(\alpha^2, A)$ instead of (46). Here a given α^2 determines the frequency σ *and* the amplitude–some measure of the size of $y_{\alpha^2}(\theta)$. As $\alpha^2 \to 0$, $H(\alpha^2) \to \sigma_0$, the limit cycle frequency, and indeed (45) for $\alpha^2 = 0$ and $\sigma = \sigma_0$ is just the equation for the limit cycle of the chemical kinetics. Of course (45) is a system of twice the order of the system of kinetic equations, and small α^2 is thus a singular perturbation. The method used in **[20]** to show the existence of plane waves near the limit cycle for small α^2 is a slight variant of the Flatto-Levinson theorem.

3. There is one simple class of reaction-diffusion equations for which the plane wave solutions can be written down explicitly in terms of formulas. These have c two dimensional, and can be written most simply by replacing the vector c with the complex function $c_1 + ic_2$, which we shall also call c. Then these special systems are

$$c_t = [\lambda(|c|) + i\omega(|c|)]c + K\nabla^2 c \tag{47}$$

where here K is a positive number and λ and ω are functions qualitatively like those graphed in Figure 18. $\lambda(r)$ is positive for $0 \leqslant r < r_*$, and negative for $r > r_*$, and we take ω to be decreasing for a reason to be mentioned later. Evidently the 'kinetics' $c_t = (\lambda + i\omega)c$ has a limit cycle whose orbit is the circle $|c| = r_*$. The plane wave solutions are most readily seen by setting $c = re^{i\theta}$ and rewriting (47) in terms of the polar coordinates as

$$r_t = r(\lambda(r) - K|\nabla\theta|^2) + K\nabla^2 r,$$

$$\theta_t = \omega(r) + Kr^{-2}\nabla\cdot(r^2\nabla\theta). \tag{48}$$

Evidently then a solution is $r = r_0$, $\theta = \sigma t - \alpha x$, where $\sigma = \omega(r_0)$ and $K\alpha^2 = \lambda(r_0)$, and this corresponds to a plane wave with frequency σ and wave number α. The dispersion relation (46) is here

$$\sigma = \omega(\lambda^{-1}(K\alpha^2)) \equiv H(\alpha^2).$$

We get such a plane wave solution for all values of r_0 for which $\lambda \geqslant 0$,

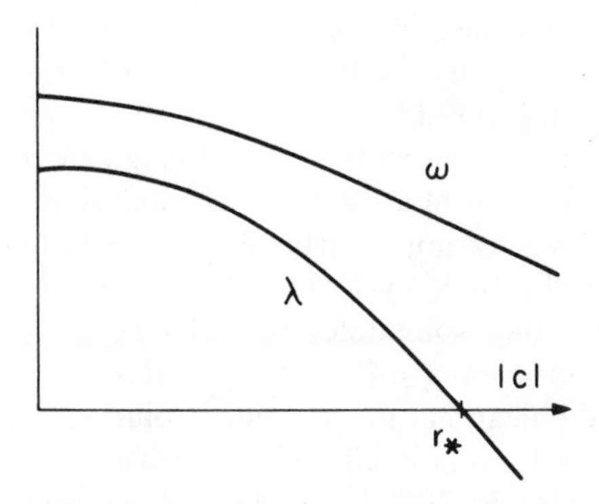

FIGURE 18.

i.e., $0 \leqslant r_0 \leqslant r_*$, and so plane waves exist with wave numbers in the range $0 < \alpha < \alpha_c$, where $K\alpha_c^2 = \lambda(0)$. The waves which occur in the real Belousov reaction seem to have the property that the shorter ones have the higher frequency. This is also the case with these "λ–ω" examples provided $\omega(r)$ is a decreasing function, which is the reason for this choice. Of course these special models are not very 'chemical'. The fact that the c's are sometimes negative conflicts with their interpretation as concentrations, but this can be readily fixed by interpreting them as deviations of the concentration from an unstable equilibrium state (since only a limited part of the c-plane is really of interest). But it does not seem to be possible to obtain equations of this form from any even vaguely plausible model of chemical reactions. The main advantage of these models is their convenience for making explicit detailed calculations, not only for the plane wave solutions but for other more complex ones as well. For example in [**20**] the *stability* of the plane wave solutions as solutions of the partial differential equation (44) was investigated; while this question can be formulated in a general way, in the case of the λ–ω models it can also be explicitly answered rather easily. In the general case it seems that rather elaborate numerical calculations are required to do this. On a number of occasions we have found results obtained for λ–ω models to be convenient and fairly reliable guides to what subsequently has turned out to be more generally true. Thus we see explicitly in the λ–ω case that plane waves exist for a finite range, $0 \leqslant \alpha < \alpha_c$, of wave numbers. The general result mentioned above shows that waves always exist for small enough α, and one can also show generally that they will *not* exist if α is sufficiently large; we must consequently expect that in general there is a 'cut-off wave number' α_c beyond which plane wave trains no longer occur. The stability investigation in the λ–ω case showed that the plane wave trains were stable on an interval $0 < \alpha < \alpha_s$, where α_s is definitely less than α_c, though positive. We conjecture that this is generally true, at least when K is near enough to a scalar matrix; i.e., that sufficiently long waves are stable, but as α is increased this stability is lost before the waves cease to exist. This has not been proved yet, although there are some partial results [**19**]. If K is far from a scalar matrix–has great disparities among its eigenvalues–this may not be true. There are for instance examples which show that the limit cycle ($\alpha = 0$), stable as a solution of the kinetic equations, may sometimes be unstable as a solution of (44) provided K has sufficiently different eigenvalues. This is analogous to the diffusive destabilization of a stationary solution (critical point of the kinetic equations) first pointed out by Turing, and related to the example given above in §IV. It should perhaps be remarked that the wave patterns we are talking about here do not depend on K being far from a scalar matrix and are not of the same nature as the so-called 'dissipative structures'–nonlinear developments of the Turing instability–which do. The latter may well be important in certain biochemical systems involving substances of very different molecular weights, or in ecological problems concerned with interactions of species of quite different mobilities (trees and insects, say). But there seems no reason to expect great differences in the diffusivities of the substances involved in the Belousov reaction.

4. Some aspects of the patterns seen in the Belousov reaction appear to be describable fairly well in terms of the idea of 'slowly varying waves', functions which *locally*, i.e. over distances of a number of wavelengths and times of a number of periods, are quite close to a particular plane wave but such that to maintain this closeness over greater distances and/or times the parameters of the plane wave (angular frequency and wave number vector especially, but also wave form) must be gradually altered. This idea has been successfully exploited in a number of areas of applied mathematics to extend the usefulness of plane wave solutions, which though not always easy to come by themselves are still much more accessible than more general solutions of the full initial value problem. The most extensive reference on this topic is the book of Whitham [**21**]. Whitham's approach is particularly well adapted to systems which are essentially conservative, and we have found a somewhat different formulation more suitable to the case of reaction-diffusion equations, but the basic idea is the same. Here is one formalization of this idea: we introduce a small parameter ε, thought of as a measure of the slowness of variation, and set $X = \varepsilon x$, $T = \varepsilon t$ (with a suitable choice of origin) for the 'large scale' or 'slow' variables. At first we consider just the idea of a slowly varying wave (SVW), neglecting the question of what it is supposed to be a solution of. Then by an SVW we mean a family (parameterized by ε) of functions $c(x, t, \varepsilon)$ expressible in the form $c(x, t, \varepsilon) = Y(\theta(x, t, \varepsilon), X, T, \varepsilon)$ where:

(1) $Y(\theta, X, T, \varepsilon)$ is a C^2 periodic function of θ with least period 2π.

(2) $Y = Y^0(\theta, X, T) + o(1)$ (or else Y has an asymptotic expansion in powers of ε up to some order) uniformly for X and T in some compact region.

(3) $\Theta(X, T, \varepsilon) \equiv \varepsilon\theta(x, t, \varepsilon) = \Theta^0(X, T) + \varepsilon\Theta^1(X, T) + o(\varepsilon)$ (or else has a corresponding asymptotic expansion to one higher order than Y).

REMARKS. (1) If θ is expanded about a particular point X_0, T_0 we have $\theta = \varepsilon^{-1}\Theta(X_0, T_0, \varepsilon) + \Theta_X(X_0, T_0, \varepsilon)(x - x_0) + \Theta_T(X_0, T_0, \varepsilon)(t - t_0)$ + terms small if $X - X_0$ and $T - T_0$ are small like, say, $o(\sqrt{\varepsilon})$, which may still be large in terms of the fast variables. This shows the appropriateness of the term 'SVW', and of the terminology 'local wave number' for $-\Theta_X = -\theta_x$ and 'local frequency' for $\Theta_T = \theta_t$.

(2) It is helpful to add to the above three conditions a *normalization condition* on the periodic functions $Y(\theta, X, T, \varepsilon)$ and Y^0 (and other terms of the asymptotic expansion if any). This is to make definite the meaning of the zero of phase, and might be done for instance by requiring that the last component of all periodic functions involved have a fixed value k at $\theta = 0$. (This precise condition may not always be convenient but something like it is; for details see **[19]**.)

With the phase normalization condition one can show that the representation of a slowly varying wave in the form $c = Y(\theta, X, T, \varepsilon)$ is then *asymptotically unique*, in spite of the fact that Y and Θ together seem to give one more dependent variable than is involved in c. This means that if X is a slowly varying wave and $X \equiv Y(\Theta/\varepsilon, X, T, \varepsilon) \equiv \overline{Y}(\overline{\Theta}/\varepsilon, X, T, \varepsilon)$, then Y and $\overline{Y}$, and Θ and $\overline{\Theta}$, have the same asymptotic expansions, respectively. A proof of this and a more complete discussion of the whole topic can be found in **[19]**.

5. We turn now to slowly varying wave *solutions*: When can an SVW be a solution of our system of reaction-diffusion equations $c_t = F(c) + K\nabla^2 c$? We suppose that these equations have a family of plane wave solutions $c = y_{\alpha^2}(\theta)$, $\theta = H(\alpha^2)t - \alpha \cdot x$ parameterized by the magnitude squared of the wave number vector α. (This is true, for instance, when the kinetic equations have a limit cycle solution, and it may also be true otherwise, although then the allowable range of α^2 does not extend down to zero.) We also suppose that for all relevant α^2 the periodic function $y_{\alpha^2}(\theta)$ is normalized by the same condition adopted for the SVW's. Thus the pair $(y_{\alpha^2}(\theta), \Theta)$, with $\Theta = H(\alpha^2)T - \alpha \cdot X$, is itself an SVW (*very* slowly varying) and also gives a solution of the reaction-diffusion equations.

Now by substituting the Y, Θ representation into the reaction-diffusion equations one obtains (suppressing the variables X, T, ε):

$$Y_\theta(\Theta/\varepsilon)\Theta_T - F(Y) - KY_{\theta\theta}(\Theta/\varepsilon)\Theta_X^2 + o(1) = 0. \tag{49}$$

Let

$$G(\theta, X, T, \varepsilon) \equiv Y_\theta(\theta, X, T, \varepsilon)\Theta_T(X, T, \varepsilon) - F(Y(\theta, X, T, \varepsilon)) \\ - KY_{\theta\theta}(\theta, X, T, \varepsilon)\Theta_X^2 + \cdots,$$

a function of the four independent variables θ, X, T, ε obtained by replacing Θ/ε where it appears as argument of Y in the left-hand side of (49) by θ. Then it is easy to see that (G, Θ) constitutes an SVW, and (49) says that if it is also a *solution*, then it must be the *zero* SVW. From the uniqueness theorem mentioned above we then conclude that the asymptotic expansion of $G(\theta, X, T, \varepsilon)$ must be zero and consequently that

$$Y_{\theta\theta}^0(\theta, X, T)\Theta_T^0 - F(Y^0(\theta, X, T)) - KY_{\theta\theta}^0(\theta, X, T)(\Theta_X^0)^2 = 0. \tag{50}$$

Here θ, X, T are regarded as independent variables, and we recognize (50), for fixed X and T, as the equation for plane waves, with Θ_T^0 as σ and $(\Theta_X^0)^2$ as α^2. Thus our SVW (Y, Θ) can be a solution only when

$$Y^0(\theta) = y_{(\Theta_X^0)^2}(\theta) \tag{51}$$

and

$$\Theta_T^0 = H\left((\Theta_X^0)^2\right) \tag{52}$$

(of course this does not prove that there really are any SVW's that are solutions, other than the plane waves, but we hope for the best).

Now the Hamilton-Jacobi equation (52) makes possible some definite predictions about the patterns which one can try to compare with observation, for it describes how the phase function evolves with time; if we know Θ as a function of X initially, assuming things are slowly varying, then we know $\Theta^0(X, 0)$ approximately, and by solving the initial value problem for (52) we can predict Θ and so the pattern of waves at later times. (This assumes that what is seen in the photographs or with the eye—the light absorption by the ferroin—really corresponds to the phase; because of the large excursion in Ce^{IV} concentration associated with the most marked color change this is probably very nearly true with say a definition of phase = 0 as where the Ce^{IV} concentration is increasing through a value near the geometric mean of typical largest and smallest values.) To attack the initial value problem for (52) we may use the method of characteristics. For this problem, this goes as follows: (see Courant and Hilbert [**22**, vol. II]). First one introduces the variables $p = -\Theta_X$ and $q = \Theta_T$, and considers the 'strip space' of X, T, H, p and q in which for the moment p and q are thought of as new independent variables (we here drop the superscript 0 for simplicity). In this space we solve the ordinary differential equations:

$$\frac{dX}{ds} = \frac{\partial}{\partial p} H(p^2) = 2pH'(p^2),$$
$$\frac{dT}{ds} = 1,$$
$$\frac{d\Theta}{ds} = q - 2p^2 H'(p),$$
$$\frac{dp}{ds} = 0,$$
$$\frac{dq}{ds} = 0, \tag{53}$$

with the initial conditions $X = X_0$, $T = 0$, $\Theta = \Theta_0$, $p = p_0$, $q = q_0$ at $s = 0$. In the present case this is very easy to do and the integral curve–the *characteristic* through the initial point–is evidently a straight line along which p and q are constant, and X, T and Θ vary linearly with s as indicated by (53). Returning now to the initial value problem for the Hamilton-Jacobi equation (52) for which we have, say, the initial data $\Theta^0(X, 0) = \hat{\Theta}(X)$, the prescription of the method of characteristics is that we take for the initial conditions on (53) an arbitrary X_0, and with it set $\Theta_0 = \hat{\Theta}(X_0)$, $p_0 = -\hat{\Theta}_X(X_0)$, $q_0 = H(p_0^2)$. We thus get from the corresponding solution of (53) X, T, Θ, p and q as functions of X_0 and s. We then solve the equations $X = X(X_0, s)$, $T = T(X_0, s)$ for $X_0(X, T)$ and $s(X, T)$, which is always possible near $s = 0$, and substitute the results into H, p and q to express these as functions of X and T. One then finds that not only initially but also thereafter we have $q = H(p^2)$ (not surprising since p and q are constant along the characteristics) and *also* (which *is* surprising with the present extremely brief description of the method) that we always have $p = -\Theta_X$ and $q = \Theta_T$. Thus $\Theta(X, T)$ solves (52) with the given initial data. It is not usually possible to carry this process through in terms of formulas. However, conceptually (and numerically) it gives an entirely satisfactory solution to the initial value problem *except* for the fact that the equations for X and T in terms of X_0 and s, which can be written down for arbitrary X_0 and s (because the characteristics are merely straight lines in the strip space) cannot always, or even usually, be solved for X_0 and s in terms of X and T if T is too great. The process of solving these equations—or trying to—may conveniently be described geometrically by projecting the characteristics from the strip space down to the (X, T) space. The projections are again straight lines, of course, with slope $dX/dT = 2pH'(p^2) \equiv c_g$, the "group velocity" vector of the plane wave with wave number vector p. X_0 is X at $T = 0$, and s is essentially the same as T; to solve for X and T in terms of X_0 and s means that for a point in (X, T) space we take $s = T$ (easy enough) and then follow back "the" projected characteristic through this point to $T = 0$ and find the point X_0 from which it came. The trouble arises when, after some time, there is more than one projected characteristic through (X, T), so that no unique X_0 can be found. This is not a weakness of the method of characteristics, but an indication of a more fundamental fact: when two projected characteristics approach one another, the local wave number vector $-\Theta_X$, which is constant along each at two different values, is becoming more and more nearly discontinuous, and thus is ceasing to be slowly varying. The hypothesis of slow variation, on which the whole of this approximate theory is based, is breaking down and it is impossible to continue beyond such a point without more careful consideration of the full equations or at any rate without introducing something beyond the SVW picture.

6. Putting aside for the moment further discussion of what happens when the projected characteristics intersect, we can still make some interesting comparisons with the real patterns in the Belousov reaction. First of all, it is clear that the *centers* of the targets, and the fairly sharp boundaries between adjacent targets (or spirals) are regions of rapid variation of wave number for which the SVW theory is not appropriate, but between these that description seems plausible. As mentioned before, the frequency of waves in the real reaction seems to be higher the shorter the waves are, which means that $H'(p^2) > 0$; thus the group velocity vector $c_g = 2pH'(p^2)$ has the same direction as the wave number vector p. Whatever is going on at a target center, it is spatially localized, and waves propagate *outward* from it, so c_g must also point outward. Since c_g gives the direction in which small variations in frequency and wave number propagate, the SVW theory indicates that the centers should appear to act like pacemakers. Given the observed fact that they oscillate with a well defined and relatively constant frequency, then we should expect them to be surrounded by a region also oscillating with this frequency, at least after some time has elapsed. This seems to agree well with what one sees. In principle the function H which describes the dispersion relation could be computed, given a sufficiently good model of the chemical kinetics. The only attempt I am aware of in this direction is that of Stanshine **[23]** using the Oregonator model and a rather complicated asymptotic method. Unfortunately the Oregonator model does not appear to be close enough to the real Belousov kinetics to make a quantitative comparison with experiment possible, and besides the experimental situation is not very satisfactory yet, the only serious attempt being that of Tatterson and Hudson **[24]**.

However if an effective method of driving a plane wave at a specified frequency could be developed, which seems quite possible, it should be possible to measure the dispersion relation carefully and also to, say, produce a train of waves and then alter the frequency of the wavemaker somewhat and thus compare the resulting evolution with the prediction of the slowly varying wave theory. This seems a feasible although nontrivial experiment, but it has not been attempted (at least successfully) yet. Finally, it should be noted that the boundary between two targets is a region toward which the characteristics of the SVW theory are converging, and is indeed a very thin region across which there is a near-discontinuity in wave number and often also in frequency. Like a center it is a region outside the scope of the SVW theory, but it differs essentially from a center in being a region of converging rather than diverging characteristics. Its observed persistence suggests that this is what happens when characteristics intersect, and we shall see in the next section that by somewhat extending the SVW theory the propagation of these boundaries can be fairly well understood. A theoretical justification for this extension, based on the full reaction diffusion equations, will also be presented.

VIII. Boundaries between regions of slow variation.

1. The breakdown of the method of characteristics when different projected characteristics intersect is familiar in gas dynamics, where it is associated with the formation of *shocks*. The latter, localized very thin regions of almost discontinuous change in pressure, density, temperature, etc. which propagate through the gas, are well known as real phenomena and have been carefully studied experimentally. The continuum theory of inviscid non-heat-conducting ('nondissipative') gas dynamics, generally very satisfactory as a description of gas flows, 'predicts' shocks in the sense that it indicates the development, in a finite time, of discontinuities from some kinds of smooth initial conditions, and this theory accurately describes the development of real shocks. Although the theory of characteristics for the system of hyperbolic equations of gas dynamics is more complicated than that of the Hamilton-Jacobi equation sketched above, the development of these discontinuities also comes about from the convergence and eventual intersection of the projected characteristics. Now in gas dynamics, the usefulness of the nondissipative theory has been successfully extended beyond the point at which a shock first forms by accepting discontinuous solutions, other considerations being introduced to fix the location of the discontinuities in space-time. These other considerations are (1) the *Rankine-Hugoniot conditions* which express the continuity of certain quantities across the shock, and (2) the *entropy condition*, which states that entropy of gas flowing across the shock must always increase. These conditions may be derived from the basic laws of conservation of mass, momentum, etc. (on which the differential equations of gas dynamics are based) and from the second law of thermodynamics. An alternative is to regard the equations of nondissipative gas dynamics as

approximations to a more complete theory based on the compressible Navier-Stokes equations, and show that the latter has genuine solutions which, for small viscosity and heat conductivity, are well approximated by solutions of the nondissipative equations except near the places where the Rankine-Hugoniot and entropy conditions predict shocks. In gas dynamics this has (more or less) been done.

These facts about gas dynamics suggest that the slowly varying wave theory discussed in the last section can perhaps also be extended beyond the first intersection of characteristics, if we can find the appropriate analogues of the Rankine-Hugoniot and entropy conditions. In our chemical wave context we do not have such firm foundations as conservation of mass and the second law of thermodynamics to base this on, but it is not difficult to guess the answer. Verification of this guess depends on returning to the full reaction diffusion equations, in much the same way as in gas dynamics one can study the *structure* of a shock on the basis of the Navier-Stokes equations.

The entropy condition in gas dynamics is equivalent to the statement that the characteristics on the two sides of the shock converge on it, rather than diverge from it. This immediately suggests the qualitative part of the guess for the chemical wave 'shock condition', namely the same statement about the characteristics for the Hamilton-Jacobi equation. Since the group and phase velocities in the Belousov reaction are in the same direction, a 'shock' for the SVW theory is then a thin region toward which the wave crests move; we thus wish to identify as *shocks* the boundaries between adjacent target patterns. (The centers, whatever they are, are something else.) Of course it might happen for some other system exhibiting chemical waves (if there are any!) that the phase and group velocities were *oppositely* directed; in that case a shock would be a thin region from which the wave crests diverge, while a center–something which acts like a pacemaker–would be a place toward which the waves (but not the characteristics) converge.

Now it was already noticed by Zaiken and Zhabotinskii **[17]** that these boundaries where waves meet move, and do so in such a way that the phase, as judged by the color, is continuous across them. Thus the quantitative part of the guessed shock condition (the analogue of the Rankine-Hugoniot conditions) is that the phase function Θ is continuous across a shock, though of course its derivatives, the local wave number and frequency, are not. With this condition it is easy to show that a shock propagates normal to itself at a speed given by $\Delta\sigma/\Delta\alpha$, the ratio of the differences in frequency and wave number across it. They really do seem to propagate this way, and with these shock conditions the SVW theory can be extended to permit the calculation of the evolution of patterns including shocks, provided what happens at centers if any is given. Example calculations can readily be made graphically, and one is presented in **[19]**. While no really careful experimental study has been made, the qualitative comparison seems good.

2. In order to provide a more definite basis for the 'phase continuity rule' and 'convergence of characteristics' shock conditions it is desirable, at least in some idealized case, to study this kind of problem using the full reaction-diffusion equations. This also would clarify the *inner structure* of the shock: the details of the way the transition from one wave number to another is made. In **[19]** we have attempted to do this by considering an 'ideal shock-structure'. The idealization has two aspects: (1) we look for a solution to the reaction-diffusion equations corresponding to a shock connecting one (semi-) infinite train of plane waves with another parallel one; (2) we also suppose that the shock structure has 'settled down' to a permanent form. In gas dynamics these two idealizations would correspond to seeking a plane shock-structure solution of the Navier-Stokes equations connecting one constant state at, say, $-\infty$ with another at $+\infty$, and which was time independent in a system moving with the shock. In our case the 'constant states' are replaced by plane wave trains, which cannot both be time independent in the same system. The two wave numbers at $\pm\infty$ are unchanged when observed in a moving system, but the two frequencies are altered because of the Doppler effect. In fact there is one moving system in which the two frequencies become the same (and this is actually the system moving at the shock speed given above) so the idealization of 'permanence' here becomes the requirement that the whole solution, in the moving system, should be *periodic in time*, with the same period everywhere. We now look at some of the details of this ideal shock-structure problem, taking for simplicity the λ–ω model. We shall also consider only the case that the two wave trains at $\pm\infty$ have the *same* frequency σ and wave numbers $\mp\alpha$, corresponding to 'amplitude' r_*, say; thus $\sigma = \omega(r_*)$ and $\alpha^2 = \lambda(r_*)$. We take the limit cycle radius to be 1, $\lambda(1) = 0$; thus $r_* < 1$. Our equations are, as before,

$$r_t = r(\lambda - \theta_x^2) + r_{xx},$$
$$\theta_t = \omega + r^{-2}(r^2\theta_x)_x, \tag{54}$$

and we are seeking a solution of this which corresponds to concentrations which are everywhere periodic in time with period $2\pi/\sigma$ and which has $r \to r_*$, $\theta_x \to \pm\alpha$ as $x \to \pm\infty$. It is consistent with the form of (54) to seek such a solution in the form $r = r(x)$, $\theta = \sigma t - \int_0^x a(x')\,dx'$; this Ansatz gives the equations:

$$r(\lambda - a^2) + r'' = 0,$$
$$\sigma = \omega - a' - 2r'r^{-1}a, \tag{55}$$

which can conveniently be written as the third order autonomous system:

$$\begin{aligned} r' &= ru, \\ u' &= a^2 - u^2 - \lambda(r), \\ a' &= \omega(r) - \omega(r_*) - 2au. \end{aligned} \tag{56}$$

We want to know if this system has a solution with $r \to r_*$, $u \to 0$, $a \to \mp \alpha = \mp \sqrt{\lambda(r_*)}$ as $x \to \pm \infty$. This system has $(r_*, 0, \mp\alpha)$ as two critical points, so with these variables the shock-structure problem becomes that of finding a *connecting trajectory* joining the two critical points–exactly the same kind of problem as the gas dynamics shock-structure problem. The possibility of this reduction is another illustration of the exceptional simplicity of λ–ω models. In the general case it is also possible *formally* to look at the problem as a connecting trajectory problem, but only in an infinite dimensional (rather than three dimensional) space, and there are very serious difficulties in making this formal picture into a mathematically rigorous one. This is discussed in [**19**]; we shall not go into it further here.

If one examines the linearization of (56) at the two critical points, it turns out (assuming λ' and ω' to be negative) that the one with $a = +\alpha$ has a two-dimensional unstable manifold and a one-dimensional stable manifold, while that with $a = -\alpha$ has a one-dimensional unstable manifold and a two-dimensional stable manifold. Since a connecting trajectory must be in the unstable manifold of the critical point it comes *from*, and in the stable manifold of the critical point it goes *to*, and since two surfaces in 3-space in general intersect in a curve while two curves in 3-space in general do not intersect at all, there is some hope of finding a connecting trajectory going from $(r_*, 0, \alpha)$ to $(r_*, 0, -\alpha)$ but one going the other way could occur only exceptionally (e.g., only for special values of α, if at all). It is for this reason that the problem was (in anticipation) stated above with $a \to \mp \alpha$ as $x \to \pm \infty$. This observation thus indicates that we cannot in general expect to find any shock structure solution except when the waves propagate *toward* the shock, so verifying the 'entropy condition' of the extended SVW theory. (When $\omega' > 0$, the situation is reversed and the waves would propagate away from the shock, but the group velocities are still toward it. The dimensions of the stable and unstable manifolds depend on the direction of the group velocity.) The exceptional case does however sometimes occur, and there is then a qualitatively different kind of connecting trajectory. We should like to regard this as a model for the 'center structures', as will be discussed later.

One way to show that equations (56) do really have a connecting trajectory going from $(r_*, 0, \alpha)$ to $(r_*, 0, -\alpha)$ is to use a bifurcation method centered on the limiting case of coalescence of the two critical points, i.e., $r_* \to 1$, $\alpha \to 0$. Such a method was used by Foy **[25]** in the gas dynamics shock-structure problem. It may be regarded as a 'weak shock' method; this can be done in the present case, and the details are presented in **[19]**. A formal generalization of this approach to the case of general chemical kinetics is also given there, and is formally successful though, as mentioned above, obtaining these results in a mathematically rigorous way for other than λ–ω models seems definitely nontrivial.

It turns out that the shock structure solution produced by the bifurcation method has a 'shock thickness', the characteristic scale of the transition layer, which is inversely proportional to the jump in wave number across the shock. Thus weaker shocks are thicker; this suggests that a modification of the SVW theory should be possible which would directly produce the first approximation results of the bifurcation method applied to the ideal shock-structure problem. Furthermore, such a modification should also be applicable (as a weak shock approximation) to other solutions of the reaction-diffusion equations, which correspond to 'nonideal shock structures', for instance solutions evolving toward an ideal (weak) shock structure. This modification of the SVW theory is given in **[19]**; it leads to an equation for the phase function which, in the original (fast) variables, has the form

$$\theta_t - H(\theta_x^2) = K_1(\theta_x^2)\theta_{xx}. \tag{57}$$

K_1 is in general a function of the local wave number which is determined in a somewhat complicated way by the solution to the plane waves problem; for λ–ω models however it is merely the diffusivity constant. This equation is similar to the integrated form of the Burgers equation, and in the simplest case of a λ–ω model with, say, $\lambda = 1 - r^2$, $\omega = a - br^2$, it is exactly the integrated Burgers equation. Since the initial value problem for the latter can be explicitly solved, various examples of (weak) shock development, interaction, etc., can be readily constructed.

Some numerical calculations of ideal shock structures using both the exact system (56) and the approximate equation (57) show that (57) gives a fairly good representation up to moderately strong shocks, but not for really strong shocks. The latter may have an oscillatory approach to the plane waves at $\pm\infty$, which is never given by equation (57). Some information about strong shocks can be obtained without numerical calculation by a different bifurcation method, also discussed in **[19]**. However, sufficiently strong shocks always put at least one of the plane waves at ∞ outside the stable range, and so presumably never really occur.

IX. Group velocity, 'spiral' and 'center' wave patterns.

1. A few more remarks about propagating waves and fronts are in order here. One should keep in mind the special nature of idealized solutions like plane wave trains or shock structures which are usually found from the partial differential equation by inserting a special Ansatz so as to simplify the problem, mostly by reducing it to an ordinary differential equation. Typically the special form of the Ansatz is not possessed by neighboring solutions of the partial differential equation, so for instance, its stability properties as a solution of the ODE are not normally relevant to questions of stability as a solution of the PDE. Also, there may be solutions of the PDE having the special form but not really relevant to the original problem as a whole: PDE plus initial conditions, for instance. As an example, consider the Fisher equation

$$u_t = u(1 - u) + k^2 u_{xx} \tag{58}$$

which has been used as a model for the propagation of an epidemic or of an advantageous gene. At $k = 0$ this equation has the solution

$$u = \tfrac{1}{2}\left(1 + \tanh\left(\tfrac{1}{2}(t - t_0(x))\right)\right)$$

for any function $t_0(x)$, since x does not enter explicitly at $k = 0$. If we take $t_0 = cx$ we get a permanent travelling wave front, for *any* propagation speed c. This is really a purely *kinematic* wave, obtained by judiciously selecting initial conditions. Now wavefront solutions of the full equation are solutions of the ODE

$$k^2 u'' + cu' + u(1 - u) = 0, \tag{59}$$

obtained by putting $u = u(x - ct)$ into (58). (59) has monotone (hence $\geqslant 0$) solutions if and only if $c^2 \geqslant 4k^2$ (and thus for all $c \geqslant 2k$), but *only* at $c = 2k$ can this solution be approached by a solution of the PDE whose initial conditions are of compact support. The underlying reason for this is that fronts with $c > 2k$ are a sort of mixture of kinematic and diffusive waves, and require that initial conditions be carefully set up on the whole line in order to be realized. (There are also fronts with $c < 2k$, but these are oscillatory, require u negative somewhere, and are more obviously irrelevant to the original problem in which u is intrinsically nonnegative.) The SVW ideas are often helpful in giving proper interpretation to idealized solutions like wave trains or fronts, because they enable one, at least approximately, to examine nearby solutions of the PDE.

2. We turn now to the question of solutions resembling the centers and spirals seen in the Belousov reaction. Examples of solutions with circular symmetry have been given by Ortoleva and Ross [**26**] and by Greenberg [**27**]. These can be made to look like centers, with phase

velocity directed outward by making $\omega' > 0$ ($\lambda' < 0$ as usual, to fit in with the limit cycle of the kinetics). But then the *group* velocity at ∞ is inward. If $\omega' < 0$, both phase and group velocities are inward. Thus these "center" solutions would not act like pacemakers, and are more properly described as circular shocks. They were obtained essentially by bifurcation-type methods, using the wave number as a small parameter and having r (in the λ–ω case) everywhere near 1. [Greenberg gives a constructive existence proof in the λ–ω case. Ortoleva and Ross treat the general case formally, making a small wave number approximation from the start. This is essentially equivalent to using the axisymmetric form of equation (57) above, taking K_1 to be constant and H to be $a + b\theta_x^2$, which would also be obtained with a λ–ω model. The solution is then obtained by using the logarithmic transformation that converts the integrated Burgers equation into the heat equation.]

A spiral solution must, by symmetry, have $r = 0$ (in the λ–ω case) at the center; such are accordingly very difficult to obtain by bifurcation methods, even in the case of a 'spiral shock' with incoming group velocity. Cohen, Neu and Rosales, in a forthcoming paper, apparently have an example, but it requires a very special (and somewhat peculiar) form of λ. It gives a logarithmic spiral, approaches the limit cycle at ∞, and there (because of the special form of λ) has *infinite* group velocity. It seems to be a sort of borderline case neither exactly a 'spiral shock' nor a 'spiral center'.

Examples of spiral centers in spatially *discrete* models have been computed by Winfree [**28**] and more generally by Hastings, Greenberg and Greene [**29**]. [The latter work was described by Hastings at this seminar.]

3. Kopell and I have found some one dimensional solutions, for λ–ω models, which may be regarded as analogues of centers and in some cases spirals. These are solutions to the equations (56) with $r \to r_*$, $u \to 0$, and $a \to \pm\alpha = \pm\sqrt{\lambda(r_*)}$ (assuming $\omega' < 0$) corresponding to a 'backwards' connection between the two critical points, i.e. a structure at $x = 0$ from which waves propagate outwards toward $\pm\infty$ and, more importantly, have there *outward* directed group velocity. Unlike the shock structure joining the critical points, which exists for all sufficiently small α, these center solutions exist only for certain special values of α. Since the latter do not accumulate at $\alpha = 0$, bifurcation methods based on bringing the critical points together ($\alpha \to 0$) do not seem to be helpful here, and we have instead used a different kind of singular perturbation method. These center structures appear to have no analogue in gas dynamics. The small parameter which we shall exploit is introduced by setting $\omega(r) = \varepsilon\Omega$ and supposing that Ω–or really its derivative–is $O(1)$. Making ε small may be regarded as treating a case in

which the limit cycle is very stable; since this also corresponds to enlarging the range of stable wave numbers for the plane wave problem, it seems a reasonable case to consider. Introducing this into (55) and replacing the parameter σ by $\varepsilon\Omega(r_*)$ one gets

$$r_{xx} - a^2 r = -r\lambda(r),$$
$$(r^2 a)_x = \varepsilon r^2\left[\Omega(r) - \Omega(r_*)\right], \tag{60}$$

and we are seeking a value of r_* ($= r_*(\varepsilon)$) such that these equations have a solution with $r \to r_*$, $a \to \pm\sqrt{\lambda(r_*)}$ as $x \to \pm\infty$. Now at $\varepsilon = 0$, (60) may be regarded as the (polar) equations of motion of a particle in a central force field (x being regarded as time) and as such have the energy and angular momentum integrals

$$e = \tfrac{1}{2}\left(r^2 a^2 + r_x^2 + f(r)\right), \qquad j = r^2 a,$$

where $f' = 2r\lambda(r)$ and we take $f(1) = 0$, $r = 1$ being the limit cycle value where $\lambda = 0$. Making use of these first integrals we can describe the solutions to (60) at $\varepsilon = 0$ quite explicitly, and this makes possible their investigation for small ε. In doing this we have found the following change of variables convenient:

$$R = r^2,$$
$$J = \frac{j}{\varepsilon} = R\frac{a}{\varepsilon},$$
$$E = \frac{e}{\varepsilon^2} = \frac{1}{2\varepsilon^2}\left(\frac{j^2}{R} + \frac{R_x^2}{4R} + F(R)\right),$$

where $F(R) = f(r)$. In these terms the equations become:

$$R_{xx} + 2(RF)' = 4\varepsilon^2 E, \tag{61}$$
$$J_x = R\left[\Omega(R) - \Omega(R_*)\right], \tag{62}$$

where the prime denotes differentiation with respect to R. Using the definition of E,

$$4\varepsilon^2 E = \left(2\varepsilon^2 J^2 + \tfrac{1}{2}R_x^2\right)/R + 2F \tag{63}$$

one also finds

$$E_x = J\left[\Omega(R) = \Omega(R_*)\right]. \tag{64}$$

Instead of using the third order system (61) and (62) with E defined by (63), one may imbed it in the fourth order system (61), (62) and (64). It is easily shown from these equations that this fourth order system has

$$4\varepsilon^2 ER - 2\varepsilon^2 J^2 - \tfrac{1}{2}R_x^2 - 2RF = \text{const.}$$

as invariant manifolds; we are actually interested only in the one for which the constant is zero. [For numerical purposes it is much better to use the fourth order system, being sure to start on the right invariant manifold. The reason for this is that the solutions of interest pass through or near $R = 0$ and the third order system then becomes very sensitive to numerical errors because of the R in the denominator in (63). No such difficulty occurs with the fourth order system, and the equations automatically keep one on the correct invariant manifold.] The two critical points which are to be approached as $x \to \pm \infty$ are then at:

$$R = R_*,$$

$$R_x = 0,$$

$$J = \pm J_* \equiv \pm R_* \sqrt{F'(R_*)/\varepsilon}\ ,$$

$$E = E_* \equiv \big(R_* F'(R_*) + F(R_*)\big)/2\varepsilon^2. \tag{65}$$

It turns out that the values of R_* for which backwards connecting trajectories exist approach 1 as $\varepsilon \to 0$ and differ from 1 by $O(\varepsilon^2)$. To incorporate this scaling we set

$$\Omega(R_*) = \Omega(1) + s\varepsilon^2 \tag{66}$$

replacing the parameter R_* by s. For small ε the critical points then have

$$J_* \cong \frac{1}{\varepsilon}\sqrt{F''(1)(R_* - 1)} \cong \sqrt{F''(1)s/\Omega'(1)}\ ,$$

which thus approaches a finite value as $\varepsilon \to 0$ (assuming s does, as our scaling implies); the same is true of E_*, which approaches $1/2J_*^2$. [Since $F(1) = F'(1) = 0$ and $F'' < 0$, a 'representative' function $F(R)$ is $-\frac{1}{2}(R - 1)^2$; similarly one may think of $\Omega(R) - \Omega(R_*)$ as something like $R_* - R$.]

4. We can now describe the 'singular solutions', that is, the case $\varepsilon = 0$. In this limit, (61) and (62) uncouple, and (61) can be solved by itself. It has the first integral

$$\tfrac{1}{2}R_x^2 + 2RF = \text{const.} = I, \quad \text{say.} \tag{67}$$

This looks like the energy equation of a nonlinear oscillator with potential energy function $2RF$. Recalling that F looks something like $-\frac{1}{2}(R - 1)^2$, we can sketch the potential energy curve, and from it immediately see the qualitative character of the phase portrait in the (R, R_x) plane. These are shown in Figure 19. Evidently the saddle point at $R = 1$ (which is the projection into this plane of both critical points of (61), (62) at $\varepsilon = 0$) is joined to itself by the homoclinic trajectory $\frac{1}{2}R_x^2 + 2RF = 0$.

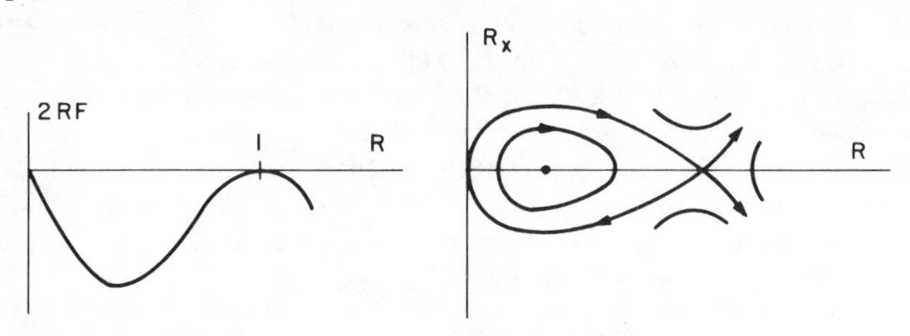

FIGURE 19.

Now consider (62) at $\varepsilon = 0$, namely

$$J_x = R[\Omega(R) - \Omega(1)].$$

As (R, R_x) goes around the homoclinic trajectory, J changes by an amount

$$\Delta J = \int_{-\infty}^{\infty} R[\Omega(R) - \Omega(1)]\, dx$$

(which converges since $R \to 1$ exponentially on both ends). Thus in order to get a trajectory joining the two critical points as we go around the homoclinic trajectory in the (R, R_x) plane, we must (and can) choose s so that if J starts at $-J_*$ it will just reach $+J_*$ as $x \to \infty$, i.e.,

$$\Delta J = 2J_* = 2\sqrt{\frac{F''(1)s}{\Omega'(1)}}\ .$$

From the symmetry we see that

$$\tfrac{1}{2}\Delta J = \int_{\infty}^{-\infty} R[\Omega(R) - \Omega(1)]\, dx$$

(if the origin of x is placed where $R = R_x = 0$) and using (67) with $I = 0$ we can change this x-integral to an R-integral, thus obtaining the value of s for the connecting trajectory at $\varepsilon = 0$:

$$\sqrt{\frac{F''(1)s_0}{\Omega'(1)}} = \int_0^1 \frac{R[\Omega(R) - \Omega(1)]\,dR}{2\sqrt{-RF(R)}}. \tag{68}$$

An attempt at a sketch of the connecting trajectory in the (R, R_x, J) space is shown in Figure 20.

It seems reasonable to expect that for small ε there will be a value of s close to s_0 which gives a connecting trajectory close to this singular one.

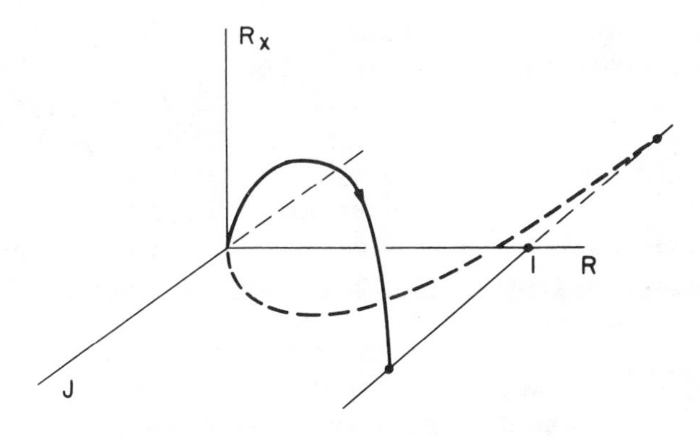

FIGURE 20.

X. Singular and actual center patterns. Finite regions.

1. It is of interest to contrast the singular center trajectory just described with the shock trajectory which goes from $+J_*$ to $-J_*$. In the $\varepsilon = 0$ picture the latter is indistinguishable from the line segment joining the two critical points. To describe it approximately by formulas, in terms of the variables we have been using, set $R = 1 + \varepsilon^2\rho$ and $x = [\varepsilon^{-2}F''(1)/\Omega'(1)]\xi$. Inserting these in equations (61)–(63), and retaining only the largest terms, gives

$$2F''(1)\rho = 2J^2, \tag{69}$$

$$J_\xi = F''(1)\left(\rho - \frac{s}{\Omega'(1)}\right), \tag{70}$$

which imply

$$J_\xi = J^2 - J_*^2. \tag{71}$$

The relevant solution of this equation, taking $\xi = 0$ where $J = 0$, is

$$J = -J_* \tanh \xi. \tag{72}$$

Since $F''(1)$ and $\Omega'(1)$ are both negative, ξ increases as x does and this solution indeed goes from $+J_*$ to $-J_*$ rather than the opposite way. We also find from this that

$$R = R_* - \frac{\varepsilon^2 s}{\Omega'(1)} \operatorname{sech}^2 \xi \tag{73}$$

so along the shock trajectory R is always (slightly) greater than R_*, while along the center trajectory, when it exists, it is less. Note that here we have no restriction on s in order to get the approximate connecting trajectory of the shock type.

Returning now to the 'singular center solution', suppose that we had chosen s so that $J_* = \Delta J$, instead of $\frac{1}{2}\Delta J$ as in (68). Then after (R, R_x)

had traversed the homoclinic trajectory, J would have advanced from $-J_*$ to 0, rather than to $+J_*$. What should be expected of a similar solution for small $\varepsilon > 0$? It seems plausible that for J_* near ΔJ the solution starting at the critical point at $-J_*$ (the unstable manifold there) would eventually reach $J = 0$, and do so near $R_x = 0$. If we choose the zero point of x to be the (first) point at which the unstable manifold crosses $R_x = 0$ from below, then the singular solution would have J approaching 0 as $x \to \infty$, but for small $\varepsilon > 0$ there is no longer a critical point on $J = 0$, so instead one would expect that J would reach 0 at a large but finite positive value of x. Now if it is really possible to adjust J_* near ΔJ so that this happens say at $x = x_1$, together with $R_x = 0$, then it is clear from the symmetry of the problem that this solution, continued on to $x \to +\infty$, would approach the critical point at $J = +J_*$; if the origin of x were shifted to x_1, R would be even and J odd. We may thus expect a second kind of backwards connection, occurring for a J_* near ΔJ rather than near $\frac{1}{2}\Delta J$: a sort of 'second mode'. The singular representation of this second mode consists of the homoclinic trajectory in the (R, R_x) plane traversed *twice*, the first time around carrying J from $-J_*$ to 0 and the second time from 0 to $+J_*$. An attempt at a sketch of such a trajectory for small ε is shown in Figure 21.

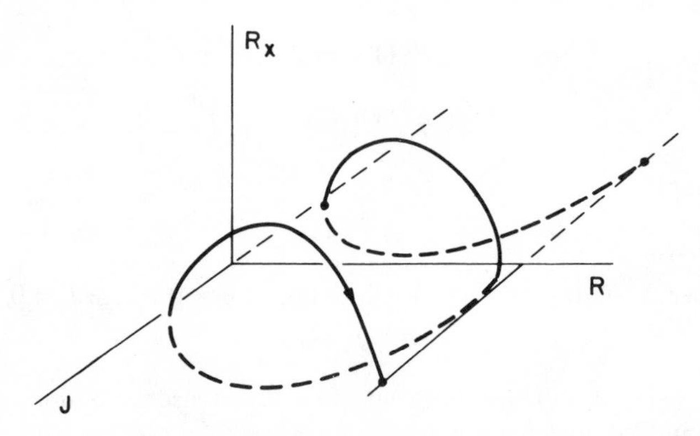

FIGURE 21.

It is scarcely necessary to stop here: if we can traverse the homoclinic orbit twice, we can do it 3, 4, 5, ... times. It is only necessary, in the singular solutions, to take $J_* = \frac{3}{2}\Delta J$, $2\Delta J$, $\frac{5}{2}\Delta J$, ..., in order to obtain candidates for approximations to third, fourth, fifth, ... modes, respectively. If this is correct for $\varepsilon > 0$, however, as one goes up in mode number the wave number of the wave trains at ∞ also increases and after some finite number of steps one must pass out of the range in which the wave trains are stable. In fact we first found an example of a center structure of this type by a numerical calculation, and in that

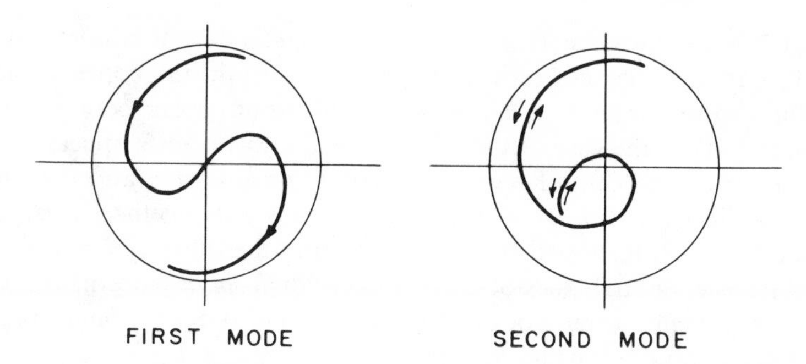

FIGURE 22.

particular case the mode we found (which was actually a second mode) was already in the unstable range. Since then, however, by making ε smaller, we have found examples in which several modes have wave trains at ∞ which are in the stable range.

In terms of the original concentration variables the structure of these center solutions is somewhat different in the case of the odd modes from what it is for the even ones. The variations of concentrations with x at a fixed time may be described by thinking of the 'motion' of the phase point in the (c_1, c_2) plane, x playing the role of the 'time' variable. For large negative x this point is going around a circle of radius $r_* = \sqrt{R_*}$ at a uniform rate in the sense corresponding to leftward propagating waves. As the center of symmetry, say $x = 0$ (the point at which J passes through 0), is approached, the phase point in the case of the *first* mode moves inward toward the origin, and passes through it at $x = 0$. At this moment the sense of rotation reverses, and the point then moves outward, approaching once again the circle $r = r_*$, but now traversed in the opposite sense, as $x \to \infty$. In the case of the *second* mode, the description starts as before (with a somewhat smaller circle), but then the phase point moves in toward the origin before x reaches zero, passing close to it (but not through it except at $\varepsilon = 0$) and then going out again to larger values, still preserving its original sense of rotation. Then at $x = 0$ it comes to a stop, reverses, and goes back over the same path, ending up going around the original circle in the opposite sense. Sketches of these paths in the (c_1, c_2) plane are given in Figure 22 for the first and second modes. The descriptions for the higher odd and even modes are similar, except that the 'turning around' process is somewhat more complex. Now these descriptions of the variation with x are the same at all times t, except that the pictures are rotated about the origin through the angle σt. It is evident from these pictures that in the case of the first mode (and also the higher odd modes) two points at equal distances on opposite sides of $x = 0$ have at all times values of θ differing by π, whereas for the even modes the values of θ are the same. (For the odd modes, θ is thus discontinuous at $x = 0$, but this does not reflect any discontinuity in the concentration variables being merely a consequence of using polar coordinates.) Consequently in the case of the odd modes the wave crests appear to be sent out from the central region alternately on one side and then the other, while with the even modes they come out together. It seems appropriate then to regard the first mode, and the other odd modes, as one-dimensional analogues of a two-dimensional *spiral* pattern, and the second and higher even modes as analogues of two-dimensional centers 'emitting' circular wave patterns.

2. Here is a sketch of the proof that for small enough $\varepsilon > 0$ there really is a first mode solution close to the singular solution given above. We start at the critical point at $R = R_*$, $J = -J_*$ for $x \to -\infty$ and follow its one-dimensional unstable manifold W_ε^u in (R, R_*, J) space (the branch of it for which R decreases); using either the three-dimensional system (61), (62) with E defined by (63) or else the four-dimensional system (61), (62), (64) with $E_* = (RF)'_*/2\varepsilon^2$. W_ε^u is close to W_0^u in some neighborhood of the critical point, as follows for instance from the iterative construction of these manifolds and the closeness (for small ε) of the critical points and unstable eigenvectors. It then remains close to W_0^u on any additional finite x interval by continuous dependence on parameters and initial conditions. We follow this manifold until it intersects $R_x = 0$, which it will do in some additional finite x interval since W_0^u does, transversely. Now it is clear from the four-dimensional system that E is then finite; hence we reach $R_x = 0$ with $R \geqslant 0$; otherwise R would have become 0 before, and E (cf. (63)) would have diverged. (E does not blow up as $\varepsilon \to 0$ either.) Let $J_1(s, \varepsilon)$ be the value of J when W_ε^u intersects $R_x = 0$ for the first time. Then

$$\lim_{\varepsilon \to 0} \left(J_1(s, \varepsilon) + J_*(s, \varepsilon) \right) = k,$$

where k is a constant independent of s, in fact the value given by the integral in (68). Thus

$$J_1(s, \varepsilon) = k - \sqrt{\frac{F''(1)s}{\Omega'(1)}} + o(1) \quad (\text{in } \varepsilon),$$

so for small ε there are values of s near s_0 which make $J_1(s, \varepsilon)$ positive and negative. We then show, using the variational equations, that $\partial(J_1 + J_*)/\partial s = o(1)$ in ε and conclude that there is a unique $s = s(\varepsilon)$ near s_0 which makes $J_1(s, \varepsilon) = 0$. Since $F(0) \neq 0$ and E is bounded it follows that for this value of s we also have $R = 0$, where $R_x = J = 0$. The trajectory can then be extended by symmetry and will go to the other critical point as $x \to +\infty$.

For the higher modes the argument is somewhat more complicated but along the same lines. For further details I must refer to a forthcoming paper by Kopell and myself. It should perhaps be mentioned however that we have not been able to show the existence of infinitely many modes for any positive ε. In order to make this kind of argument work, more and more stringent restrictions on the size of ε are required as the mode number increases and the best we could do was show that for any integer N there is an $\varepsilon_1(N)$ such that N modes exist for $\varepsilon < \varepsilon_1(N)$. But it may be that $\varepsilon_1 \to 0$ as $N \to \infty$ and that for any particular ε there is only a finite number of these center-structure solutions. Of course, as already mentioned, in any case at most finitely many could be stable.

3. For a particular λ–ω model, i.e. for a particular choice of Ω and F, such trajectories can be satisfactorily computed numerically as follows. Having selected a trial value of R_* or s, one calculates the unstable eigenvector at $(R_*, 0, -J_*)$ in (R, R_x, J) space. Then select an initial point on the line in this direction through the critical point, with an R coordinate slightly less than R_*, e.g., $0.98R_*$. Calculate the approximate initial value of E from equation (63), and then integrate (61), (62) and (64) forward until $R_x = 0$. For the first mode, one stops here; for the second, continue until $R_x = 0$ for the second time, and similarly for higher modes. At this point examine the value of J. This gives a function $J_1(R_*)$, and the desired solution is determined by adjusting R_* (for instance by the method of false position) so that J_1 is zero. The singular solution value s_0, given by (68) for the first mode, is naturally a good starting point. Although the initial point is not exactly on the unstable manifold, it is near it and the trajectory soon approaches it very closely since there is only one unstable eigenvalue at the critical point. Accuracy in this regard can be checked by repeating the calculation with a different initial point, such as one with $R = 0.99R_*$. Accuracy of the numerical integration can be checked by standard methods, such as reducing the step size. A rough check can be made by verifying that (63), or R times it, continues to be accurately satisfied. The numerical integration using the fourth order system is not difficult: we have made satisfactory preliminary calculations using a programmable calculator, although a computer is better for more extensive exploration, especially with the higher modes.

4. c_x vanishes at the center of an ideal (symmetrical) shock-structure solution. Thus one could imagine placed there a wall through which no chemicals pass. The same is true of any of the symmetrical (even mode) center structures just described in the λ–ω case. A symmetrical shock-structure solution can thus be interpreted as corresponding to a semi-infinite train of waves impinging on an impermeable wall, at which they appear to be absorbed. Since a center structure, centered at $x = 0$, say,

looks essentially like a wave train propagating to the right for large x, and a wave train can always be joined to a wall by a (localized) half-shock structure, it appears likely that there are (sometimes) periodic solutions of reaction-diffusion equations which look like a center structure for x near 0, but terminate in two half-shock structures at walls ($c_x = 0$) placed at $x = \pm L$, at least for large enough L. On the other hand, it has been shown by Othmer [**6**], and by Conway and Smoller [**30**] in a different way, that in a sufficiently small domain with $c_x = 0$ on the boundary (or in a fixed domain for sufficiently large diffusivities) *all* solutions of reaction-diffusion equations must tend to a spatially uniform one. It is thus of interest to try to verify the above conjecture that for a large enough but finite domain, periodic spatially heterogeneous solutions do sometimes exist. By a modification of the above construction of center structures for λ–ω models with small ε we have been able to do this. For values of R_* near but somewhat below the value corresponding to, say, the first mode center structure, there is a *periodic* solution of (61) and (62) (with E given by (63)) which passes through the R axis at a value of R near 1 going from positive J toward negative, follows a path similar to a shock-structure solution into the neighborhood of the critical point at $(R_*, 0, -J_*)$, then follows a path much like that of the center solution back through the R axis (at $R = 0$) and on to the neighborhood of the critical point at $(R_*, 0, J_*)$, from whence it returns to the crossing of the R axis near $R = 1$ where it started, again near half of the shock trajectory joining the two critical points. The period of this solution in x gives the length ($2L$) of a segment on which this solution may be regarded as representing a periodic heterogeneous structure which looks like a one-dimensional 'spiral' at the center sending out waves alternately in either direction. These waves seem to be 'absorbed' by impermeable walls at the ends. A similar periodic solution exists which can be interpreted as a structure in which waves are *symmetrically* emitted at the center, and propagate outward to impermeable walls at L. Since in this case we also have $c_x = 0$ at the center, half of this solution can be regarded as corresponding to a situation in which waves appear at a wall at $x = 0$ and propagate to the right toward another wall, where they disappear.

Our proof shows the existence of these periodic solutions for a certain range of values of R_* below the value (R_*^∞, say) corresponding to the ideal center structure (antisymmetrical or symmetrical as the case may be) with L fairly large and increasing toward ∞ as R_* increases toward R_*^∞. We have also computed a number of examples of these structures, and attempted to follow them down as R_* decreases from R_*^∞. [The efficient computation of these periodic solutions is somewhat more complicated than is the case for the centers in an infinite domain, and cannot be described here.] In the case of the structure which resembles the first mode (antisymmetrical) at the center, the situation seems to be that as R_* is reduced, the trajectory in (R, R_x, J) space decreases in size (and changes significantly in form when it starts to get fairly small) and finally shrinks down to the origin. The length $2L$ of the finite domain structure continually decreases as R_* does, but tends to a nonzero value as the trajectory shrinks down to the origin. This suggests that these structures come into existence as L is increased from the small values at which only spatially homogeneous solutions can persist by a *bifurcation from the unstable critical point* at $r = 0$. Indeed it is easy to see that such a bifurcation does take place. For this purpose it is more convenient to return to the form (47) of the λ–ω reaction-diffusion equations. The formal calculation goes as follows.

The linearization of (47) about $c = 0$ is

$$c_t = (\lambda(0) + i\omega(0))c + Kc_{xx} \tag{74}$$

for which we have boundary conditions $c_x = 0$ on $x = \pm L$. The eigenfunctions of the operator on the right of (74), with these boundary conditions, are multiples of $\cos m\pi(x + L)/2L$, with corresponding eigenvalues $\lambda(0) - Km^2\pi^2/4L^2 + i\omega(0)$ for $m = 0, 1, 2, \ldots$. In real form the equations are the real and imaginary parts of (74), which are linear combinations of (74) and its complex conjugate; thus the eigenvalues are actually those just cited, together with their complex conjugates. We thus see that there is always a conjugate pair $\lambda(0) \pm i\omega(0)$ in the right half-plane, corresponding to the instability of $c = 0$ to perturbations independent of x. If $2L < \pi\sqrt{K/\lambda(0)}$, all other eigenvalues are in the left half-plane, but as L increases through $L_1 \equiv \frac{1}{2}\pi\sqrt{K/\lambda(0)}$ we get a Hopf bifurcation with the eigenvalue pair corresponding to $m = 1$ crossing the imaginary axis. Calculation of the amplitude expansion shows that the corresponding periodic solution exists for $L > L_1$, and is thus 'supercritical'. However it will not (at least at first) be stable because the *other* pair of unstable eigenvalues for the critical point implies an unstable pair of Floquet exponents for the periodic solution. Because of this instability, results obtained by such bifurcation methods are not of particular physical interest, but it *is* just at this size $2L_1$ that spatially heterogeneous periodic solutions first become possible. [Remark. A similar bifurcation method can be applied with general kinetics and a scalar or near-scalar diffusivity, given an unstable spiral critical point.]

One might suppose that the *symmetrical* center structures we found for R_* near 1 by the singular perturbation method would similarly arise at small amplitude via a Hopf bifurcation at the unstable critical point corresponding to $m = 2$, i.e. at length $2L_2 \equiv 2\pi\sqrt{K/\lambda(0)}$. This however seems not to be the correct description. There is indeed such a

Hopf bifurcation, but it is not really different from the $m = 1$ case: the 'symmetrical center structure' periodic solution which arises for $L > L_2$ is really just two antisymmetrical ($m = 1$) ones placed back-to-back. (The same is true for the higher bifurcations with $m > 2$.) This kind of structure might be better described as a 'double antisymmetrical center structure'. The numerical calculations show that the symmetrical center structures found by singular perturbation for R_* near 1 arise by a *secondary bifurcation* from the double antisymmetrical structure as R_* increases from small values toward 1. At this bifurcation a pair of symmetrical structures split off from the double antisymmetrical one. One of these is the symmetrical structure described before with a 'center' in the middle and two 'half-shocks' on the ends. The other is a reinterpretation of the same trajectory in (R, R_x, J) space, which has a full shock in the middle and two 'half-centers' on the ends.

REFERENCES

1. A. A. Andronov and C. E. Chaiken, *Theory of oscillations*, 1st ed., English transl., Princeton Univ. Press, Princeton, N. J., 1949. Also: A. A. Andronov, A. A. Vitt and S. E. Khaiken, *Theory of oscillators*, 2nd ed., English transl., Pergamon Press, London, 1966.

2. S. Kauffman, *Dynamic models of the mitotic cycle*: *Evidence for a limit cycle oscillator*, Some Mathematical Questions in Biology. VIII, Lectures on Mathematics in the Life Sciences, vol. 9, Amer. Math. Soc., Providence, R. I., 1977, pp. 87–122.

3. I. Prigogine and R. Lefever, *Symmetry breaking instabilities in dissipative systems*. II, J. Chem. Phys. **48** (1968), 1695–1700.

4. E. Hopf, *Abzweigung einer periodischen Lösung von einer stationären Lösung eines Differentialsystems*, Berichte Math.-Phys. Kl. Sächs. Akad. Wiss. Leipzig **94** (1942), 3–32.

5. J. E. Marsden and M. McCracken, *The Hopf bifurcation and its applications*, Applied Math. Sci., vol. 19, Springer-Verlag, New York, 1976.

6. H. G. Othmer, *Current theories of pattern formation*, Some Mathematical Questions in Biology. VIII, Lectures on Mathematics in the Life Sciences, vol. 9, Amer. Math. Soc., Providence, R. I., 1977, pp. 57–85.

7. A. M. Turing, *The chemical basis of morphogenesis*, Philos. Trans. Roy. Soc. London Ser. B **237** (1952), 37–72.

8. J. D. Cole, *Perturbation methods in applied mathematics*, Ginn and Co., Boston, 1968.

9. R. FitzHugh, *Impulses and physiological states in theoretical models of nerve membrane*, Biophys. J. **1** (1961), 445–466. J. Nagumo et al., *An active pulse transmission line simulating nerve axon*, Proc. I.R.E. **50** (1962), 2061–2070.

10. R. Casten, H. Cohen and P. A. Lagerstrom, *Perturbation analysis of an approximation to the Hodgkin-Huxley theory*, Quart. Appl. Math. **32** (1975), 365–402.

11. G. A. Carpenter, *A geometric approach to singular perturbation problems with applications to nerve impulse equations*, J. Differential Equations **23** (1977). See also: C. Conley, *On travelling wave solutions of nonlinear diffusion equations*, Lecture Notes in Physics, vol. 38, Springer-Verlag, New York, 1975; S. P. Hastings, *On travelling wave solutions of the Hodgkin-Huxley equations*, Arch. Rational Mech. Anal. **60** (1976), 229–257.

12. L. Flatto and N. Levinson, *Periodic solutions of singularly perturbed systems*, J. Rational Mech. Anal. **4** (1955), 943–950.

13. W. Wasow, *Asymptotic expansions for ordinary differential equations*, J. Wiley, New York, 1965. Also: N. Fenichel, *Geometric singular perturbation theory for ordinary differential equations*, Preprint, Math. Dept., Univ. British Columbia, 1978.

14. R. J. Field and R. M. Noyes, *Oscillations in chemical systems*. IV. *Limit cycle behavior in a model of a real chemical reaction*, J. Chem. Phys. **60** (1974), 1877–1884.

15. S. P. Hastings and J. D. Murray, *The existence of oscillatory solutions in the Field-Noyes model for the Belousov-Zhabotinskii reaction*, SIAM J. Appl. Math. **28** (1975), 678–688.

16. R. J. Wolfe, *Temporal oscillations in a kinetic model of the Belousov-Zhabotinskii reaction* (to appear). Also, Thesis, SUNY at Buffalo, 1976.

17. A. N. Zaiken and A. M. Zhabotinskii, *Concentration wave propagation in two-dimensional liquid-phase self-oscillating system*, Nature **225** (1970), 535–537.

18. A. T. Winfree, *Spiral waves of chemical activity*, Science **175** (1972), 634–636. See also: Scientific American **230** (1974), 82–95.

19. L. N. Howard and N. Kopell, *Slowly varying waves and shock structures in reaction-diffusion equations*, Studies in Appl. Math. **56** (1977), 95–145.

20. N. Kopell and L. N. Howard, *Plane wave solutions to reaction-diffusion equations*, Studies in Appl. Math. **52** (1973), 291–328.

21. G. B. Whitham, *Linear and nonlinear waves*, Wiley-Interscience, New York, 1974.

22. R. Courant and D. Hilbert, *Methods of mathematical physics*, Interscience, New York, 1962.

23. J. A. Stanshine, *Asymptotic solutions of the Field-Noyes model for the Belousov reaction*. II. *Plane waves*, Studies in Appl. Math. **55** (1976), 327–349.

24. D. F. Tatterson and J. L. Hudson, *An experimental study of chemical wave propagation*, Chem. Engng. Comm. **1** (1973), 3–11.

25. L. R. Foy, *Steady state solutions of hyperbolic systems of conservation laws with viscosity terms*, Comm. Pure Appl. Math. **17** (1964), 177.

26. P. Ortoleva and J. Ross, *On a variety of wave phenomena in chemical reactions*, J. Chem. Phys. **60** (1974), 5090–5107.

27. J. Greenberg, *Axi-symmetric, time-periodic solutions to λ–ω systems*, SIAM J. Appl. Math. **34** (1978), 391–397.

28. A. T. Winfree, *Rotating solutions to reaction/diffusion equations in simply connected media*, SIAM-AMS Proceedings, vol. 8, Amer. Math. Soc., Providence, R. I., 1974.

29. J. M. Greenberg, C. Greene and S. P. Hastings, SIAM J. Appl. Math. (to appear). See also: J. M. Greenberg and S. P. Hastings, *Spatial patterns for discrete models of diffusion in excitable media*, SIAM J. Appl. Math. **34** (1978), 515–523; and J. M. Greenberg, B. D. Hassard and S. P. Hastings, *Pattern formation and periodic structures in systems modelled by reaction-diffusion equations*, Bull. Amer. Math. Soc. **84** (1978), 1296–1327.

30. E. Conway and J. Smoller, *A comparison theorem for systems of reaction-diffusion equations*, Comm. Partial Differential Equations **2** (1977), 679–697. See also: SIAM J. Appl. Math. **35** (1978), 1–16.

DEPARTMENT OF MATHEMATICS, MASSACHUSETTS INSTITUTE OF TECHNOLOGY, CAMBRIDGE, MASSACHUSETTS 02139

Lectures in Applied Mathematics
Volume 17, 1979

Studies of the Ear[1]

Charles R. Steele

ABSTRACT. From the many studies of perception and electrophysiology, the auditory system has a bewildering variety of nonlinearities in response to a physical stimulus. The question is, where do these nonlinearities originate? Helmholtz (1877) decided it was in the middle ear, which was later found not to be the case. In this presentation the gross mechanical response of the inner ear (the "first filter") is discussed in some detail, which seems to be free of any interesting nonlinearity in the physiological range. Some "micromechanics" of the organ of Corti are discussed, which may provide the elusive, highly nonlinear, "second filter" between the basilar membrane and the receptor cell stimulation.

1. Introduction–auditory periphery. The parts of the human outer, middle, and inner ear and their function are summarized in Figure 1. An attractive "pop science" introduction to this subject is in the book by Stevens and Warshofsky **[1965]**, while Yost and Nielsen **[1977]** provide an excellent introductory textbook. Virtually the same arrangement as shown in Figure 1 is found in all mammals, with the exception of the sea mammals who have lost the pinna and have a greatly modified middle ear. The function of the periphery is to transform the external mechanical motion due to air (or water) conducted sound or to motion of the head into the proper neural excitation, which is transmitted by the bundles of individual nerve fibers comprising the vestibular and cochlear nerves to the central nervous system. Nerves can apparently transmit direct frequency information in the range 20–300 Hz. For lower frequencies, the information is lost in the spontaneous, random firing of the nerve, while the nerve is unable to regain its electrochemical charge quickly enough to follow the higher frequencies. For this reason, the inner ear consists of two mechanical devices, the semi-circular canals and the cochlea.

AMS (MOS) subject classifications (1970). Primary 92A05, 76Z05.

[1]Work supported by a grant (NS12086) from the National Institute of Neurological and Communicative Disorder and Stroke.

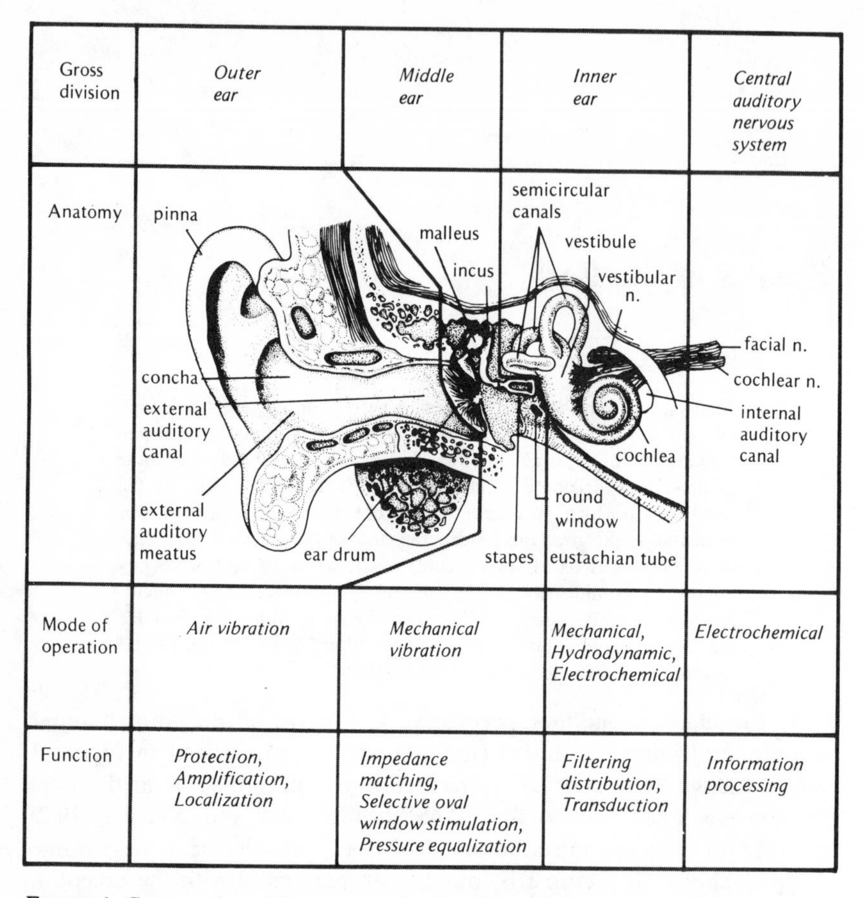

FIGURE 1. Cross section of human ear, showing divisions into outer, middle, and inner ears. Below are listed the predominant mode of operation of each division and its suggested function. (*Adapted from Ades and Engstrom*, 1974; *Dallos*, 1973) [2]

For the low frequencies of body motion, the fluid in the semi-circular canals will lag behind the motion of the bone walls. The relative displacement of fluid causes a displacement of cilia attached to receptor cells (hair cells), which causes the excitation of the nerve synapses attached to the cells. When the displacement of the cilia is in the opposite direction, instead of excitation, an inhibition of the normal random firing of the nerves is caused. Thus, at the receptor cell, there is a basic nonlinearity in the information transformation.

The cochlea is the device which provides the neural stimulation necessary for discrimination of the high frequencies. Much is known

[2] From *Fundamentals of Hearing*: *An Introduction* by William A. Yost and Donald W. Nielsen. Copyright © 1977 by Holt, Rinehart and Winston. Reprinted by permission of Holt, Rinehart and Winston.

about the cochlea, but the fundamental mechanism of neural excitation remains a mystery. Since the receptor cells are similar to those in the semi-circular canals, one might expect that the mechanical design of the cochlea has the purpose of obtaining the proper streaming of fluid. The coiled cochlea seen in Figure 1 is strictly a mammalian invention. Birds and the higher amphibians (and also the egg-laying mammals–spiny anteater and duckbills) have a slightly curved tube and can perceive frequencies only up to about 7 kHz. Primates go to 20 kHz, cats and guinea pigs to 50 kHz, while the Odontoceti (dolphins, sperm whales, etc.) use up to 200 kHz for echolocation. An interesting aspect of evolution is discussed by Sales and Pye [**1974**]. It seems that some insects "... found a place on wings of lace, to make an ear in haste ... ", in order to perceive the echolocation signals of bat predators and take appropriate evasive action. However, these insect "ears" are relatively simple resonators, which are sensitive to only a few frequencies. The mammalian cochlea is the most highly developed sound receptor, and it is surprisingly invariant across species. Take, for example, the size. In comparison with man, the cochlea in mouse in only one-fifth as long, while that in elephants and sperm whales is just twice as long. In detail, however, there is a substantial species dependence, much of which has been correlated by Fleischer [**1973**] to the frequency range of the animal.

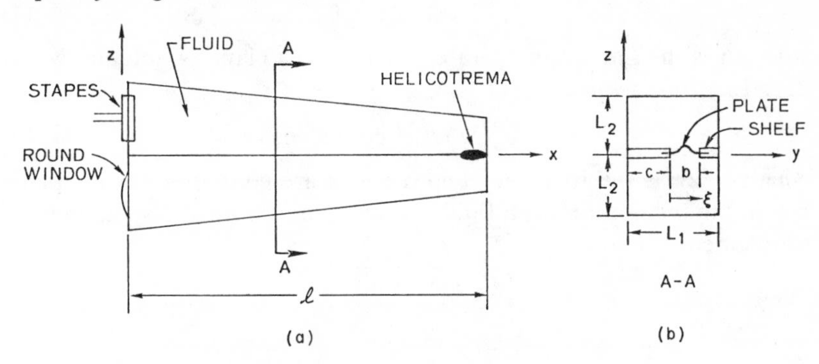

FIGURE 2. Cochlear model (a) side view, (b) cross-section.

Now we will open the cochlea and show by some reasonably simple calculations some main features of the mechanical design. In Figure 2 is shown the typical model of the cochlea used for mathematical analysis and for experiment. The distinctive coiling of the mammalian cochlea is the first feature to be disregarded in such a straight box model. What is of primary significance, at least for the gross response, are the two fluid ducts bounded by essentially rigid walls, except for the flexible plate (basilar membrane) in the partition shown in Figure 2(b).

For low frequencies of excitation, corresponding to the lowest tones on the piano, the fluid in the upper fluid duct, the scala vestibuli (SV), simply follows the motion of the stapes, which is proportional to the air presure exerted on the eardrum. The fluid flows through the orifice, the helicotrema, at the apex of the cochlea into the lower fluid duct, the scala tympani (ST), causing a displacement of the flexible membrane covering the round window. For the higher frequencies, however, the flexibility of the plate in the partition becomes significant.

2. Gross design of cochlea. The very significant feature of the cochlea is the gradient in stiffness of the basilar membrane (the plate in Figure 2(b)). The membrane is narrow and thick near the stapes, giving a high value of stiffness, and wide and thin near the helicotrema. Except for very specialized cochleas, such as in the bats who use pure tone echolocation signals, the stiffness is quite "slowly varying". Figure 2(a) does not show the true proportion; the length of the cochlea is generally about 30 times the cross-sectional dimension. Therefore, a perturbation method is useful in the analysis of the motion.

A. *Spatial decomposition of frequency*. The easiest demonstration of the mapping of frequency to "place" on the basilar membrane comes from consideration of the vibration modes in cylindrical coordinates of a plate on the $z = 0$ plane immersed in an infinite, incompressible, inviscid fluid. The fluid velocity potential φ satisfies Laplace's equation

$$\Delta\varphi + \varphi_{zz} = 0 \tag{1}$$

where Δ is the Laplacian operator for the (r,θ) plane. A solution which decreases for $z \to +\infty$ is

$$\varphi = J_n(\lambda r)e^{-\lambda z} \cos n\theta \cos \omega t \tag{2}$$

where $\omega/2\pi$ is the frequency and λ and n are constants. On the plane $z = 0$, the relation between fluid pressure p and normal displacement w is obtained

$$p = -\rho\varphi_t, \tag{3}$$

$$\dot{w} = \varphi_z, \tag{4}$$

$$p = -w\rho\omega^2/\lambda = -w\rho\omega^2 h_{\text{eq}}, \tag{5}$$

where ρ is the fluid density. From (5) one obtains the interpretation that, as far as the plate is concerned, the fluid appears as a mass with the equivalent thickness $h_{\text{eq}} = \lambda^{-1}$. For an elastic, homogeneous, isotropic plate of constant thickness, the equation is the biharmonic

$$D\Delta\Delta w = -2p \tag{6}$$

in which the pressure loading is doubled for fluid on both sides of the plate. The plate bending stiffness is

$$D = Eh^3/9 \tag{7}$$

in which E is the elastic modulus and h is the thickness. Using (2)–(5) shows that both (1) and (6) are satisfied if

$$D\lambda^5 = 2\rho\omega^2. \tag{8}$$

From (2) the deflection of the plate is seen to be zero on equally spaced rays from the origin. One sector between two such rays can be considered as the basilar membrane, for which the taper is typically so small that n is very large, i.e., $n \sim 300$. For large n the Bessel function in (2) is exponentially small for $r < r_T$ and sinusoidal for $r > r_T$ where r_T is the "transition" point

$$r_T = n/\lambda. \tag{9}$$

Thus, (8) gives the relation between the location of this transition point and frequency

$$Dn^5/r_T^5 = 2\rho\omega^2. \tag{10}$$

For high frequency ($\omega \to \infty$), the transition point is in the narrow, stiffer region of the plate sector ($r_T \to 0$), while for low frequency the transition point is in the wider, more flexible region of the plate. Experiments and more details of the analysis are given by Cole and Chadwick [**1977**].

For large n the Bessel function (2) can be approximated by the Airy function, as is well known. When the plate-fluid system is subjected to a driving force, the approximate equation governing the behavior near the transition point is

$$-\eta''(\zeta) + \zeta\eta(\zeta) = 1 \tag{11}$$

where the nonhomogeneous term arises from the driving force. The unique solution which is bounded and satisfies the radiation condition at $\zeta \to +\infty$ is algebraic on one side of the transition point and oscillatory on the other

$$\eta(\zeta) \to \begin{cases} \zeta^{-1} & \text{for } \zeta \to +\infty, \\ \pi^{1/2}|\zeta|^{-1/4} \exp -i\left(\dfrac{2}{3}|\zeta|^{3/2} + \dfrac{\pi}{4}\right) & \text{for } \zeta \to -\infty. \end{cases} \tag{12}$$

Thus, the plate strip will vibrate "quasi-statically", in phase with the driving force in the stiffer side of the transition point. On the more flexible side, traveling waves are generated which propagate away from the transition point. This is the behavior observed in the cochlea and in experimental models of the cochlea by Békésy [**1960**].

Light damping has little effect on the solution for $\zeta > 0$, but causes a substantial decay of the traveling waves for $\zeta < 0$. Thus, the transition point is near the maximum of the plate amplitude. As shown in Steele [**1976**], the value of D can be obtained from the static, point load tests on the actual basilar membrane by Békésy [**1960**]. The estimate equation

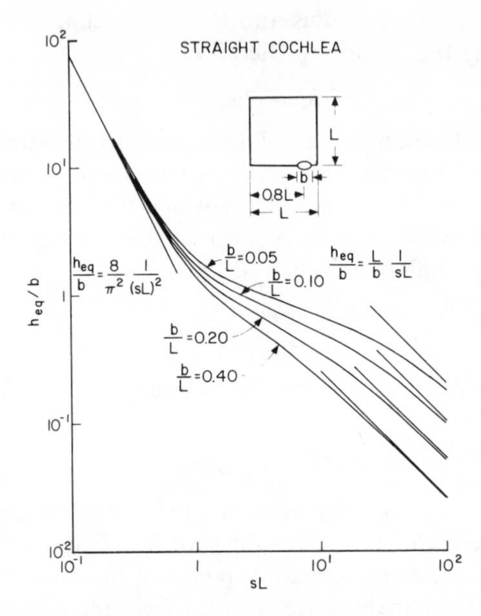

FIGURE 3. Impedance of incompressible, inviscid fluid in duct (Scala Vestibuli) against prescribed wavelength of displacement of plate (Basilar Membrane). (s = wave number)

(10) does come close to the relation between locus of maximum response and frequency observed in the cochlea by Békésy.

It is interesting that the more realistic model in Figure 2 with rigid walls on the fluid ducts can be handled more easily than the plate strip in an infinite fluid. The confinement of the fluid produces traveling waves along the entire cochlea. A "transition point" is not so easily defined; instead, the wavelength is "long" in the stiffer region near the stapes and rapidly becomes "short" in the region of maximum deflection of the basilar membrane. However, the entire response can be calculated from a "WKB" approach, since the longer wavelengths are still sufficiently short in comparison with the distance over which the basilar membrane stiffness varies significantly.

The most efficient first approximation is obtained by an adaptation of the method of Whitham [**1974**] used in Steele and Taber [**1978** a,b]. The displacement of the basilar membrane (the plate in Figure 2(b)) is assumed in the form of a traveling wave in the x-direction and a half sine wave (for hinged edges) in the y-direction

$$w(x, y, t) \sim W(x)e^{i\theta} \sin \pi(y - c)/b \tag{13}$$

where the phase is

$$\theta = \omega t - \int_0^x \lambda(x; \omega)\, dx \tag{14}$$

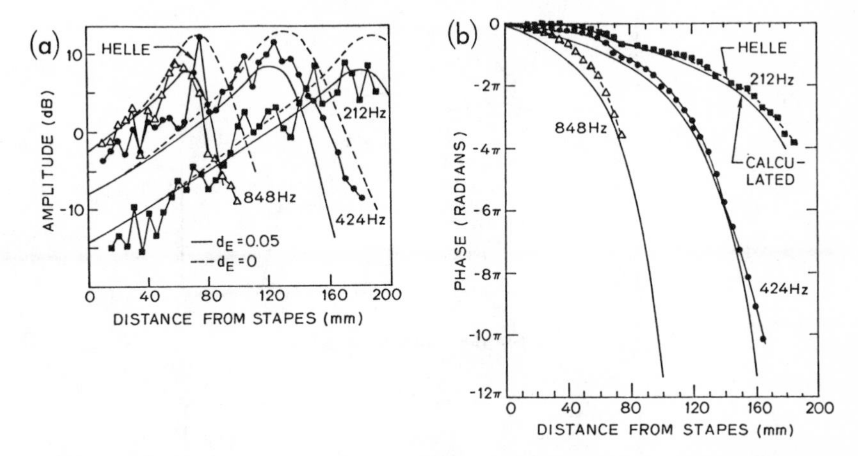

FIGURE 4. Amplitude and phase, relative to stapes, of plate vibration in Model K measured by Helle [1974] compared to calculations. Plate viscosity parameter d_E has little effect on phase.

in which λ is the local wave number. It is assumed that λ is sufficiently slowly varying that the fluid motion in a particular region can be computed with λ constant. The result from this integration over the fluid cross section, a straightforward boundary value problem which can be handled with a Fourier series expansion in the y-direction, is the fluid impedance shown in Figure 3.

For short wavelength the equivalent thickness of fluid mass is similar to that of the plate strip in an infinite fluid (5). For long wavelength, the velocity of the fluid is uniform in the cross section, i.e. the fluid flow is essentially one-dimensional.

The time-averaged Lagrangian density is

$$\mathfrak{L} = W^2(x)F(\lambda(x, \omega), x, \omega) \tag{15}$$

where

$$F = 2\omega^2 \rho h_{\text{eq}} b - \pi^4 D b^{-3}\left[1 + (\lambda b/\pi)^2\right]^2. \tag{16}$$

After the time average, Hamilton's principle is reduced to

$$\delta \int_0^L \mathfrak{L}\, dx = 0. \tag{17}$$

Since W and θ are the unknown functions, the Euler equations of the variational problem are

$$\frac{\partial \mathfrak{L}}{\partial W} = 0, \tag{18}$$

$$\frac{d}{dx}\frac{\partial \mathfrak{L}}{\partial \lambda} = 0, \tag{19}$$

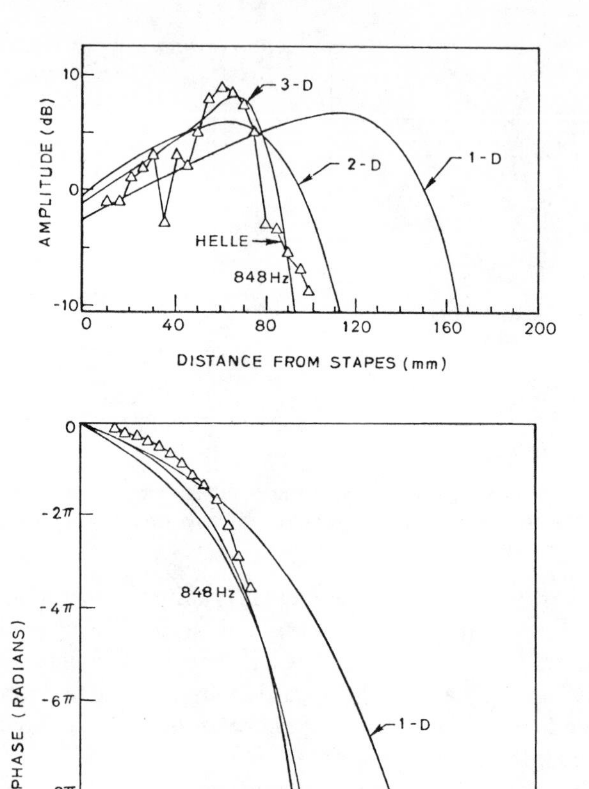

FIGURE 5. Comparison of 1- and 2-dimensional fluid approximation with full 3-dimensional calculation for Model K at 848 Hz.

the first of which gives the eikonal equation

$$F = 0 \tag{20}$$

and the second of which is the transport condition which gives the amplitude function

$$W(x) = \left(\frac{\partial F}{\partial \lambda}\right)^{-1/2} \times \text{const.} \tag{21}$$

For a given frequency ω, the solution of (20) for λ at each of a set of values of x is obtained numerically. From this, the amplitude function equation (21) is easily evaluated and a numerical integration yields the phase, equation (14). The damping due to fluid viscosity and plate viscoelasticity causes only a slight modification of the formulation, but yields a small imaginary part of λ, equation (14), which causes a strong damping of the waves, equation (13).

Comparison with a finite difference solution for 2-dimensional fluid motion by Neely [**1978**], i.e. a tall narrow cross-section, shows that this first approximation solution is reasonably accurate and reduces computer time by a factor of 100 or so (Steele and Taber [**1978**a]). Computer time for a finite difference calculation of the 3-dimensional fluid motion in a square cross section would probably be exorbitant. The use of equations (20) and (21) requires only about one second of CPU time (on the IBM 370/168 computer) for the calculation of displacement envelope and phase function ($\int \lambda dx$) for a given frequency (Steele and Taber [**1978**b]). A comparison of the experimental model results of Helle [**1974**] and calculations is shown in Figure 4. All the material and geometric parameters of the model were measured by Helle and were used for the calculation. No free parameters were available for adjustment. Thus, the agreement shown in Figure 4 is strong evidence that equations (20) and (21) do represent the physical reality with a satisfactory degree of accuracy.

Figure 5 shows that accuracy is lost by a 2-dimensional approximation of the fluid and that the 1-dimensional approximation is seriously in error. The vast majority of attempts to analyze the wave motion in the cochlea have made use of the 1-dimensional fluid approximation (Dallos [**1973**]). With such an improper assumption, correlation with observation of the dynamic response in the cochlea can be obtained only with gross distortion of the parameters. Since the values of stiffness, etc., in the cochlea are not known with much precision, investigators have felt free to make such distortions. Now, with the capability for economical and accurate 3-dimensional calculations, a stringent attempt must be made to use only realistic parameters.

Figure 6 shows such calculations for the box model of Figure 2, but with our best available estimates of basilar membrane stiffness and mass. The flexibility of the bony shelf is also included, the stiffness of which was calculated from the geometry of the shelf shown by Fleischer [**1973**]. No viscosity of the basilar membrane is included, but from Figure 4, one concludes that this will not change the slopes. It is interesting that decreasing the basilar membrane and shelf stiffness in the x-direction to zero ($k_x = g_x = 0$) does steepen the high frequency drop-off (Figure 6, left). However, Békésy [**1960**] reports a deformation pattern that would indicate $k_x > 0.5$. The conclusion, at this time, is

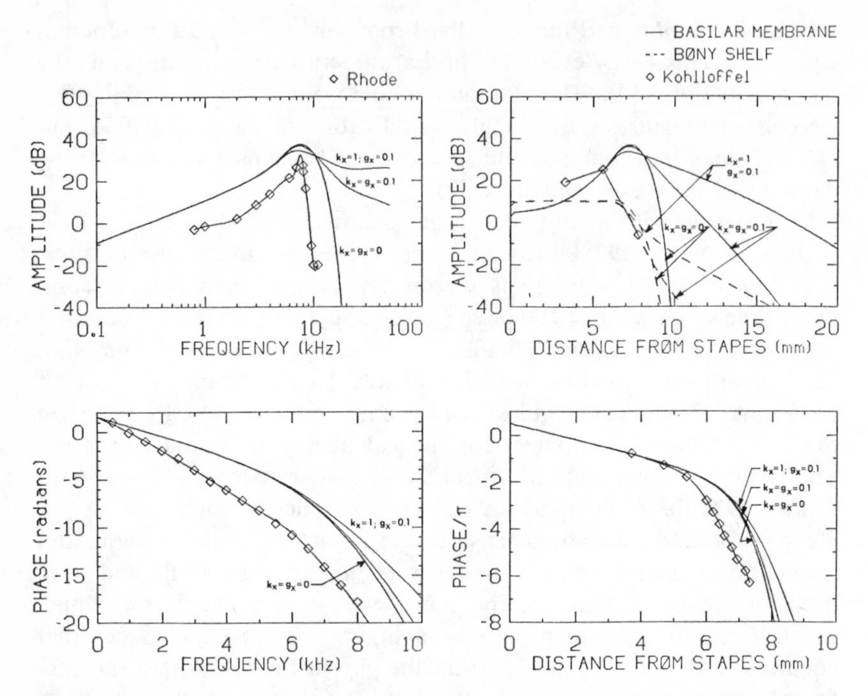

FIGURE 6. Calculations for (3-dimensional) model of guinea pig cochlea including flexibility of bony shelf (right) compared to *post mortem* measurements of Kohllöffel **[1972]** for a fixed frequency of 10 kHz and (left) compared to *in vivo* measurements of Rhode **[1973]** at a fixed point in the squirrel monkey. For model, best estimates for stiffness of shelf and basilar membrane are used. Curves show the deflection at the center of the pectinate zone for a basilar membrane which is isotropic ($k_x = 1$), with a slight stiffness in the axial direction ($k_x = 0.1$), and for membrane and shelf without axial stiffness ($k_x = g_x = 0$).

that the box model Figure 2 does represent the traveling wave phenomenon in the actual cochlea fairly well, but that some essential features of the real cochlea are omitted. Results very similar to Rhode's for the *in vivo* basilar membrane response have been obtained independently by other investigators using the same technique and by Wilson **[1974]** and co-workers using a capacitive probe. So, there is no reason to doubt the sharpness of the experimental results shown in Figure 6

B. *Transient time*. The solution for the response to a pure tone given by equation (13) can be used through a Fourier transform to obtain the transient response in the form

$$w(x, y, t) \sim \int_{-\infty}^{\infty} g(\omega) W(x, \omega) e^{i\theta} \, d\omega \sin \pi(y - c)/b \qquad (22)$$

in which $g(\omega)$ depends on the form of the excitation of the stapes. Since the phase θ is generally "rapidly varying", the main contribution to the integral (for fixed x and t) comes from the point of stationary phase, at which

$$\frac{d\theta}{d\omega} = 0. \tag{23}$$

From (14) this gives the relation

$$t = t(x, \omega) = \int_0^x \frac{\partial}{\partial \omega} \lambda(x, \omega)\, dx. \tag{24}$$

For another interpretation, this equation provides for a fixed frequency the time of travel t from the stapes ($x = 0$) to the station x of the energy in the input signal associated with the frequency $\omega/2\pi$. The local group velocity is $(\partial\lambda/\partial\omega)^{-1}$. This approach has been utilized by Dotson **[1974]** to obtain the response of a 2-D model to a variety of input signals. Of interest is that the one parameter family of curves given by $t(x, \omega)$ has two envelopes. For an impulse of pressure on the stapes, which in terms of psychoacoustics is a short "click", the function g in (22) is constant, and all the significant response occurs in the region of the (x, t) plane between the two envelopes. Thus, the one envelope gives the arrival of the disturbance front, which reaches the helicotrema in about 4 msec while the second envelope gives the disturbance tail which reaches the helicotrema in about 12 msec.

Thus, the box model does resolve the paradox of the resonance theory, most elegantly put forth by Helmholtz **[1877]**, in which the damping of the tuned resonators must be small to provide quick response and sharp frequency discrimination, but must be large to prevent prolonged ringing after the cessation of excitation. For the tapered membrane in the box of fluid, the transients at the onset and end of, say, a pure tone pulse both propagate to the helicotrema and are dissipated in the same time, around 20 msec.

C. *Unresolved features*. Since the coiling is such a distinctive feature of the mammalian cochlea, one would expect it to have a significant effect on the gross response of the basilar membrane. The solution for the fluid impedance (Figure 3) for a toroidal box has been worked out. A distortion of the long wavelength limit does occur. This distortion is most significant for the upper turns of the cochlea where the ratio of radii to the inner and outer wall of the fluid duct is largest. For the first turns, particularly for toothed whales, where the very sensitive high frequencies are localized, this ratio is so close to unity that the distortion of Figure 3 is negligible. So, the reason for the coiling remains unresolved.

The cross-sectional area of the fluid ducts varies in a systematic way, which, according to Fleischer (personal communication) from his comparative study of virtually all mammals, must be deliberate. Calculations with the box model of Figure 2 with various tapers of the cross

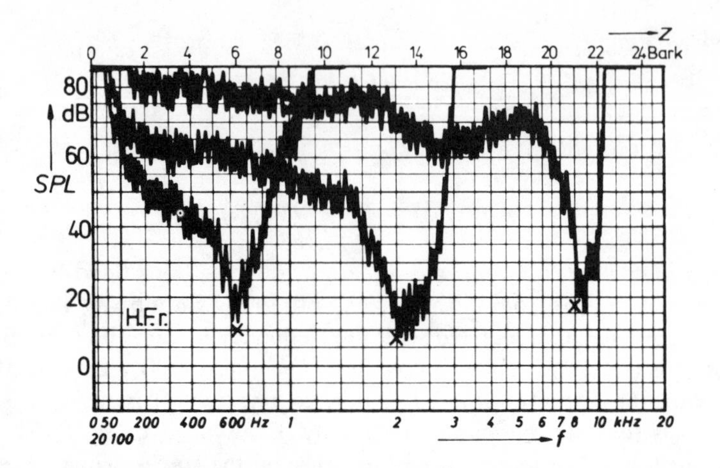

FIGURE 7. Original recordings of psychoacoustical tuning curves (SL = 5dB) for characteristic frequencies of 630 Hz, 2 kHz and 8 kHz. abscissa: frequency f (lower scale) and critical band rate (upper scale), respectively. (From Zwicker [1974].)

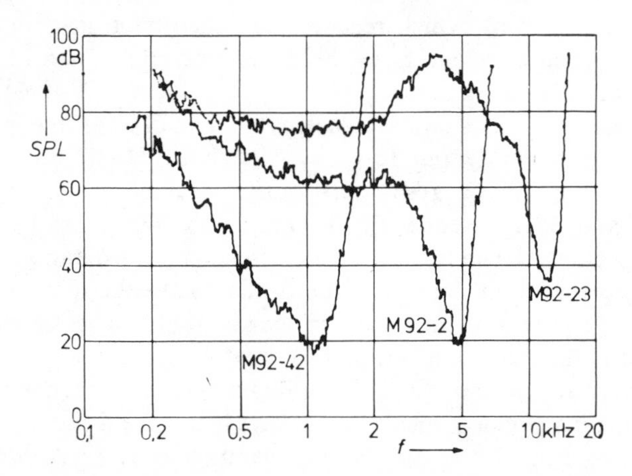

FIGURE 8. Computer processed tuning curves from single auditory nerve fibres (cat) of different characteristic frequencies (Kiang and Moxon [1974].) (From Zwicker [1974].)

section have, however, so far failed to show any interesting effect.

So far, the reason for the two fluid ducts and the gradient in stiffness of the basilar membrane, with the stiffer end at the stapes and the more flexible end near the orifice seems very clear from the analysis of the pure tone and transient response of the straight box model (Figure 2). Figure 6 shows that much more is to be considered in the actual cochlea.

The major mystery concerns the contrast in basilar membrane response and the output of the cochlea. Figure 7 shows the "psychoacoustic tuning curves" for a human subject from the work of Zwicker [**1974**]. The threshold curves for single auditory nerve fibers from the work of Kiang and Moxon [**1974**] in Figure 8 are remarkably similar. There is no anatomical evidence for neural processing in the cochlea, so the tuning curves in Figure 8 must be due to nonneural activity in the cochlea. The agreement with the psychological tuning curves indicates that the frequency discrimination does take place in the cochlea at the periphery. The higher levels of the auditory neural pathway have more subtle functions. If the neural output were proportional to basilar membrane displacement, then the tuning curves of Figure 8 would be the reciprocal of the basilar membrane response in Figure 6 (left). Instead, the tuning curves are an order of magnitude sharper in the usual range of intensity 20–70 db SPL.

3. Organ of Corti. The analysis of the mechanical behavior of the organ of Corti, which contains the receptor cells and is attached to the basilar membrane, is in a primitive stage. The necessity for such an analysis can at least be indicated.

A. *Nonlinearities in neural, electrical, and psychoacoustic response*. A wealth of recent literature shows a remarkable progress in the variety and quality of measurements of the response of single fibers in the auditory nerve and the activity of the electric field in the fluid and cells of the cochlea. In addition, a large number of workers are involved in psychoacoustic studies relating the physical sound stimuli to perception both in human subjects and in animals through behavioral studies. In all of these measurements there occurs an overwhelming variety of nonlinear effects which I will not attempt to itemize, much less explain. On the other hand, the basilar membrane response (Figure 6) seems remarkably linear. Wilson [**1974**] and others have found no evidence of nonlinearity in the physiological range of intensity in the guinea pig. Rhode [**1973**] did find nonlinear behavior beginning at around 70 db SPL in the squirrel monkey, which caused a flurry of excitement, but in later tests on the guinea pig also found only linear response. The type of phase and amplitude envelopes are certainly independent of nonlinearities. Helle [**1974**] repeated the measurements of his model at several intensity levels and duplicated every wiggle of the curves in Figure 4. Thus, one concludes that these wiggles are merely due to construction irregularities.

In Figure 9 is an interesting psychoacoustic result from Zwicker [**1977**]. The subject adjusts the intensity of a test tone SL_T^* of frequency $f_T = 4$kHz and duration $T_T = 15$ msec to hold the perception level of the test tone constant. Meanwhile a very slow, periodic, and continuous fluctuation of air pressure $p(t)$ takes place. The time within the period of $p(t)$ at which the test tone occurs is given by Δt. The curves for SL_T^*

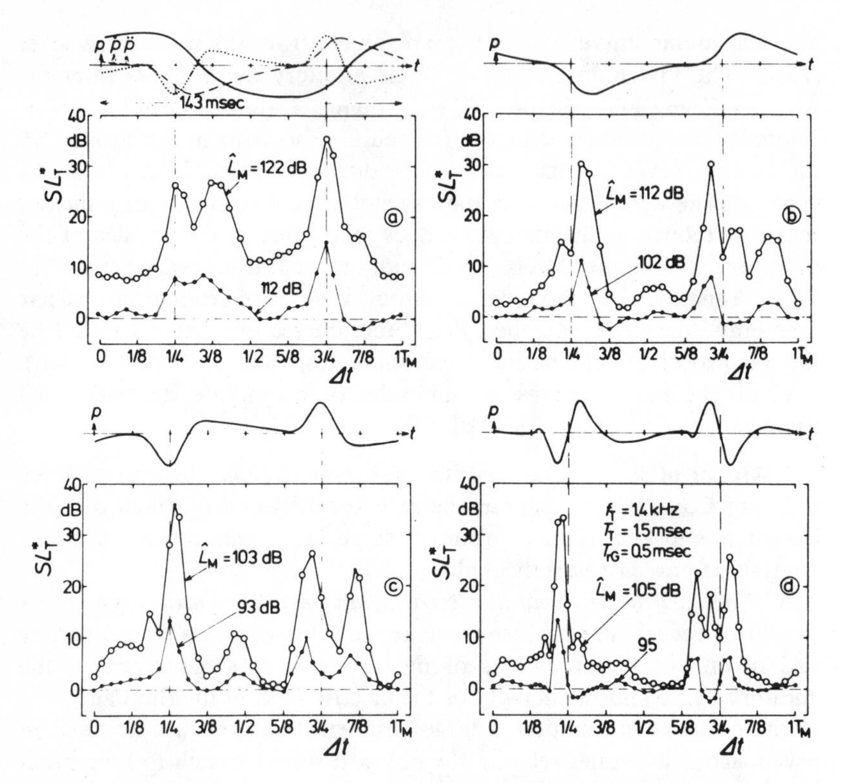

FIGURE 9. Masking-period patterns produced by maskers, the sound pressure time function $p(t)$ of which–in (a) together with its first derivative $\dot{p}(t)$ (dashed) and second derivative $\ddot{p}$ (dotted)–is indicated on top. The peak value of the masker's time function, expressed in level $\hat{L}_M$ is parameter. The sensation level SL_T^* of the test tone burst is shown as a function of its temporal spacing Δt within the masker's period T_M. Test tone frequency f_T is indicated together with test tone duration T_T and Gaussian rise time T_{rG}.[3]

show that it is most difficult to hear the test tone when $\ddot{p}(t)$ has maximum values, whether positive or negative. As previously discussed, for the very low frequencies, $\ddot{p}$ is proportional to the pressure difference in the scali

$$\ddot{p}(t) \sim C(p_{\mathrm{SV}} - p_{\mathrm{ST}}) \tag{25}$$

where C is a positive constant. From additional studies Zwicker postulates that a quasi-static, downward (toward ST) displacement of the basilar membrane causes a suppression of neural activity while an upward (toward SV) displacement causes an excitation, both of which make it more difficult to distinguish the test tone. The displacement

[3] Reprinted with permission from E. Zwicker, *Masking period patterns and hearing theories,* Psychophysics and Physiology of Hearing, E. F. Evans and J. P. Wilson, eds., Academic Press, London, 1977.

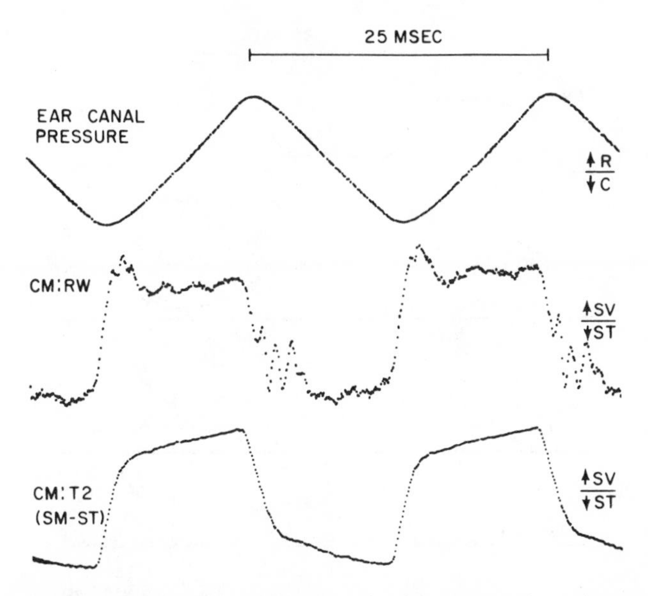

FIGURE 10. Upper record shows the 40 Hz triangular time pattern of sound pressure in the ear canal. Middle record shows the resulting time pattern of potential recorded near round window. Lower record shows the resulting time pattern of cochlear microphonics recorded differentially (SM–ST) in the second turn. R, C, SV, ST: rarefaction, condensation, scala vestibuli, and scala tympani, respectively. (From Sokolich et al. [**1976**].)

toward ST has a somewhat larger effect. According to Zwicker [**1977**], frequencies between 50 Hz and 1 kHz cause masking proportional to the first derivative $\dot{p}(t)$. In this range, the traveling wave develops so that the pressure in scala vestibuli is proportional to $\dot{p}(t)$.

This is evidently the situation in Figure 10 from Sokolich, et al. [**1976**]. The oscillatory electrical potential (cochlear microphonic (CM)) recorded from the round window (RW) and in the second turn of the cochlea is generally considered to be proportional to basilar membrane deflection for low frequencies. However, no simple behavior of individual nerve fibers is indicated by Figure 11. The spontaneous activity of one fiber (U 23-3, bottom trace), which has the characteristic (i.e., most sensitive) frequency CF of 6.1 kHz, suffers an inhibition of the spontaneous activity when the basilar membrane is displaced in either direction and an excitation when the acceleration is in either direction (next-to-last line). The remaining two fibers with lower CF and presumably originating from points further from the stapes have quite different responses.

An entirely different single fiber response can be found in an animal which has been treated with the ototoxic drug kanamycin, as shown in Figure 12. A strong nonlinearity is evident, since the pattern of excitation and inhibition changes with intensity level.

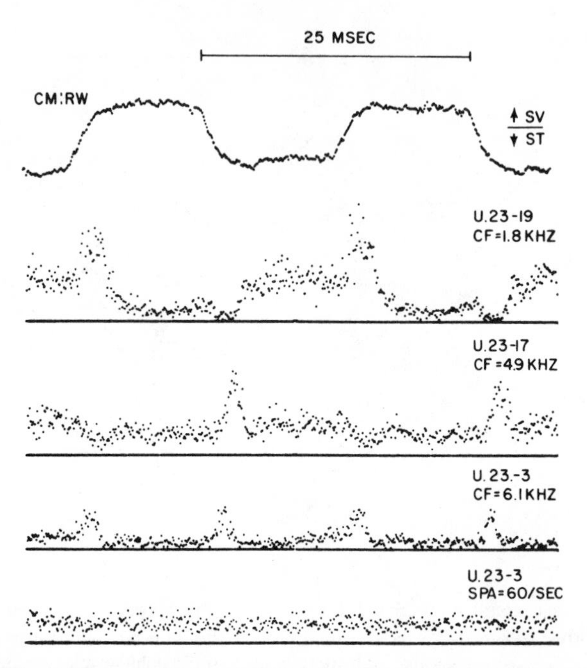

FIGURE 11. Single-fiber responses produced by trapezoidal wave pattern: untreated gerbils. Upper record shows averaged round-window microphonic with suppressed whole-nerve responses and basilar-membrane deflections toward scala vestibuli (SV) and scala tympani (ST). Middle records are PST histograms (0.1-msec bins; 2000 sweeps) illustrating the three typical response patterns associated with low, medium, and high CF fibers in untreated gerbils. SPL = 85 dB. (From Sokolich et al. [**1976**].)

The existence of such nonlinearities has motivated a search for the source. The obvious mechanical nonlinearities in the box model (Figure 2) have been investigated by Hallauer [**1974**] and others, with the conclusion that they become significant only at levels of intensity exceeding the threshold of pain. From Figure 13 one concludes that the basilar membrane support is designed to eliminate the cubic nonlinearity due to stretching.

Figure 14 indicates the structure of the organ of Corti. When the tectorial membrane is peeled back and the reticular lamina viewed from above, the arrangement of hair cell cilia shown in Figure 15 can be seen. The cilia of the inner hair cells (IHC) are not attached to the tectorial membrane. Furthermore, each IHC has some 20 afferent nerve synapses. The recordings of Figures 8, 10–12 are presumably from these nerves since the number of nerves passing through the arches and making synaptic contact with the outer hair cells is relatively small. So it seems that the IHC are the primary receptors. Russell and Sellick [**1977**] have recently accomplished the formidable task of making intracellular recordings from inner hair cells and have found a DC potential change,

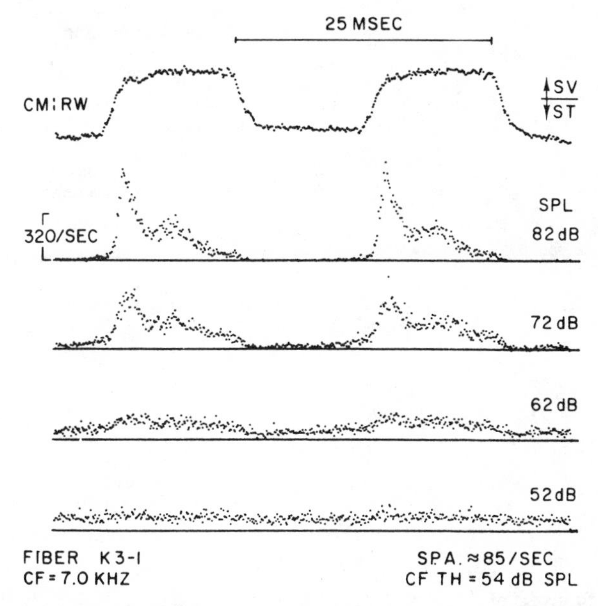

FIGURE 12. Single fiber response from gerbil treated with kanamycin. Fiber is presumably from region of normal inner hair cells but absent outer hair cells. (From Sokolich et al. [1976].)

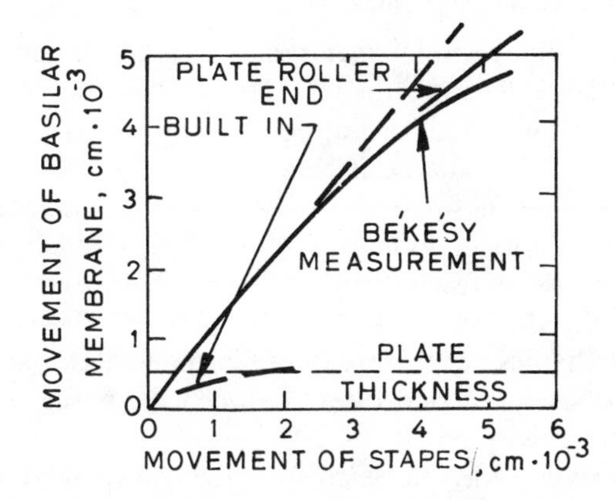

FIGURE 13. Observation by Békésy [1960] of nonlinearity of basilar membrane displacement at a point 30 mm from stapes in human cochlea. For beam (or plate) with "built-in" ends, i.e. constrained from tangential displacement, stretching nonlinearity is significant when the displacement is the order of magnitude of the thickness. For "roller" supported ends, this occurs when the displacement approaches the order of magnitude of the span. From this comparison, it seems that the spiral ligament is designed to permit tangential displacement.

called the intracellular summating potential (SP), which is as sharply tuned as the auditory nerve fibers in Figure 8. Since the IHC cilia are

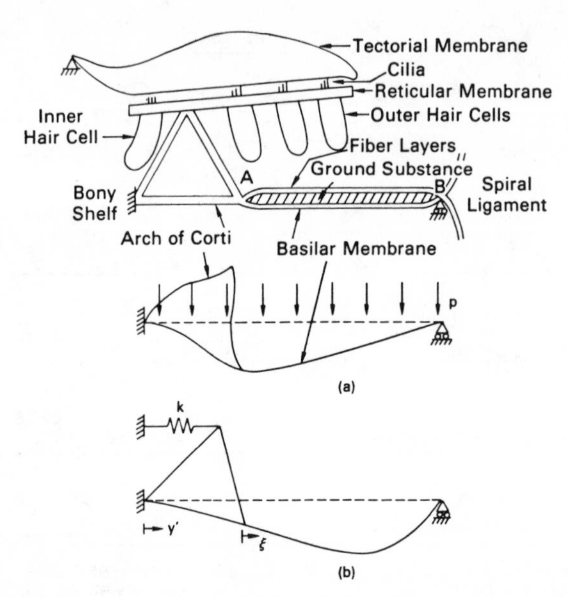

FIGURE 14. Structure of organ of Corti, which contains the receptor cells. In the pectinate zone of the basilar membrane the fibers form a "sandwich" which greatly increases bending stiffness. When the outer hair cells are missing, the significant distortion under static loading is in the arches. Normal attachment of OHC cilia provides effective spring k, and main deflection is in the pectinate zone.

not attached to the tectorial membrane, it must be the motion of the fluid which exerts pressure on the IHC cilia causing the intercellular potential change, which according to Russell and Sellick [**1977**], is large enough to cause the neural excitation.

So the question is, how can the exceedingly small amplitude of basilar membrane oscillatory motion (the "first filter") which is quite linear, induce a response of the organ of Corti which causes a strong DC streaming of the fluid through the inner hair cell cilia? In my opinion, the "micromechanics" of the organ of Corti and tectorial membrane do provide the missing "second filter" between the basilar membrane and the neural excitation.

B. *Experimental models for micromechanics*. Several large scale models have been constructed in the Institute of E. Zwicker at the Technische Hochschule München, in which the organ of Corti and tectorial membrane are simulated. Helle [**1974**] molded a rectangular shaped "organ" with his basilar membrane (the plate in Figure 2(b)) and attached a flap of softer material to the shelf for the "tectorial mebrane". No attachment between the "tectorial membrane" and the "organ" was made, so his model might be similar to the actual cochlea with the OHC removed from ototoxic drugs or noise damage. Results show that for pure tone excitation the nonlinear flapping does induce a DC streaming of the

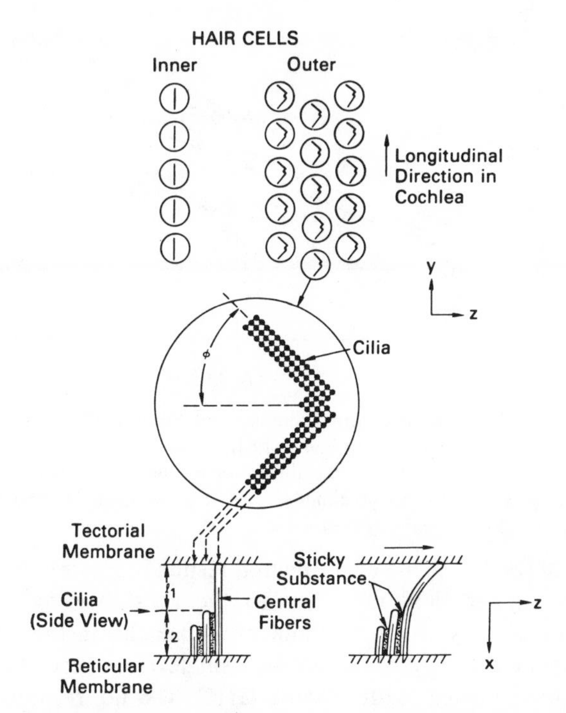

FIGURE 15. Form of IHC and OHC cilia. From Flock [1977] individual cilia are attached forming a cohesive bundle. Only the tallest row of OHC cilia are attached to the tectorial membrane.

fluid in the inner spiral sulcus, the fluid duct to the left of the inner hair cell shown in Figure 14, and a strong outflow through the layer between the tectorial membrane and recticular lamina which is more sharply concentrated than the basilar membrane response. The approach seems very promising. The satisfactory mathematical analysis for Helle's model is, however, yet to be worked out.

For the normal cochlea, the tectorial membrane is attached to the OHC cilia, so the flapping of the membrane as in Helle's model would not occur. The asymmetry of shape of the OHC cilia is intriguing. In order to determine whether or not this might have some interesting effect on the fluid flow, models ($\sim 400\times$) were constructed at Stanford and the effective permeability measured. The results are shown in Figure 16. As shown in Frommer and Steele [1978], the experimental results for linear behavior can be confirmed by theoretical considerations. As pointed out by Acrivos (personal communication), an increase in permeability with an increase in flow rate can be easily proven to be impossible, if the flow velocity distribution at the inlet and outlet of the test section is constant. The flow visualization in Figure 17 shows that the outlet flow velocity may not be constant, since long "tails" form

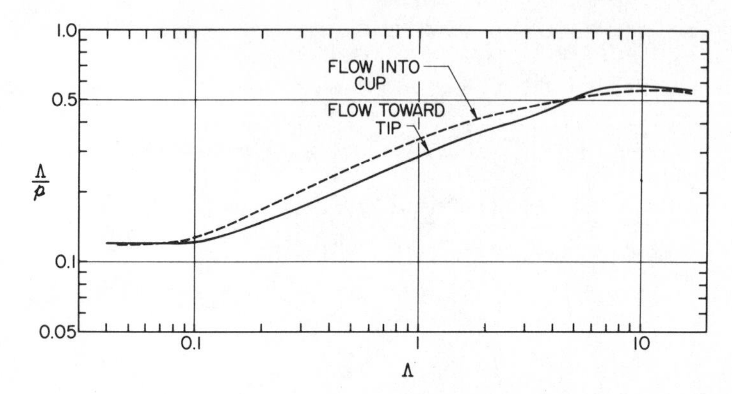

FIGURE 16. Experimental results from Frommer and Steele [**1978**] for flow through a model (400 ×) of the outer and inner hair cell cilia bundles. The dimensionless flow rate (Reynold's number) is Λ, and p is the dimensionless pressure drop. The *increase* in permeability Λ/p with the onset of nonlinearities at $\Lambda \sim 0.1$ was unexpected. For high flow rates $\Lambda > 10$, the permeability decreases.

behind the obstacles. If the experimental results in Figure 16 are correct, then at the higher flow rate $\Lambda \sim 10$ the resistance to flow of the obstacles is relatively small. Without any obstacles in the channel, the permeability factor Λ/p is unity. The single fiber recordings from a region of the cochlea with absent OHC has no response for low intensities of sound ($\lesssim 70$ db SPL), but is similar to the normal fibers shown in Figure 8 for high intensity. Is it a coincidence that according to Figure 16, the presence or absence of OHC cilia has little effect on the fluid flow for high intensity? By one calculation in Frommer and Steele [**1978**], $\Lambda = 0.1$ may correspond to 80 db SPL in the cochlea. By another equally plausible calculation, however, the nonlinear effect in Figure 16 occurs at 160 db SPL, out of the physiological range.

C. *Nonlinear equations*. Several investigators have studied the nonlinear behavior of the auditory periphery as well as higher levels with the use of mathematical models based on conjecture rather than physics. For example, Duifhuis [**1977**] postulates a "second filter" with an essential nonlinearity and obtains several interesting correlations, but also several contradictions. I will close by mentioning that such an essential nonlinearity may have a physical basis. Without attachment to the cilia, and without axial flow of fluid, i.e. for very low frequency, then a simplified physical model for the flapping of the tectorial membrane in Figure 14 is governed by an equation

$$J\ddot{g} + \gamma g^{-3}\dot{g} + \alpha g = F(t) \tag{26}$$

where J is the inertia, α is the spring constant, and γ is proportional to the fluid viscosity. The dependent variable g is the gap distance between the tectorial membrane and reticular lamina, and $F(t)$ is the forcing

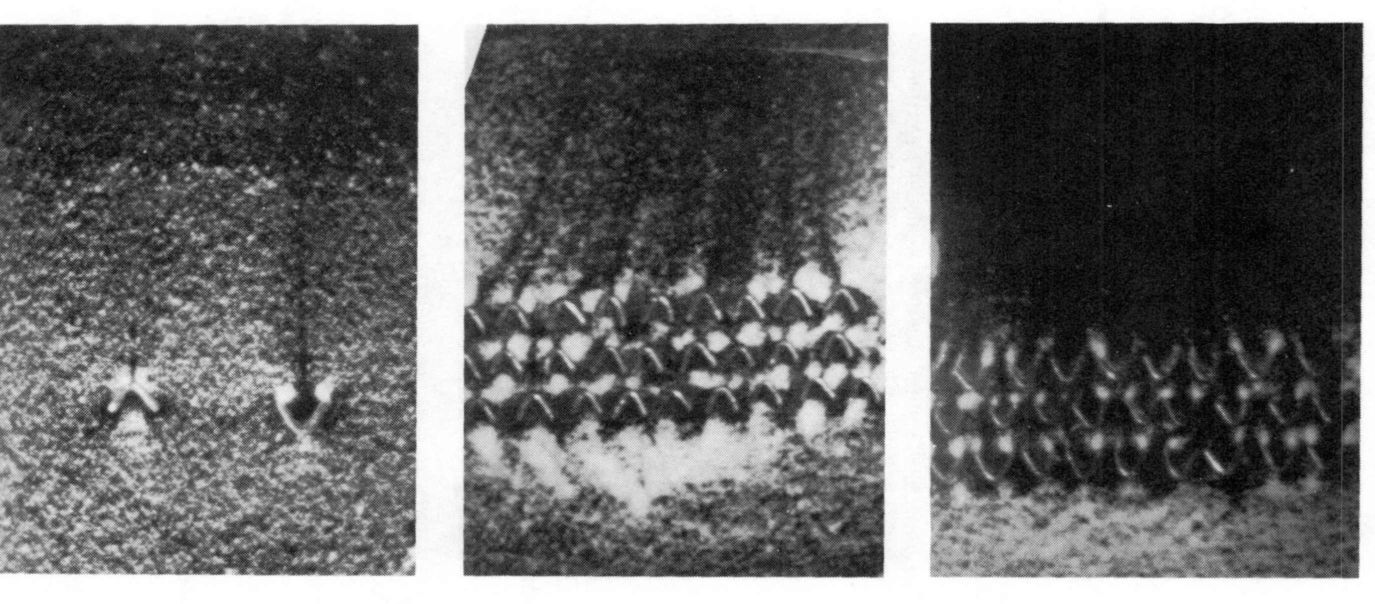

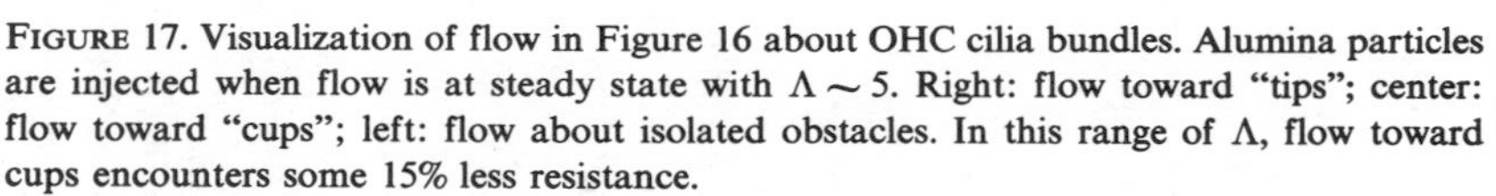

FIGURE 17. Visualization of flow in Figure 16 about OHC cilia bundles. Alumina particles are injected when flow is at steady state with $\Lambda \sim 5$. Right: flow toward "tips"; center: flow toward "cups"; left: flow about isolated obstacles. In this range of Λ, flow toward cups encounters some 15% less resistance.

function which can be related to basilar membrane displacement.

This equation has an essential nonlinearity and several interesting properties. For example, if the gap is initially closed $g(0) = 0$, then an infinite impulse is required to initiate motion. In one of Zwicker's models with an equilibrium position of $g = 0$, the gap remained closed over much of the distance and opened only near the region of maximum basilar membrane displacement. Another interesting aspect of equation (25) is that no periodic (nonzero) solution for a periodic forcing function $F(t)$ exists, unless $F(t)$ has a positive average value.

A model of the organ of Corti including IHC cilia can explain Figure 12. For low intensity (62db), the gap is closed and the cilia are bent outward by the static pressure in the inner sulcus caused by a displacement of the basilar membrane toward the scala vestibuli. For displacement toward scala tympani, the inner sulcus pressure is negative causing the cilia to be bent inward and a resulting inhibition. For higher levels of intensity, the gap opens, allowing a flow of fluid as the discontinuity of pressure propagates along the cochlea.

From Figure 11 for the normal cochlea with attached OHC cilia, the response is more complex and has not yet been explained by several more elaborate versions of equation (25) including the motion of fluid in the axial direction.

So we end at this point in an unsatisfactory state of confusion about this fundamental mechanism of perception. With the conclusive evidence of Russell and Sellick [**1977**] that the sharp tuning does occur at the hair cell, the micromechanics of the organ of Corti should be pursued with vigor.

REFERENCES

Békésy, G. von (1960). *Experiments in hearing*, McGraw-Hill, New York.

Cole J. D. and R. S. Chadwick (1977). *An approach to mechanics of the cochlea*, Z. Angew. Math. Phys. **28**, 785–804.

Dallos, P. (1973). *The auditory periphery*, Academic Press, New York.

Duifhuis, H. (1977). *Cochlear nonlinearity and second filter–a psychophysical interpretation*, in Psychophysics and Physiology of Hearing, E. F. Evans and J. P. Wilson, eds., Academic Press, New York, pp. 153–164.

Dotson, R. D. (1974). *Transients in a cochlear model*, Ph.D. Thesis, Stanford University.

Fleischer, G. (1973). *Studien am Skelett des Gehörorgans der Säugetiere, einschliesslich des Menschen*, Säugetierkundl. Mitt. **2**, **3**, 131–239.

Flock, A. (1977). *Physiological properties of sensory hairs on the ear*, Psychophysics and Physiology of Hearing, E. F. Evans and J. P. Wilson, eds., Academic Press, New York, pp. 15–26.

Frommer, G. H. and C. R. Steele (1979). *Permeability of fluid flow through hair cell cilia*, J. Acoust. Soc. Amer. **65**, 759–764.

Hallauer, W. L. (1974). *Nonlinear mechanical behavior of the cochlea*, Ph.D. Thesis, Stanford University.

Helle, R. (1974). *Beobachtungen an hydrodynamischen Modellen des Innenohres mit*

Nachbildung von Basilarmembran, Corti-Organ, und Deckmembran, Dr. Ing. Dissertation, Technische Universität, München.

Helmholtz, H. (1877). *Sensations of tone*, Dover, New York, 1954.

Kiang, N. Y. S. and E. C. Moxon (1974). *Tails of tuning curves of auditory-nerve fibers*, J. Acoust. Soc. Amer. **55**, 620–630.

Kohllöffel, L. U. E. (1972). *A study of basilar membrane vibrations*. II, *The vibratory amplitude and phase pattern along the basilar membrane* (*post mortem*), Acustica **27**, 66–81.

Neely, S. T. (1977). *Mathematical models of the mechanics of the cochlea*, E. D. Thesis, California Institute of Technology.

Rhode, W. S. (1973). *An investigation of post-mortem cochlear mechanics using the Mössbauer effect*, in Basic Mechanism in Hearing, A. R. Møller, ed., Academic Press, New York, pp. 49–68.

Russell, I. J. and P. M. Sellick (1977). *The tuning properties of cochlear hair cells*, Psychophysics and Physiology of Hearing, E. F. Evans and J. P. Wilson, eds., Academic Press, New York, pp. 71–88.

Sales, G. and D. Pye (1974). *Ultrasonic communication by animals*, John Wiley, New York.

Sokolich, W. G., R. P. Hamernik, J. J. Zwislocki and R. A. Schmiedt (1976). *Inferred response polarities of cochlear hair cells*, J. Acoust. Soc. Amer. **59**, 963–974.

Steele, C. R. (1973). *A possibility for sub-tectorial membrane fluid motion*, in Basic Mechanisms in Hearing, A. R. Møller, ed., Academic Press, New York, pp. 69–94.

______. (1976). *Cochlear mechanics*, in Handbook of Sensory Physiology, Vol. V, Part 3, W. D. Keidel and W. D. Neff, eds., pp. 443–478.

Steele, C. R. and L. A. Taber (1979a). *Comparison of "WKB" and finite difference calculations for a two-dimensional cochlear model*, J. Acoust. Soc. Amer. **65**, 1001–1006.

______. (1979b). *Comparison of "WKB" calculations and experimental results for three dimensional cochlear models*, J. Acoust. Soc. Amer. **65**, 1007–1018.

Stevens, S. S. and F. Warshofsky (1965). *Sound and hearing*, Life Science Library, Time-Life Books, New York.

Whitham, G. B. (1974). *Linear and nonlinear waves*, John Wiley, New York.

Wilson, J. P. (1974). *Basilar membrane vibration data and their relation to theories of frequency analysis*, in Facts and Models in Hearing, E. Zwicker and E. Terhardt, eds., Springer, Berlin, pp. 56–63.

Yost, W. A. and D. W. Nielson (1977). *Fundamentals of hearing–An introduction*, Holt, Rinehart, and Winston, New York.

Zwicker, E. (1974). *On psychoacoustical equivalent of tuning curves*, in Facts and Models in Hearing, E. Zwicker and E. Terhardt, eds., Springer, Berlin, pp. 132–141.

______. (1977). *Masking-period patterns and hearing theories*, in Psychophysics and Physiology of Hearing, E. F. Evans and J. P. Wilson, eds., Academic Press, New York, pp. 393–402.

DEPARTMENT OF MECHANICAL ENGINEERING, STANFORD UNIVERSITY, STANFORD, CALIFORNIA 94304

Lectures in Applied Mathematics
Volume 17, 1979

24 Hard Problems about the Mathematics of 24 Hour Rhythms

Arthur Winfree

A. Introduction. I have chosen to seize this opportunity for something a little bit out of the ordinary. Instead of trying to *tell* you something (e.g., how some organism works and how smart I am to have found out), I am trying here to *ask* you something. Specifically, I am asking for mathematical insight into a number of structured dynamical situations commonly encountered by living organisms. My emphasis is on the questions rather than on the biological context or experimental evidence. The questions I pose all concern the so-called "circadian" rhythms, the internal "biological clocks" supposed to underlie the persistent rhythms of physiological activity in most kinds of living organism, even when isolated from known external time cues.

B. Circadian rhythms. The following seem to me reasonable and widely accepted suppositions grounded in most of the biological clocks literature:

(1) In an ideally constant environment, the organism's "clock" has an autonomous kinetics. That is, the rate of change $\dot{x}_i$ of each of many variables of state, x_i (e.g., biochemical concentrations) is given as a function of all those variables. In other words we deal with an ordinary differential equation, $\dot{\mathbf{x}} = \mathbf{R}(\mathbf{x}, \mathbf{P})$, where $\mathbf{P}$ represents the departure of external parameters from some standard configuration.

(2) That ordinary differential equation has a unique attracting limit cycle and returns to it swiftly following most kinds of transient perturbation.

(3) For many purposes it is commonly considered reasonable to suppose that the clock is always sufficiently close to its attracting cycle so that its kinetics can be written in terms of a single "phase" variable,

---

AMS (MOS) subject classifications (1970). Primary 92–XX.

$0 < \phi < 1$. Phase increases steadily along the attracting cycle from an arbitrarily chosen origin: $\dot{\phi} = f(\phi, \mathbf{P})$, and $f(\phi, \mathbf{0}) \equiv 1$.

(4) A stimulus (e.g., exposure to visible light or cooling during the night) is a change of the parameters, **P**, from **0**. In many laboratory situations, $\mathbf{P}(t)$ is zero up to t_0, is constant for duration M, and then is again zero.

(5) The observable output of a biological system is rarely **x** or even any function $\mathbf{F}(\mathbf{x})$. More commonly we observe variables O_i indirectly affected by state variables x_i. Given regular rhythmicity of **x**, it is not necessary that **O** be regularly rhythmic. Nonetheless it often *is*, after a few cycles of transient irregularity. The dynamics of transients have been almost universally ignored. But to my mind it is the essence of temporal adaptation and the foremost target of evolutionary change. This is because circadian rhythms function autonomously only in the laboratory. In nature they are perpetually driven by the succession of days and nights, so their overt behavior is little more than a continually renewed sequence of "transients".

(6) $\mathbf{O}(t)$ is believed to have a unique waveform uniquely phase-related to $\mathbf{x}(t)$ after transients subside. A priori, this strikes me as unlikely. No counterexamples have been published, but this might be because no one has deliberately looked for counterexamples experimentally. In general, rhythmically driven systems do entrain in more than one stable mode while rhythmically driven. For a myriad of intriguing examples see Hayashi [**1964**].

(7) Sometimes we do not even observe $\mathbf{O}(t)$ directly, because the organism is a mosaic of many similar biological oscillators (e.g., in many separate cells) whose collective behavior alone is directly observable by gross measurements on the whole organism's behavior.

(8) These many individual oscillators may not always necessarily interact in a way adequate to keep them all in strict synchrony on a common attracting limit cycle. This internal distribution of phase may be a vitally important feature of biological rhythms.

(9) Particularly in situations involving spatially extensive fields of cells, there may be spatial gradients of phase. Neighbor interactions may achieve *local* synchronization without eliminating all patterning in space.

A great diversity of models naturally arise within this context. Many of these suggest questions of a mathematical nature which have interest quite apart from their supposed biological applications. Probably some are answerable using standard techniques. Others might motivate mathematically significant explorations once properly formulated. (This has already been the case, I think, in Guckenheimer's paper [**1975**].) Here are my attempts to formulate a sampling of both kinds of problems.

C. Isochrons. In a variety of biological contexts, one needs to know the timing difference between two independent oscillators when both have come sufficiently close to a common attracting limit cycle. The "given" is only an initial state of each oscillator, in the attractor basin, but not on the attracting cycle. In Winfree **[1967]** I argued, essentially by example, that there exists a foliation of the attractor-basin into a 1-parameter family of manifolds called "isochrons" with the following property: all states in a given isochron are confluent in the sense that the rhythms ensuing from such initial conditions are eventually indistinguishable. All these initial conditions are said to have the same "latent phase" Φ. This Φ is a map from the oscillator state space $R^n \to S^1$, the circle of phases (the best summary is in Winfree **[1974]**).[1]

This notion proved useful for several years in contriving experiments with interesting outcomes, but it was not until 1971 that Graeme Mitchison (unpublished) and then Guckenheimer **[1975]** independently initiated a proper topological examination of the circumstances under which isochrons exist and have the intuitively-postulated behavior. Kawato and Suzuki **[1978]** have pursued the matter a little further. The following are a few outstanding questions.

I. HOW CAN ISOCHRONS BE COMPUTED ANALYTICALLY?

The following four examples admit analytic solution but require extravagant simplifications, namely that the dynamics involves only two variables and has exact polar symmetry.

Suppose that $\dot{\phi} = A(R)$ with the unit of time chosen so that $A(R_0) = 1$, and $\dot{R} = B(R)$ with $\dot{R} = 0$ and decreasing at some $R_0 > 0$.

This is the simplest attractor-cycle oscillator. If there is only one $R = R_0$ at which $B(R) = 0$ and $dB/dR < 0$, then R_0 is a unique attracting cycle and we may as well scale R to make $R_0 = 1$.

Since the dynamical flow has polar symmetry, the isochrons must also have polar symmetry:

$$\Phi = g(\phi, R) = \phi - f(R).$$

Now the latent phase Φ necessarily increases at unit angular velocity as the oscillator follows its kinetic equation. Thus we write

$$\dot{\Phi} \equiv 1 = \dot{\phi} - \frac{df(R)}{dR}\dot{R} = A(R) - \frac{df(R)}{dR}B(R),$$

so

$$\frac{df}{dR} = \frac{A(R) - 1}{B(R)}.$$

[1] It was brought to my attention at Salt Lake City that the notion of Φ had long been entertained under the name "asymptotic phase". See Coddington and Levinson **[1955]** and Hale **[1963]**, **[1969]**.

This is a differential equation for $f(R)$. We can integrate it (by parts if A and B are polynomials) and then obtain $g(\phi, R)$.

EXAMPLE 1. Suppose (Figure 1(a))

$$A(R) \equiv 1,$$
$$B(R) = 5(1 - R)R.$$

Then $df/dR \equiv 0$ so $f(R) = \text{const}$, chosen so $\Phi = 0$ at $R = 1$, $\phi = 0$, and we have (Figure 1(b))

$$\Phi = g(\phi, R) = \phi.$$

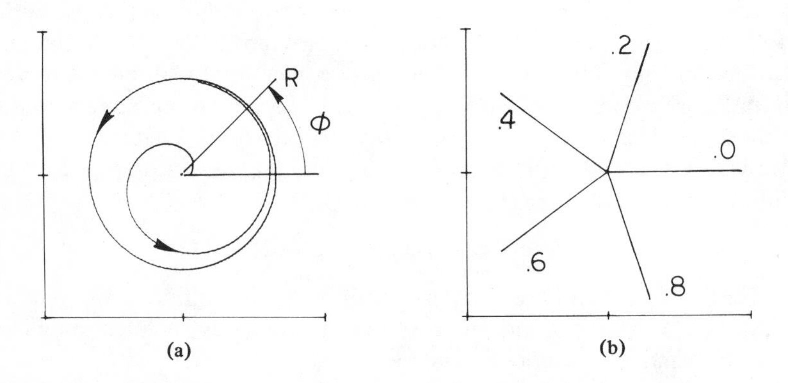

FIGURE 1. (a) A single trajectory of the dynamical scheme described in polar coordinates in Example 1. The origin is placed at Cartesian point (1, 1), as though it were a repelling chemical equilibrium. The unit circle is an attracting limit cycle.

(b) Isochrons of the same dynamic, marking off intervals of 1/5 cycle along all trajectories in the attractor basin of the cycle. Dots started on any isochron reappear on it at unit intervals ot time, closer to the point $R = 1$.

EXAMPLE 2. Suppose (Figure 2(a))

$$A(R) = R,$$
$$B(R) = 5(1 - R)R^2.$$

Here the trajectories have the same geometry but the state traverses these trajectories at speeds R times faster than in Example 1. In this case the angular velocity $A(R)$ is no longer independent of amplitude, but increases in proportion to R.

Thus,

$$\frac{df}{dR} = \frac{R - 1}{5R(1 - R)R} = -\frac{1}{5R^2},$$

and so

$$f(R) = \frac{1}{5R} + \text{const},$$

and (Figure 2(b))

$$\Phi = g(\phi, R) = \phi - \frac{1}{5R} + 0.2.$$

Thus the isochron structure close to the repelling focus of this 2-variable kinetics consists of tight spirals.

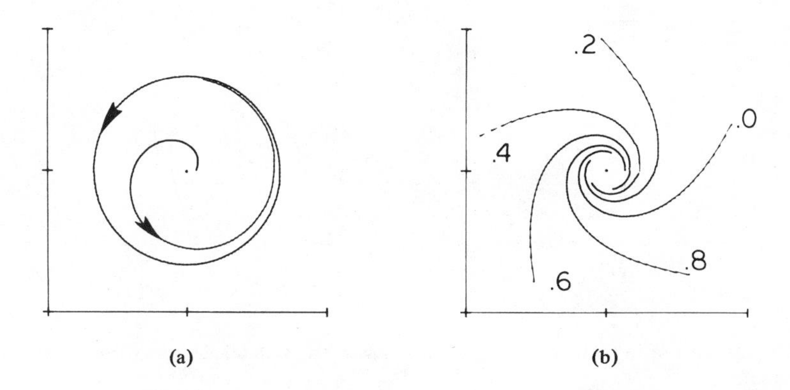

FIGURE 2. (a) As in 1(a) except both components of velocity are multiplied by a scalar field, leaving directions unaltered.

(b) The changed isochrons reflect the radial dependence of angular velocity.

EXAMPLE 3. Suppose (Figure 3(a))

$$A(R) = 1 + \varepsilon(1 - R),$$
$$B(R) = (1 - R)R.$$

Now,

$$\frac{df}{dR} = \frac{\varepsilon}{R}.$$

It follows that

$$f = \varepsilon \ln R + \text{const}$$

and (Figure 3(b))

$$\Phi = \phi - \varepsilon \ln R.$$

This example shows that the isochron spiral can turn either way relative to trajectories, depending on the sign of ε.

EXAMPLE 4. Suppose (Figure 4(a))

$$A(R) = 1 + \varepsilon(1 - R),$$
$$B(R) = 5(1 - R)\left(R - \tfrac{1}{2}\right)R.$$

The point here is to examine a self-sustaining oscillator that is *not* also self-exciting, as in the circadian clock model of Kalmus and Wigglesworth [**1960**] and in Best's computation of isochrons for a pacemaker

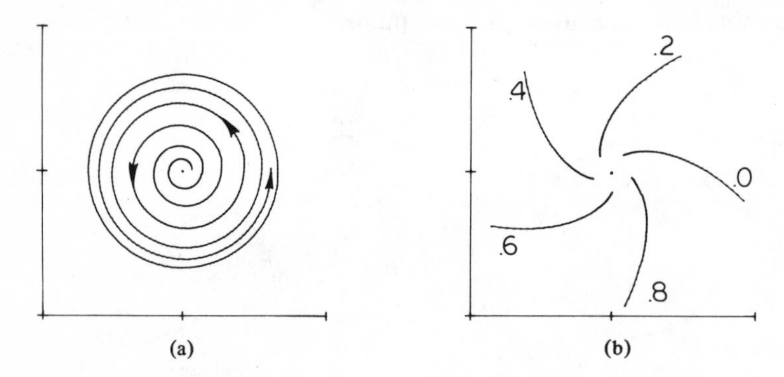

FIGURE 3. (a) As in 1(a) but radially 5 times slower, and azimuthally somewhat dependent on radius.

(b) The isochrons make exponential spirals into the origin, in contrast with the hyperbolic spirals of 2(b).

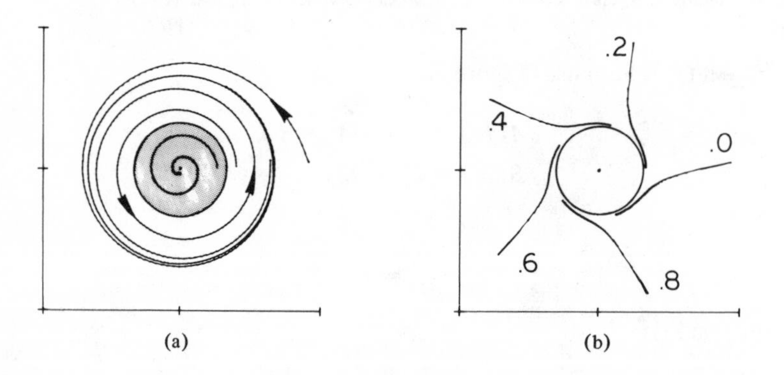

FIGURE 4. (a) The azimuthal equation is as in 3(a), while the radial equation is modified to create a repelling limit cycle at $R = \frac{1}{2}$. Trajectories inside that radius wind into the *attracting* equilibrium.

(b) The phaseless manifold is the disk $R = \frac{1}{2}$. Along its border all isochrons converge (as they did in 3(b) and 4(b)) but with $R = 0$.

nerve (Best [**1978**]):

$$\frac{df}{dR} = \frac{\varepsilon}{5R\left(R - \frac{1}{2}\right)} = \left(\frac{1}{R - \frac{1}{2}} - \frac{1}{R}\right)\frac{2\varepsilon}{5}.$$

For $R < \frac{1}{2}$, Φ will go undefined.

For $R > \frac{1}{2}$,

$$f = \frac{2\varepsilon}{5} \ln \frac{R - \frac{1}{2}}{R} + \text{const}$$

so (Figure 4(b))

$$\Phi = \phi - \frac{2\varepsilon}{5} \ln\left(2 - \frac{1}{R}\right).$$

The phase singularity in this case is not a point but the whole circular border of a "black hole".

EXAMPLE 5. In less symmetric cases in the plane, no analytic solutions for the isochrons have been proposed. Even within the linear range near equilibrium, the isochrons behave most peculiarly, according to the only numerical calculation I have done. This example is taken from FitzHugh [**1961**]. FitzHugh sought a useful simplification of the Hodgkin-Huxley equation. His model uses only two variables to mimic a pacemaker neuron with a repelling steady-state and an attracting cycle. I computed its isochrons as follows:

(1) The attracting cycle was found by integrating the rate equations numerically. A point on the cycle is chosen arbitrarily to be "phase 0". The cycle period is taken as unit time.

(2) An arbitrary initial state was chosen and once again the rate equations were integrated by iterating the difference equation in steps Δt until the state comes within a small tolerance of phase 0 on the cycle.

(3) Backing up along that trajectory, the latent phase was noted at each step as $\Phi = -N\Delta t$ modulo 1, N being the step number.

(4) This was done for many different initial states.

(5) The isochrons were sketched through states of equal latent phase at intervals of one-tenth cycle. From states on one isochron, the time to phase 0 is the same, modulo 1. All trajectories pass from one isochron to the next at intervals of one-tenth cycle.

The results near the steady state are shown in Figure 5. For purposes of convenience a linear transformation of variables was effected in plotting the trajectories and isochrons. This was done in order to display trajectories near the steady state as exponential spirals and to give the whole diagram polar symmetry. The isochrons approach the steady state in a distinctly complicated way.

Further out, the trajectories approach their attracting cycle and the isochrons behave as one might expect intuitively (Figure 6):

(1) They are packed close together near the separatrix: a slight displacement of state where the system is in one of these states elicits a big change of phase in its eventual rhythm.

(2) A small depolarizing (excitatory) pulse during the depolarizing phase of the action potential advances phase. It delays phase during the

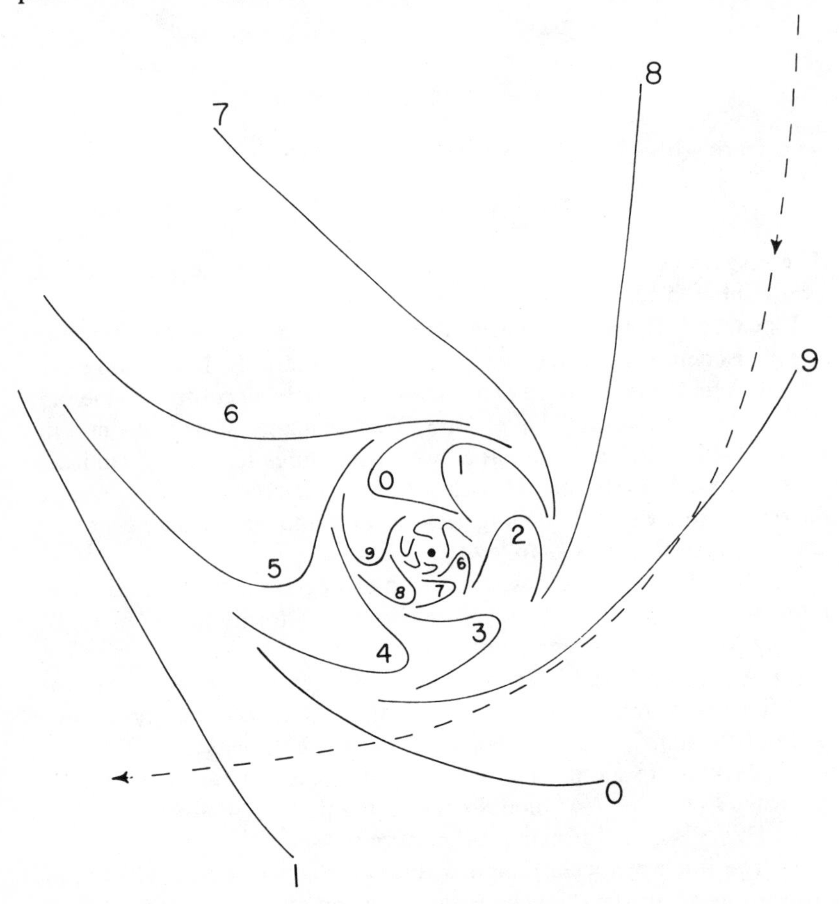

FIGURE 5. Similar in principle to parts (b) of Figures 1–4. The isochrons are calculated in the linear region near the equilibrium of FitzHugh's BVP model of pacemaker activity in nerve membrane:

$$\dot{x} = c(y + x - x^3/3 + z),$$
$$\dot{y} = -(x - a + by)/c,$$
$$a = 0.7, \quad b = 0.8, \quad c = 3., \quad z = -0.4.$$

The isochrons are interrupted where the calculation becomes too delicate, along a spiral locus. Coordinates are as in Figure 6 but the scale is greatly enlarged. All trajectories (not shown) are clockwise exponential spirals out of the equilibrium, cutting isochrons at unit rate. The dashed curve is the lower right-hand arc of the attracting cycle shown complete in Figure 6.

relatively-refractory recovery interval. Hyperpolarizing (inhibitory) pulses have the opposite effect.

However, in keeping with the polar symmetry of the dynamics near equilibrium (in suitably linearly transformed coordinates), the isochrons do exhibit polar symmetry.

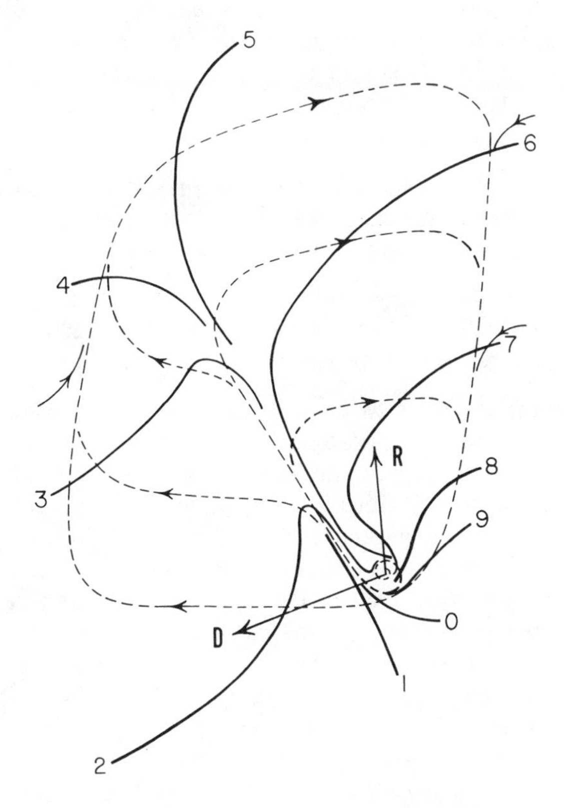

FIGURE 6. Similar in principle to parts (a) of Figures 1–4. The scale is grosser than in Figure 5. Isochrons are solid curves. Trajectories (dashed) wind out of the repelling equilibrium to diverge along a separatrix and wind onto the attracting cycle. The picture is somewhat skewed from the conventional representation in FitzHugh's Figure 5 (see depolarizing (D) and recovery (R) variables) by the linear transformation required to symmetrize the spirals near equilibrium.

II. CAN ANYTHING MORE BE SAID ABOUT THE ISOCHRONS OF ARBITRARY LIMIT-CYCLE KINETICS, AT LEAST NEAR EQUILIBRIUM?

Guckenheimer [**1975**] proved that in the generic case of a unique attracting cycle in a contractable state space, the isochron foliation *fails* along a singular locus of codimension 2 or 1, bounding a "phaseless manifold" of the same or larger dimension.

In Examples 1–3 above the phaseless manifold has codimension 2: it is a point in a 2-dimensional plane. In Example 5, it appears one-dimensional but probably is not really. The spiral-shaped locus of ostensible discontinuity is only an extension of a separatrix, along which phase is exquisitely sensitive to small changes in state transverse to that trajectory. I thank Eric Best for an example in which a phaseless manifold of codimension 1 *does* appear. It joins two repelling nodes and a saddle point, all surrounded by a unique limit cycle in R^2. In Example 4 the equilibrium is an attractor and the phaseless manifold is its attractor basin: the phaseless manifold is definitely of codimension 0 in the plane and the isochrons run amok along its boundary of codimension 1.

III. WHAT CAN BE SAID ABOUT THE PHASELESS MANIFOLD OF A "KNOTTED" ATTRACTING CYCLE IN R^3? IT WOULD APPEAR THAT THIS MAY HAVE CO-DIMENSION 1, TOGETHER WITH A RATHER PECULIAR TOPOLOGY.

IV. IS IT TRUE AS CONJECTURED, AND AS PROVED IN SOME CASES BY GUCKENHEIMER [**1975**], THAT EVERY ISOCHRON COMES ARBITRARILY CLOSE TO EVERY POINT OF THE PHASELESS MANIFOLD OR ITS BOUNDARY?

Mathematical solutions to these riddles, and others which naturally come to mind while contemplating these, would add a great deal to the qualitative understanding of the ways in which biological clocks are reset in phase by stimuli which kick the state off the attracting cycle to new initial conditions (e.g., Kawato and Suzuki [**1978**]).

It should be noted before passing on to the next topic that the definition of "isochrons" involves the words "limit" and "eventually". But it may take several unit durations (limit cycle periods) before the perturbed oscillator gets close enough to its attracting cycle to justify those adjectives. In Figure 3, for example, after the initial condition shown x_1 goes through five or six successively higher maxima before it begins to repeat monotonously. During this transient the trajectory passes any given isochron at unit intervals, but it passes through maxima or through any other locus of points in state space at other intervals. Whatever our criterion of "phase ϕ" in the $x_1(t)$ rhythm is, it necessarily coincides with the corresponding isochron ϕ on the attracting cycle. But these two criteria–one secular, one asymptotic–generally differ elsewhere.

This discrepancy is a source of grief to experimentalists who for technical reasons often cannot monitor organisms reliably for more than a day or two after perturbation of their biological clocks. Kawato and Suzuki [**1978**] show that this general discrepancy leads to spurious discontinuities in resetting curves constructed by measuring event times soon after the stimulus.

Kaus [**1976**], on the other hand, interprets transients as a consequence of signal filtering by a dynamical system intermediate between the clock and our indirect observation of its rhythmicity. Kaus' observable $O(t)$ is the first harmonic Fourier component of $x(t)$ under conditions of entrainment by an external light/dark cycle.

If Kaus is right then there may yet be some hope for experimentalists who wish to acquire more information from observations limited to the immediate aftermath of a stimulus. If Kawato is right there may be a tidy geometric way to relate transients to isochrons in state space. As noted in supposition (5) in §B above, questions about transients are usually swept under the rug, but they are all-important for understanding rhythmic physiology in a rhythmic environment:

V. DO THE VIEWS OF KAUS AND OF KAWATO SUGGEST A GEOMETRIC RECONCILIATION IN STATE SPACE, OR DO THEY REPRESENT OPERATIONALLY DISTINGUISHABLE DYNAMICAL HYPOTHESES?

D. Simple clocks. There exist other models of biological clocks, whose behavior is unlike those considered above (in §C). These include discrete-state models akin to Kauffman's switching networks [**1969**]. They also include continuous dynamical systems of a simpler sort called "simple clocks". These are restricted to a state space of a single dimension, topologically connected in a ring. The dynamical equation of such a system has the form $\dot{\phi} = f(\phi, P)$. This seems a reasonable approximation to an attracting limit cycle in a space of more degrees of freedom, in which the flow transverse to the cycle (which leads onto the cycle) is exceedingly swift in comparison with the slower component of flow parallel to the cycle ($\dot{\phi}$), and is relatively little affected by parameter P. Thus we might write as a first approximation:

$$\phi = \tau^{-1} + PZ(\phi),$$

where Z is a phase-dependent sensitivity to influence P. To complete the picture we give each oscillator a phase-dependent output of influence $X(\phi)$ and define

$$P = \sum_i X(\phi_i).$$

For applications of this model see Winfree [**1967**] and Peskin [**1975**].

With these simplifications some biologically pertinent questions about interactions among many clocks in a population become tractable, but not so tractable that all the interesting questions have been solved.

1. *Entrainment*. The first priority is to write the conditions for entrainment of such an oscillator by a rhythmic influence $P(t)$, with period τ_0. The necessary condition is that

$$\int_0^{\tau_0} \dot{\phi}\,dt = 1, \qquad \text{i.e.} \quad \langle \dot{\phi} \rangle = \frac{1}{\tau_0}.$$

This is easy if we approximate $\phi = \psi + t/\tau_0$, where ψ is just ϕ at $t = 0$.

Then $P(t)$ has the same period as $Z(\phi(t))$, so

$$\langle\dot{\phi}\rangle = \frac{1}{\tau_0}\int_0^{\tau_0}\dot{\phi}\,dt = \frac{1}{\tau_0}\int_0^{\tau_0}\left(\frac{1}{\tau} + P(t)Z\left(\frac{t}{\tau_0} + \psi\right)\right)dt = \frac{1}{\tau} + F(\psi).$$

This is a smooth periodic function of ψ, the phase difference between the influence and sensitivity rhythms. The criterion for entrainment is that

$$\dot{\psi} = \langle\dot{\phi}\rangle - \frac{1}{\tau_0} = 0, \qquad \text{so } F(\psi) = \frac{1}{\tau_0} - \frac{1}{\tau}.$$

This is feasible if and only if $F(\psi)$ has at least that amplitude, which in turns requires that Z and P have sufficient amplitude relative to the discrepancy of periods. If this equilibrium exists its stability requires $d\dot{\psi}/d\psi < 0$, so $F' < 0$.

The problem becomes interesting when we let $P(t) = \Sigma_i X(\phi_i(t))$. Now there arises an intriguing requirement of self-consistency.

2. *Mutual entrainment*. Depending on the frequency difference, an entrained oscillator must lag a certain phase behind its driver. But what if the driver *is* the collective influence of those self-same oscillators? That would seem to require a certain phase difference between the influence and insensitivity functions, and/or some adjustment of frequency. The situation has never been properly formulated for mathematical solution. In (Winfree [**1967**]) I made the only attempt I know of for simple clocks.[2] Hobbled as it was by assumptions and approximations, it nevertheless correctly predicted the empirical outcome of mutual interaction in such populations of simple clocks in computer simulations. In a nutshell, entrainment depends on the area and the moment of inertia of the loop formed by plotting $X(\phi)$ vs. $Z(\phi)$. Depending on those quantities, a population may achieve mutual synchronization at any frequency, even outside the range of natural frequencies of the individual clocks. Or it may not. It may in fact even actively resist entrainment by generating an opposing rhythm, to whatever extent some exogenous driving rhythm does impose any synchronization. These results are implicit in the following three equations (see Winfree [**1967**]):

$$P(\psi) = \oint X(\phi - \psi)p(\psi)\,d\psi, \tag{1}$$

$$F(\psi) = \oint Z(\phi + \psi)P(\psi)\,d\phi, \tag{2}$$

---

[2] At the Salt Lake meeting, my attention was drawn to Kuramoto [**1975**] and Neu [**1979**], both of which reduce chemical oscillator interactions to a simple-clock problem and derive mutual entrainment.

$$p(\psi) = -F'(\psi)N\left(\frac{1}{\tau_0} - F(\psi)\right)$$
$$= 0 \quad \text{if } F' > 0. \tag{3}$$

τ_0 may be chosen so that

$$\oint \psi p(\psi)\, d\psi = 0. \tag{3'}$$

So if $N(1/\tau)$, the distribution of autonomous frequencies, is narrow, then $1/\tau_0 \cong 1/\tau + F(0)$. (See Figure 7.)

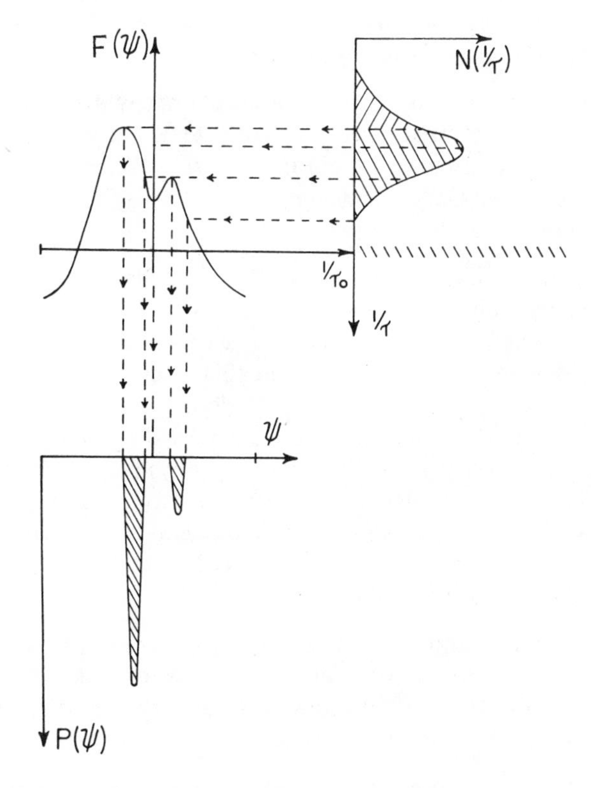

FIGURE 7. A guess at the relationship between τ_0, $N(1/\tau)$, $F(\psi)$ and $P(\psi)$. Oscillators whose $1/\tau$ is outside the range of $1/\tau_0 - F(\psi)$ remain unentrained and are ignored. The dip in $F(\psi)$ is unnecessary, but can occur. It gives us two disjoint but overlapping downslopes, thus two peaks of $p(\psi)$. (Oscillators with τ in the overlap region can occupy either peak; here I put them all in the right-hand peak.)

These equations have never been solved, except in a few very specialized approximations. It would be of great interest to solve them in order to answer the following questions:

VI. GIVEN A DISTRIBUTION OF NATURAL PERIODS $N(1/\tau)$, UNDER WHAT CONDITIONS OF X VS. Z SHAPE DOES MUTUAL ENTRAINMENT ARISE SPONTANEOUSLY FROM AN INITIALLY UNIFORM DISTRIBUTION OF PHASES?

VII. UNDER WHAT CONDITIONS DOES IT NOT, WHILE THE MUTUALLY ENTRAINED STATE IS NONETHELESS STABLE ONCE ACHIEVED?

VIII. UNDER WHAT CONDITIONS ARE THERE TWO OR MORE ALTERNATIVE PHASES OF ENTRAINMENT OF ANY INDIVIDUAL OSCILLATOR TO THE AGGREGATE RHYTHM, AND WHAT DETERMINES THE NUMBER OCCUPYING EACH OF THESE STABLE POSITIONS? (This hysteresis makes the aggregate waveform *plastic* to perturbing influences, which switch some oscillators from one stable position to another, thus altering their contribution to the aggregate rhythm.) (See Figure 7.)

IX. UNDER WHAT CONDITIONS DOES THE STEADY STATE AMPLITUDE OF THE AGGREGATE RHYTHM VARY SMOOTHLY WITH THE PARAMETERS OF THE X VS. Z LOOP? UNDER SOME CONDITIONS (see Winfree [**1967**]) MUTUAL SYNCHRONIZATION ARISES AS A FIRST ORDER PHASE-TRANSITION WHEN A CRITICAL LOOP AREA IS REACHED.

Answers to such questions would contribute substantially, I think, to an understanding of temporal organization in almost all living organisms, which are now generally accepted to be populations of coupled oscillators in respect to circadian rhythmicity as well as in respect to metabolic cycles of shorter period.

3. *The precision of some biological clocks*. I draw your attention to a stochastic question repeatedly posed by physiologists, but never yet answered by mathematicians. Suppose we have, not a population of individually precise oscillators of various periods, but rather a population of oscillators of identical statistical behavior, individually somewhat irregular of rhythmicity.

X. IF THEY DO ACHIEVE MUTUAL ENTRAINMENT, DOES THE AGGREGATE RHYTHM DRONE ON WITH THE STEADINESS ORDERS OF MAGNITUDE GREATER THAN ANY INDIVIDUAL'S, DISCIPLINING EACH INDIVIDUAL TO ADHERE TO THE COLLECTIVE RHYTHMICITY?

By analogy to more demonstrable instances (e.g., Jongsma et al. [**1975**], Clay and DeHaan [**1979**]), biologists are inclined to account in this way for the uncanny accuracy of some circadian rhythms, but so far as I am aware, the mathematical essence of such a mechanism has never been revealed.

In one limit, the problem can be viewed as a trivial departure from the deterministic case sketched above. Suppose mutual entrainment is achieved, so $P(t)$ and τ_0 are known, and so $F(\psi)$ is known. Then each oscillator rides the downslope of $F(\psi)$ at altitude $1/\tau_0 - 1/\tau$. Now suppose τ varies slowly. Then the oscillator drifts up and down the slope of F, changing its ψ as τ varies. If τ varies too fast, ψ will not always be thus in equilibrium, and the oscillator will vary above and below $F(\psi)$ (Figure 8). Such an oscillator is exposed to $\dot{\psi} = F(\psi) + 1/\tau - 1/\tau_0 \neq 0$ so it is pushed back toward the downslope. The density distribution $p(\psi)$ is not much broadened, so $P(t)$ and $F(\psi)$ are little changed. So the situation remains qualitatively the same (more quantitative analyses are available: see Stratonovich [**1967**]).

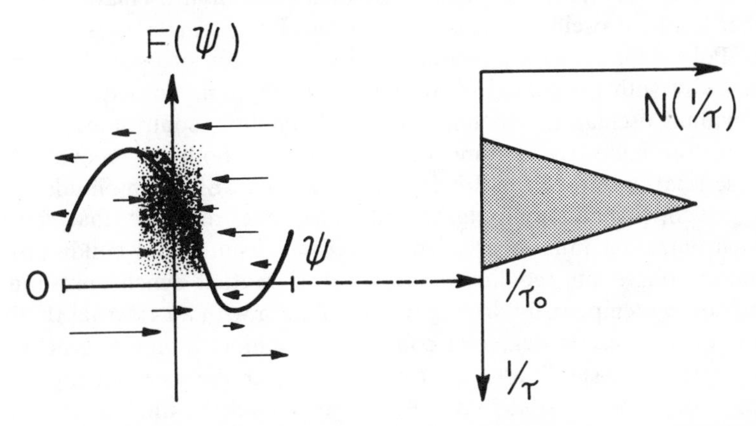

FIGURE 8. If ψ changes fast enough within the confines of $N(1/\tau)$ then the oscillator will stray from $F(\psi)$ within the confines of $p(\psi)$. The arrows indicate $\dot{\psi} = 1/\tau + (F(\psi) - 1/\tau_0)$.

It would appear that the aggregate rhythm *does* discipline each individual to enormously enhanced regularity in the sense of the number of cycles passed, though ψ may jitter noisily without ever transgressing a full cycle. Can this problem be posed in exact form and solved to learn what are the limits of validity of such handwaving?

XI. WHAT KIND OF OSCILLATOR OR WHAT KIND OF MUTUAL COUPLING BEST LENDS ITSELF TO THIS MECHANISM OF ERROR-REDUCTION?

4. *A needed improvement*. Returning to determinism, the three equations suggested above still ignore a factor that probably ought not to be ignored. They suppose that only mutually entrained oscillators contribute to $P(t)$, and that other oscillators whose native periods are too different from τ_0, which do not adopt a fixed phase relation to P, make no contribution to P. But that is not really so because even such oscillators linger longer near that ψ where F is closest to $1/\tau_0 - 1/\tau$, and zip fastest through that ψ where the opposite is true. This could be quite important, as the influence contributed by these oscillators to the rise of $P(t)$ does increase the amplitude of F, and so expands the range of native periods which can be captured to further enhance $P(t)$, etc.

XII. CAN THE THREE EQUATIONS OF MUTUAL ENTRAINMENT BE REFORMULATED USING $F(\psi)$ TO DEFINE SOME KIND OF PROBABILITY DISTRIBUTION OF OSCILLATOR PHASES,[3] FOR USE IN THE SUMMATION OF $X(\phi_i)$ INTO $P(t)$?

5. *Another needed improvement*. The whole discussion above is conducted exclusively in terms of phase alone, as though each oscillator rigidly adheres to a fixed cycle, varying only its rate. That might be a realistic assumption, e.g. in the case of a relaxation oscillator. But for other kinds of oscillator, it is not realistic. For example consider a van der Pol oscillator with ε about 0.2. Its amplitude of oscillation varies markedly with the period of an entraining rhythm. Consequently each individual oscillator, depending on its frequency, contributes more or less to the collective rhythm which entrains it, both on account of its phase relationship to the driver, *and* on account of its amplitude while so driven. In my computer simulations, one result is that mutual synchronization fails to arise spontaneously from initial conditions of random phase on the limit cycle. But if mutual synchronization is initiated by temporarily driving the population with an external rhythm, then, once a certain degree of coherence is achieved, mutual synchronization remains stable and improves even after the external rhythm is removed. Some physiologists have proposed such a principle as a basis of the mysterious periodic diseases so widespread in physiological control systems: without mutual entrainment the cycles of individual cells or organs continue parochially without sending the whole community into convulsive crashes and booms; but an initially synchronizing shock elicits pathological collective rhythmicity.

Grattarola and Torre [**1977**] recently published a mathematical criterion for mutual synchronization in populations of *identical* van der Pol oscillators. Linkens [**1974**] has given some attention to the case of

---

[3]Kuramoto [**1975**] achieves something close to this for fixed-amplitude oscillators coupled by

$$\dot{\phi}_j = 1 + a \sum_{i=j} \sin(\phi_i - \phi_j).$$

This rule of coupling can be rewritten as

$$1 + Z_1(\phi_j) \sum_{i \neq j} X_1(\phi_i) + Z_2(\phi_j) \sum_{i \neq j} X_2(\phi_i),$$

where (Z_1, Z_2) and (X_1, X_2) are orthogonal *pairs* of influences and sensitivities, perhaps representing two substances oscillating 90° out of phase. Kuramoto's more symmetric format lends itself more conveniently to analytic solutions than does Winfree's

$$\dot{\phi}_j = 1 + Z(\phi_j) \sum_{i \neq j} X(\phi_i).$$

nonidentical oscillators, especially by computer simulation. Zwanzig **[1976]** gives a very tidy analytical model of mutual synchronization in *identical* piecewise-linear oscillators with two variables. Kuramoto **[1975]** and Aizawa **[1976]** have recently studied populations of coupled oscillators of the polar-symmetric form used in Examples 1–4 above (the "λ-ω" oscillator of Kopell and Howard **[1973]** or the *AB* oscillator of Ortoleva and Ross **[1974]**, recast in the notation of complex numbers). As in Winfree **[1967]**, they use a narrow but finite distribution of native periods and couple the oscillators via the average of their states. Aizawa simplifies this by postulating interactions such that the collective period is the mean of the native periods, but generalizes by letting the *amplitude* depend on native period in the way characteristic of entrained λ-ω oscillators. Computer simulations are carried out and plotted in the same format, facilitating comparison of the two cases (e.g., compare Winfree's **[1967]** Figure 6 with Aizawa's **[1976]** Figures 1 and 2, and Winfree's Figure 7 with Aizawa's Figure 6). Aizawa's analytical approximations are a cut above Winfree's but still leave much to be desired. (Kuramoto takes a simple-clock limit in which no oscillator's amplitude can change.)

XIII. CAN THESE PAPERS BE USED TO PRODUCE CRITERIA FOR INITIATION AND STABILITY OF MUTUAL ENTRAINMENT OF POPULATIONS WITH A FINITE DISTRIBUTION OF PERIODS?

E. Oscillators communicating by delta functions.

1. *Mutual entrainment*. A quite different formulation of mutual entrainment questions would be appropriate to another common case. Suppose oscillators interact not smoothly and continuously, but by single spikes of output. This is the case in the swarms of fireflies routinely observed to synchronize their flashing so that whole trees explode with light at intervals of 6 seconds all evening. It is the case with cells of the social amoeba, which releases a pulse of cAMP every 5 minutes. It seems to be the case with cells of your heart's pacemaker, which fire an electrical spike once a second. In all these cases the consequence of receiving a signal has been measured: the rhythm of the receiving cell or animal is reset in phase. A phase transition curve, ϕ' vs. ϕ, summarizes the effect. In these measurements it turns out that near the time of pulse emission $\phi' = \phi$ with $d\phi'/d\phi < 0$. With this dependence of ϕ' on ϕ, laggards get advanced and precocious individuals get delayed each time, so mutual synchronization is stable. But this description pretends that the aggregate rhythm is a single discrete spike, while simultaneously admitting the existence of untimely individuals. How are untimely spike outputs to be reckoned in? Peskin **[1975]** does it by making the ϕ' vs. ϕ map a function of spike magnitude: $\phi' = \mu(\phi, M)$. Given this extra flexibility,

XIV. ARE THERE MORE INTERESTING MODES OF TEMPORAL ORGANIZATION IN WHICH SEVERAL SPIKES OF DIFFERENT HEIGHTS ARE FIRED OFF WITHIN THE POPULATION, BY AS MANY TEMPORAL "SUB-CULTURES" OF INDIVIDUALS?

This seems to be a problem for difference equations, but I have not the slightest notion how to formulate it.

2. *Resetting maps*. Some thought about the ϕ' vs. ϕ maps of an individual oscillator generates a fairyland of intriguing analytical consequences. First of all, such curves are most naturally drawn on a crystal lattice (wallpaper), or equivalently on the torus. If the curve is smooth then it must belong to one of the homotopy groups of mappings from $S^1 \to S^1$, which correspond to the integers. Thus there are curves which rise up through any integer number W of cycles of ϕ' while ϕ scans one cycle. Graphed on a torus, such curves link W times through its hole. In principle, limit cycle dynamical systems permit any integer winding number, though systems with only two degrees of freedom perturbed by a temporarily constant change of parameters can yield only types $W = 1$ and $W = 0$.

Unless I am mistaken (counterexamples would be welcome) the reason for this latter constraint is geometrically transparent. Consider any of the planar limit cycles in Examples 1–4, and their isochrons. Consider any fixed modification of the dynamical flow during the temporary stimulus. During that time oscillators initially on the attracting cycle follow other trajectories (Figure 9). At any moment they lie on a new ring of initial conditions homologous to the attracting cycle. This new ring, like the attracting cycle, is non-self-intersecting because dynamical flow in the plane during the stimulus is non-self-intersecting. This new ring of initial conditions either does or does not once encircle the phaseless manifold, the convergence point of the isochrons. Accordingly the winding number of phase along this ring (that is, the dependence of ϕ' on the initial ϕ) is either 1 or 0. It seems interesting in this context that all known resetting curves, measured on the greatest variety of biological clocks, are either type 1 or type 0.

This is not typically true of ordinary differential equation models involving three or more variables. It therefore comes to me as some surprise that a time delay model for circadian rhythms (an integral equation) has resetting curves only of type 1 and type 0, according to the analog computations of Johnsson and Karlsson [**1971**].

XV. IS THERE ANY ANALYTIC WAY TO DESCRIBE THE ISOCHRONS AND RESETTING MAPS OF LIMIT CYCLE OSCILLATORS WHOSE STATE SPACE IS NOT FINITE-DIMENSIONAL (FOR EXAMPLE, $\dot{x}(t) = kx(t - \tau)$)?

See Hale [**1977**], p. 242, for proof of the existence of asymptotic phase ("isochrons") for delay equations, and a method of construction near the attracting cycle.

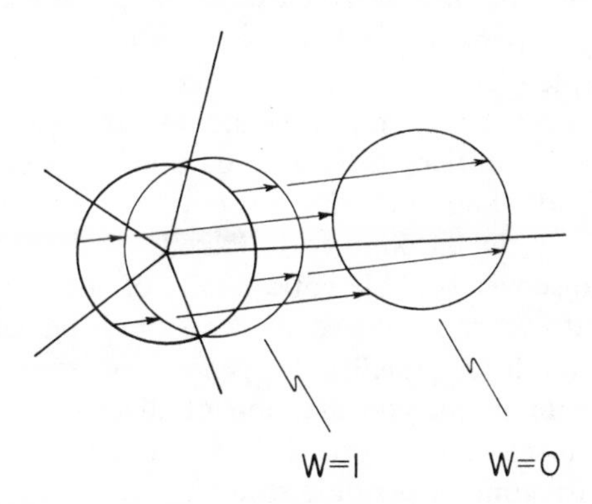

FIGURE 9. From initial conditions on an attracting cycle, trajectories during perturbation displace the ring of states from $W = 1$ to $W = 0$. With time-varying geometry of flow, or with a third dimension, other winding numbers can be attained.

Iteration of the ϕ' vs. ϕ map (as by repetitive stimuli) yields a curve of type W^m, m being the number of iterations. Only in cases $W = 0$ and $W = 1$ does the winding number remain the same. The hobby of iterating such maps has peculiar consequences unless the map has slope <1 everywhere (as only a type 0 resetting map can; but type 0 has another pathological implication, developed below). These consequences include multiple fixed points and the emergence of chaotic behavior, a result explored in another context by May [**1976**] and by Guckenheimer et al. [**1977**]. A variety of physiological systems must design around such considerations, as in the example of heart-beat synchronization, the coordination of swimming, crawling, walking, running and flying, and the synchronization by natural photoperiods of daily and seasonal metabolic rhythms such as flowering and hibernation. Though the difference equations of population biology ($R^1 \to R^1$) have of late received much attention from this point of view, the special peculiarities of $S^1 \to S^1$ maps have not yet been exploited so far as I am aware.

XVI. ARE THERE ANY?

Given a picture of any dynamical system's isochrons, it is easy to derive these resetting maps geometrically. It is only necessary to integrate the perturbed dynamical equations during a stimulus to see how initial conditions on the attracting cycle are displaced by the stimulus. Each point on the attracting cycle is displaced from one isochron to another. That change is from ϕ to ϕ'. Knowing a little about the general layout of isochrons in state space, and especially about their topological structure, can thus provide a convenient key to behavior of practical importance for timing of repetitious activity. If the oscillator can be counted on to return sufficiently close to its attracting cycle before a next impulse again perturbs it, then the ϕ' vs. ϕ curve is *all* we need to know for a complete analysis of its reaction to any sequence of stimuli. In the case of periodically repeated stimuli, the ϕ' vs. ϕ curve, appropriately iterated, suffices to answer questions about fixed points of higher period and about multiple fixed points of any given period (nonunique, hysteretic entrainment). Exactly this program was used successfully to discover bistable entrainment of the fruit fly's circadian clock under light cycles composed of two stimuli per day (e.g., sunrise and sunset (Pittendrigh [**1965**])). In general, it seems reasonable to suspect that the temporal coordination of diverse metabolic rhythms within any organism may be established in a diversity of stable ways under any given regime of periodic stimuli.

XVII. CAN THE ϕ' VS. ϕ RESETTING MAP BE USED TO PRESCRIBE CHARACTERISTICS OF AN OSCILLATOR AND ITS DRIVER THAT CAN BE EXPECTED TO RESULT IN ALTERNATIVE STABLE MODES OF ENTRAINMENT PULSES?

F. Lines and sheets of biological clocks.

1. *Intestinal peristalsis.* Up to here we have implicitly ignored any spatial considerations, supposing that every clock interacts on an equal footing with each other. But suppose instead that they are arranged in physical space, and every oscillator interacts only with its immediate neighbors. Physiologically this may be the commonest situation. For example, the successive cylindrical area elements of your small intestine are autonomous neural oscillators, each coupled to the next. Synchronization proceeds on a local basis. Because there is a substantial gradient of native period from stomach toward anus, the intestine breaks up into smaller domains of mutual entrainment, resulting in a discrete staircase distribution of frequency along its length (Figure 10). Oscillators at the boundary of adjacent domains are apparently in a temporally disorganized state, not unlike the geometrical disorganization of magnetic dipoles in the transition layer between magnetic domains. This problem has been simulated on analog computers by chains of coupled van der Pol oscillators (e.g., Brown et al. [**1975**]). But so far as I know there is no theorem about domain widths, stability, critical steepness of the frequency gradient, etc. Is this problem tractable to mathematical analysis?

2. *Quantized rhythmic pattern formation by clocks in fungi.* Gradients, though interesting, play lesser roles in contemporary thought about circadian rhythms. In fact there *is no* thought about spatial distribution

even of phase alone in circadian rhythms. This seems quite natural since circadian clocks are almost always all entrained by the rhythm of light and darkness to which most cells, regardless of their locations, may be

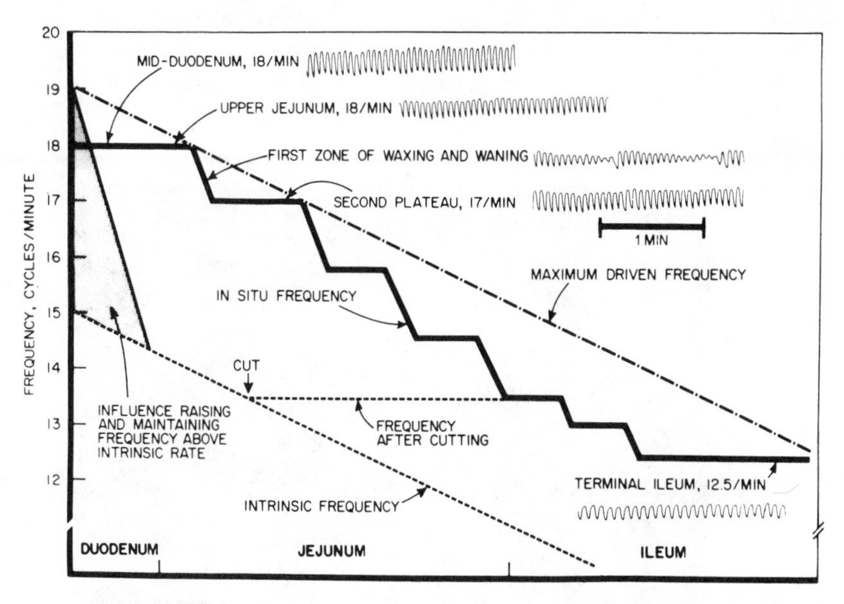

FIGURE 10. An idealized picture of the synchronization plateaus of smooth muscle along a smooth gradient of native frequencies (from Davenport [1977]).[4]

exposed to a sufficient degree to ensure entrainment. But in laboratory conditions, with the cadence caller locked outdoors, spatial patterns of phase do arise.

For example, there are a great diversity of colonial fungi which harbor internal biochemical clocks with a period around one day. This rhythm and its spatial arrangement of phase are permanently recorded as a regular banding of the fungus along its path of growth (Figure 11). The period is about a day of time, which corresponds to anywhere from a millimeter to a centimeter of space as the frontier of the colony advances along its path of growth across the nutrient substrate.

If the whole colony is exactly synchronous then the banding of a colony consists of a succession of rings, each being the image of its frontier at that past hour when clock phase passed 0. The principle here is much the same as in the formation of growth rings in trees, except that the clock is internal instead of external.

---

[4] Reproduced with permission from Davenport, H. W.: *Physiology of the Digestive Tract,* 4th edition. Copyright © 1977 by Year Book Medical Publishers, Inc., Chicago. (Adapted from Diamant, N. E., and Bortoff, A.: Am. J. Physiol. 216; 734, 1969. Electric records supplied by A. Bortoff.)

But it turns out that only about two-thirds of the colonies (in the particular species I have most worked with) do in fact make rings. About two-thirds of the remainder make a clockwise or anticlockwise spiral of the same spacing. And about two-thirds of the remainder make 2-armed spirals. Each arm has double spacing so the bands still appear at the same intervals along any growth path. A few make 3-arm spirals. Each arm has triple-normal spacing (Bourret et al. [**1969**], Winfree ([**1970**], [**1973**])).

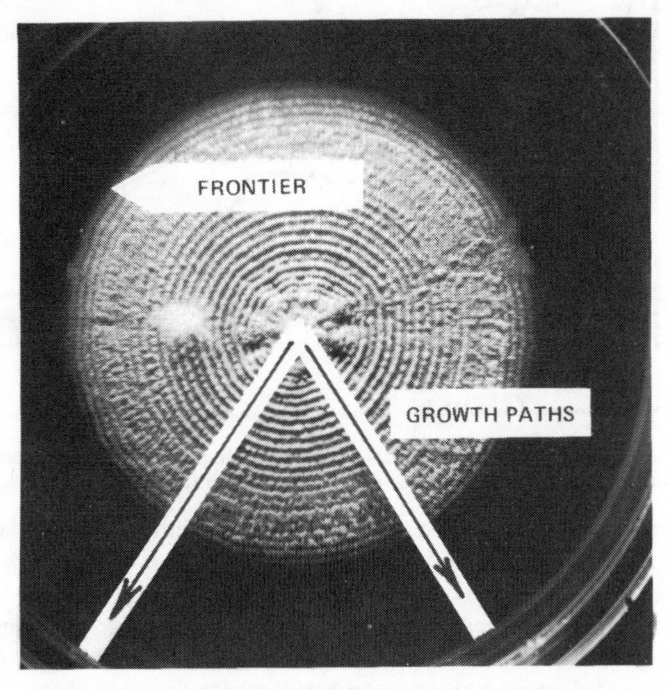

FIGURE 11. A disk of fungus expanding radially (see "growth paths") from an inoculum at the center. Growth is mainly along the frontier where the patterning is laid down. Bands are about 1mm apart, representing 16 hours' growth in this particular species.

I take these patterns to indicate that biochemical clocks lay down the rhythm along each growth path but the clocks on different growth paths are not all synchronous. In the case of single spirals, phase advances through one full cycle around the frontier of the colony. In the case of double spirals, phase advances through exactly two cycles around the frontier. How can this happen?

One easy interpretation, compatible with the existing experiments, invokes a biochemical oscillator in each volume element and ignores lateral coupling between adjacent volume elements (Winfree, ([**1970**], [**1973**])). Since volume elements are independent in first approximation, we can deal with the ordinary differential equation by mapping each point on the colony's frontier into state space, and watching it follow the trajectories of the ordinary differential equation. I assume a simple limit cycle kinetics with a repelling equilibrium, and assume that the whole fungus starts from near equilibrium when the spore germinates. However soon after germination, the various parts of the frontier have grown away from each other and have departed from equilibrium somewhat. The image of the frontier is a ring in state space lying near equilibrium and winding some integer number of times around it (Figure 12). Since all volume elements follow the local trajectories away from equilibrium, that winding number is conserved. Eventually the whole frontier is on the attracting cycle, winding around it an integer number of times. That integer W characterizes the pattern. W is a topological "quantum number" of the banding pattern.

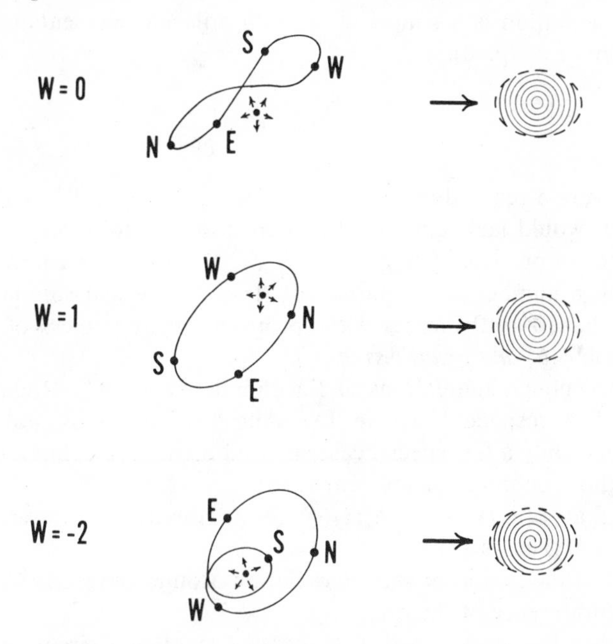

FIGURE 12. The image of the fungus' frontier is a ring in state space, initially near equilibrium. As it departs from equilibrium it acquires a winding number ($W = 0$, 1 or -2 shown here) which cannot change. Each winding number determines a different connectivity of bands on the disk.

This interpretation leads to a variety of mathematical problems. The first and least well formulated is, "What is the probability distribution $p(W)$?" Taking the curve's center of mass to lie at the equilibrium, this probability distribution is undoubtedly determined by some measure of the "wiggliness" of the closed curve (the frontier's image in state space). I suppose this wiggliness could best be described as a Fourier spectrum of amplitude of harmonics of the function $\phi(\theta)$ which assigns an angle in state space to each angle on the frontier. High harmonics, implying steep concentration gradients, would have lower amplitudes according to some plausible physical chemical rule.

XIX. GIVEN SUCH A RULE, HOW CAN A HISTOGRAM OF $p(W)$ BE CALCULATED?

3. *Waves and rings of simple clocks.* A second problem is implicit in this last assumption, that the frontier's image is only finitely wiggly, namely that the phases of adjacent volume elements are not completely independent. Once we seriously admit that fact, the character of the mathematical problem changes radically (at least if we are prepared to wait a long time and allow arbitrarily large frontiers), because now we deal with a *partial* differential equation. The simplest useful caricature of that equation is a simple clock with adjacent elements of arc along the frontier coupled:

$$\frac{\partial \phi}{\partial t} = f(\phi) + D\frac{\partial^2 \phi}{\partial \theta^2}.$$

If ϕ were a real-valued scalar and f were a map f: $R^1 \to R$, then this problem would lack some of the interest that it does in fact acquire in our case. In our case $\phi \in S^1$, $\theta \in S^1$, f: $S^1 \to S^1$. f is characterized by a winding number W. It turns out (from numerical simulations once again) that $W \neq 0$ implies wave propagation, regardless of the initial condition $\phi(\theta)$ *and regardless of* $f(\ )$.

When enough simulations had accumulated in 1973, Richard Casten (now of Aerospace Corp. in Los Angeles) undertook some analytic solutions. Only a few special cases proved manageable, but a conjecture, still without counterexample, emerged:

(A) that $\phi(\theta, t) \to W\theta + t\oint f(\phi)\, d\phi$ as the ring circumference approaches 0 (i.e., as $D \to \infty$);

(B) that the period of wave circulation around the ring increases with the circumference of the ring;

(C) that the wave speed also increases with the circumference (but only up to an upper asymptote and not fast enough to prevent the circulation time from increasing).

XX. ARE (A), (B), AND (C) TRUE OF SOME BROAD CLASS OF $f(\ )$'S?

4. *Rotating waves in media of nonsimple clock oscillators.* Circulating waves do appear in a wide variety of physiological media including heart muscle, nerve fiber, brain tissue, and the retina of the eye. They have not yet been observed in tissue parochially harboring circadian clocks, but few have looked for such spatial patterning of phase in

circadian rhythms. It would be useful to anticipate the conditions under which stable spatial patterning might arise. An interesting problem arises if we attempt this by extending the previous simple-clock model into a 2-dimensional medium. Suppose initially that we have a narrow annulus of material, at each point of which $\dot{\phi} = f(\phi) + \nabla\phi^2$. By analogy with the 1-dimensional ring problem it seems plausible that a rotating wave solution should be stable on this annulus. Now make the annulus fatter, more like a disc with its center removed. The inner rim has a shorter circumference than the outer rim and so more nearly adopts the limiting solution (A) above. But in the limit, which must be taken as we fill in the hole to make a simply connected 2-dimensional medium (a disc), all phases come arbitrarily close together. We would have a phase singularity at a pivot point of a rotating wave. But this is scarcely plausible in biochemical or biological context. So the next step is to make the oscillator model more realistic by allowing it more than one internal state variable. Kuramoto and Yamada [**1976**] undertook analytic studies and calculations of spatial patterns of phase using a very similar simple clock model. They ran into this contradiction and resolved it by numerical calculations using a 2-variable oscillator such as in Examples 1–4 above (Yamada and Kuramoto [**1976**]). With such an equation replacing $f(\phi)$ in each volume element, the pivot has a finite option: *its amplitude can go to* 0 so all phases can coexist near that point while the spatial gradient of both state variables remains finite.

Solutions of this sort have been attempted by Erneux and Herschkowitz-Kaufman [**1976**], Greenberg [**1978**], and by Cohen et al. [**1978**] in simply connected 2-dimensional continua.

Their numerical solutions verify the existence of a stable solution, but so far only one case has yielded analytic solutions of its dynamics all the way down to radius zero (the pivot). At this time it is not yet clear that this solution has physically reasonable behavior. The problem seems especially seductive on account of its many symmetries: the solution is periodic in time and has rotational symmetry in space, the local kinetics has polar symmetry in state space, diffusion can be handled symmetrically, and the solution far from the origin is a plane wave train. It is also intriguing because the pivotal volume element must be at the repelling equilibrium inside the attracting cycle. It resides there stably on account of the symmetries of the problem. From this timeless point, the rotating pattern of phase emits a spiral wave toward infinity.

XXI. GIVEN AN ORDINARY DIFFERENTIAL EQUATION $\mathbf{R}(\mathbf{x})$ WHICH HAS AN ATTRACTING LIMIT CYCLE, IS IT POSSIBLE TO SOLVE $\partial\mathbf{x}/\partial t = \mathbf{R}(\mathbf{x}) + \nabla^2\mathbf{x}$ NEAR THE CENTER OF A STABLY ROTATING 2-DIMENSIONAL WAVE WHICH BECOMES AN OUTGOING PERIODIC WAVE TRAIN AT LARGE DISTANCES FROM THE CENTER?

5. *Nonperiodic solutions*. There is another intriguing problem lurking behind this one. Numerical work shows (Gulko and Petrov [**1972**]; Rössler [**1978**]) that rotating solutions of this sort, using less symmetric kinetic schemes, need not always have a pivotal volume element. While volume elements far from the center exhibit nearly periodic activity, the center can still exhibit peculiar irregularities that have resisted description up to now. In addition to numerical observations, this effect is observed in the laboratory, not yet in circadian rhythms, but in oscillatory chemical reactions and in beating heart muscle.

XXII. WHAT PROPERTY OF THE ORDINARY DIFFERENTIAL EQUATION OF REACTION **R**(**x**) DISTINGUISHES ROTATIONALLY EXACT SOLUTIONS FROM SOLUTIONS WHICH SEEM "CHAOTIC" NEAR THE WOULD-BE PIVOT?

G. Probability distributions of phase in populations of independent clocks. The prospect of spatially organized inhomogeneity of circadian rhythm phasing has not yet been realized experimentally, but a similar problem has been and cries out for competent modelling. As noted above, it appears that living organisms are typically "communities" of clocks whose collective behavior may be dominated by mutual coupling as noted above, or might not be. There is a body of experimental evidence suggesting that interactions (in the situations involved) are not important on the short time scale of several days. Taken together with the inevitable variability of biological material, this notion has at least two striking implications, neither of which has been adequately formulated for exact mathematical inference.

1. *Phase resetting in a single simple clock*. For the first, we again conjure back the simple clock model, $\dot{\phi} = f(\phi, P)$. Suppose some stimulus $P(t - t_0)$ affects the clock, starting at an initial time $t = t_0$ when the clock is at phase $\phi = t_0$ mod 1. Let the stimulus end at some time $t_0 + M$. Taking a simple form of $f(\;)$, e.g., $f(\phi, P) = 1 + P\cos\phi$, it is easy to integrate the rate equation and so obtain the new phase $\phi = g(\phi, M)$ reached at the end of the stimulus. But *whatever* smooth bounded f we choose, the new phase ϕ will scan through + 1 cycle as the old phase ϕ scans a cycle. The resetting curve of a simple clock model necessarily has winding number $W = 1$ (see Winfree [**1975**], appendix). And a simple clock's $g(\phi, M)$ has no singularities. Its contour map consists of parallel displacements of a single smooth curve.

The behavior of real circadian clocks contrasts starkly with this picture. Already it has been 10 years since the first resetting curve of winding number $W = 0$ was recognized, and now they have been established in over a dozen different species of organisms, using diverse stimuli and laboratory procedures. Moreover in at least four circadian systems, $g(\phi, M)$ has been laboriously measured. Figure 13 shows the first example, using my data from the fruit fly *Drosophila*. Figure 14 shows the surface that fits the data of Engelmann et al. [**1973**] for the plant *Kalanchoe*. There are unmistakable singularities in these measured functions. Their level contours converge to points disposed at regular intervals on the contour map (Figure 15). So simple clock models are plainly unrealistic in some essential way.

In my opinion the essential flaw in the simple clock model is its

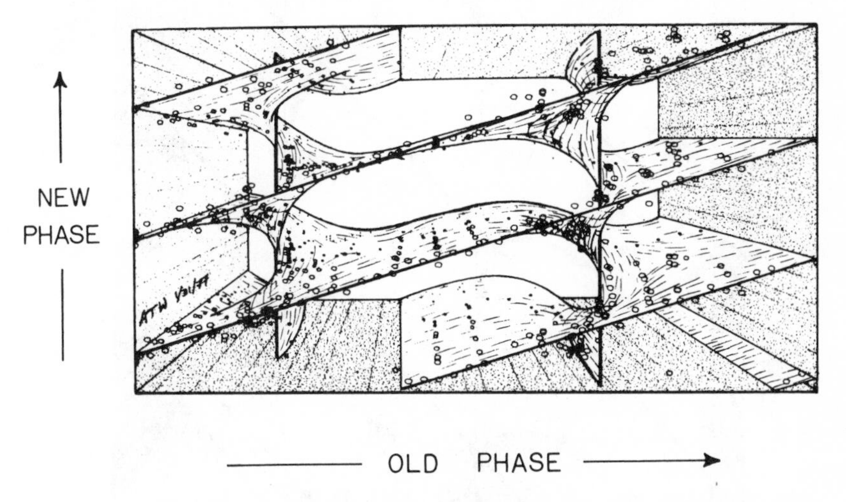

FIGURE 13. A perspective drawing of the fruit fly clocks' "time crystal" showing 6 unit cells (from Winfree [**1978**], Figure 6). Each contains one turn of a screw surface. Old phase to the right, new phase upward, stimulus magnitude increasing into the background.[5]

restriction to a cycle of fixed amplitude. Resetting surfaces like Figures 13 and 14 are typical of oscillators not so restricted. For example, take Example 1 above, with radial isochrons: Let the stimulus' effect be a decrease of x_1 by amount M. The new phase is then given by

$$\tan 2\pi\phi' = \frac{M + \sin 2\pi\phi}{\cos 2\pi\phi}$$

which is a screw surface like Figures 13 and 14. Its level contours $k \cos 2\pi\phi = M + \sin 2\pi\phi$ are arcs of sines equispaced along the ϕ axis at $M = 0$, symmetrically converging from all directions to $M = 1$, $\phi =$ any integer (Figure 15).

[5] This first appeared in New Scientist, London, the weekly review of science and technology.

2. *Phase resetting in a population of independent simple clocks.* But there remains an excluded possibility. What if the rhythm we observe in the whole organism is the *sum* of rhythmic outputs of many independent clocks? Suppose, for example that each simple clock secretes a smooth periodic function $X(\phi)$, and that these secretions are summed,

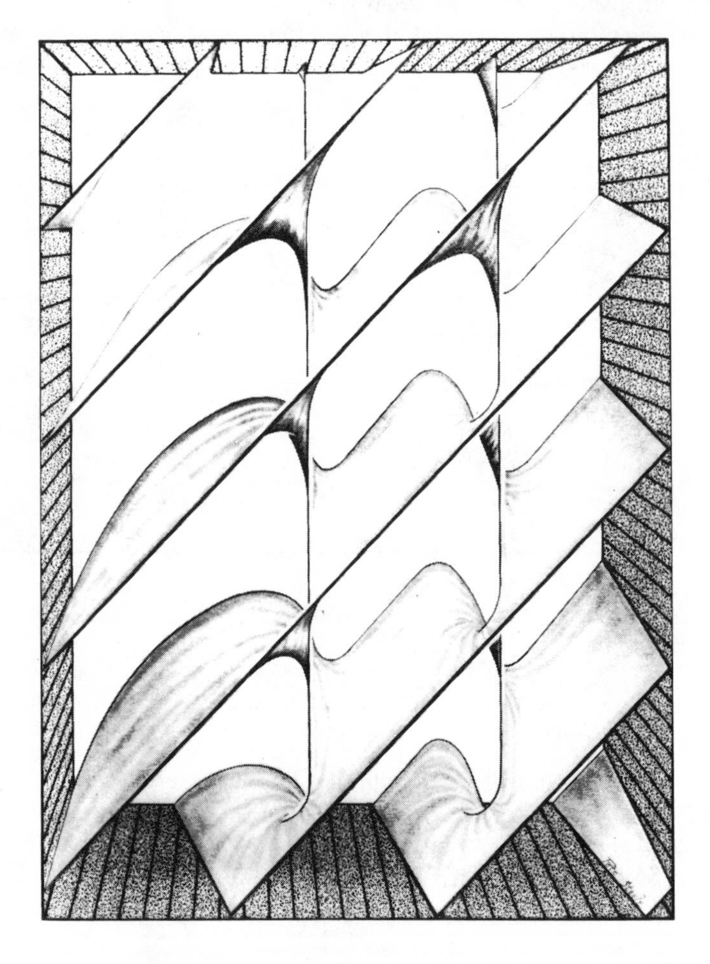

FIGURE 14. A perspective drawing of the *Kalanchoe* flower clock's "time crystal" showing 12 unit cells (from Winfree **[1978]**, Figure 7). Each contains one turn of the screw surface. Old phase to the right, new phase upward, stimulus magnitude increasing into the background. As in Figure 13, sensitivity to stimuli increases with time in the dark: sensitivity is so low in the first column of unit cells that the screw axis is outside the range of magnitudes used. Data points (from Engelmann et al. **[1973]**) are omitted for clarity.[6]

---

[6] This first appeared in New Scientist, London, the weekly review of science and technology.

and that for the phase of the superposition we take as our measure the phase of the fundamental harmonic of $\Sigma_i X(\phi_i)$.

A convenient geometrical way to compute the collective phase is shown in Figure 16. Here a circle depicts the fundamental of $X(\phi)$ and the dots are oscillators disposed at various phases around the circle. The phase is simply the geometrical angle from the top of the circle. The center of mass of any cluster of dots also lies at some angle, measured clockwise from the top of the circle, about the center. That phase is the phase of the collective rhythm $\Sigma_i X(\phi_i)$ of that cluster of clocks.

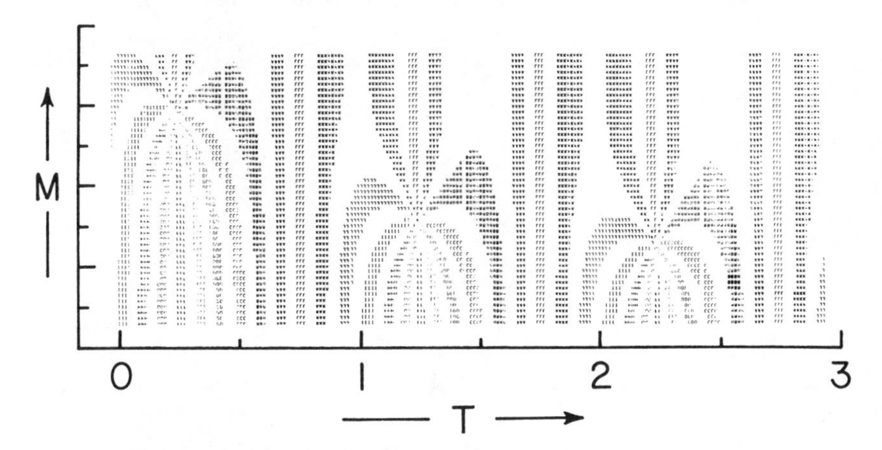

FIGURE 15. Digitized contour lines of the new phase surface of Figure 13 above the (T, M) plane: $\phi = T \bmod 1$. The three unit cells are not identical because photoreceptor sensitivity increases secularly through the first two cycles. Note that the contours are equispaced at $M = 0$ and converge to repeated singularities: the screw axes.

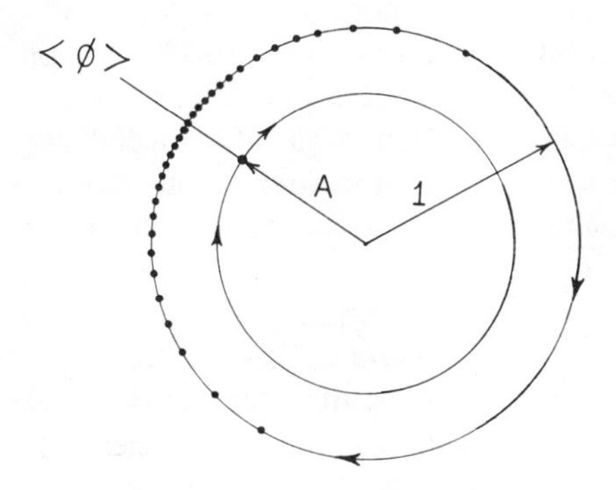

FIGURE 16. The circle represents unit amplitude of the first harmonic of any rhythm. Dots represent clocks at corresponding phases. The inner dot represents the phase and amplitude of the collective rhythm (averaged). It follows the inner circle so long as clocks are independent and unperturbed.

A "quick and dirty" argument suffices to show that a population can exhibit resetting with winding number $W = 0$. Figure 17 represents the phase circle of such a simple clock. The arrows indicate its phase velocity during a stimulus of a given intensity, which, by definition of a "simple clock", depends (smoothly) only on the phase. Let us assume clock motion is eventually arrested by prolonged exposure to the stimulus, i.e., the clock advances or delays toward a stagnation point Q.

Now consider a population of simple clocks, all endowed with the same circadian period but spanning some small range of phases. A population straddling the boundary between delaying and advancing

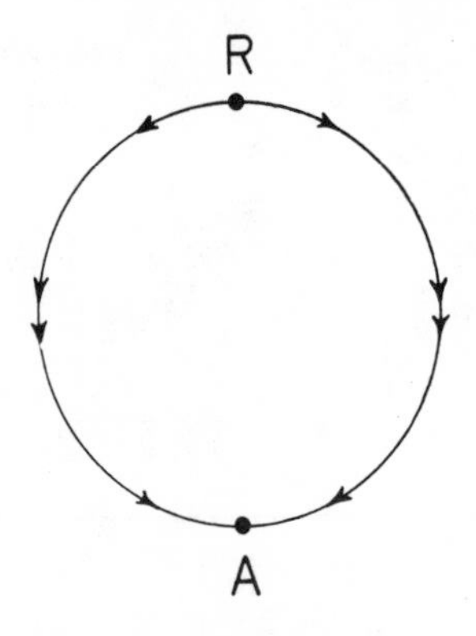

FIGURE 17. Analogous to Figure 16. Arrows on the ring indicate a simple clock's phase-dependent angular velocity during a perturbation that drives it to rest at attractor A. The corresponding repellor is at R. Phase increases uniformly clockwise in the absence of perturbation.

regions (phase R in Figure 17) at the moment the stimulus begins is split into advancing and delaying portions, leaving only some few clocks near phase zero, where phase velocity is so close to nil that they hang almost motionless during the finite-duration stimulus. Choose stimulus duration long enough so that the bulk of both half-populations moves into the bottom quarter-cycle, surrounding the stagnation phase. Thus the mean phase of the whole population (which determines the collective rhythm's phase) is also inside the bottom quarter-cycle. Under the same assumptions, a population initially lying wholly to one side or the other of phase zero also would end up in the bottom quarter-cycle. Thus the whole resetting curve lies inside that quarter-cycle range. So the resetting curve cannot scan through any full cycles, and must therefore be type 0.

On the other hand, resetting is necessarily type 1 ($\phi' \approx \phi$) in response to vanishingly small stimuli. Let us put these facts together. In Figure 18 Box $ABCD$ spans one cycle of old phase ϕ and reaches from stimulus magnitude $M = 0$ up to a magnitude sufficient to elicit type 0 resetting. Now let us traverse path $ABCDA$ while keeping track of $g(\phi, M)$. Along path AB, $\phi' = \phi$ so g moves up one cycle. Along BC some further change $\Delta\phi$ occurs. Along CD, where resetting is type 0, g just rises and falls back, adding nothing to the net change of g. DA is the same as BC but one cycle earlier and it is traversed backwards, so g changes by $-\Delta\phi$, cancelling the change along BC. The net effect of one tour of inspection around $ABCDA$ is that g rises through 1 cycle. Thus a plot of $\phi' = g$ above the (ϕ, M) plane has a *helical* boundary along $ABCDA$. The surface interior to $ABCDA$ is accordingly a screw surface, as seen in the measurements on actual organisms in Figures 13 and 14.

Thus can the principal qualitative features of resetting behavior in

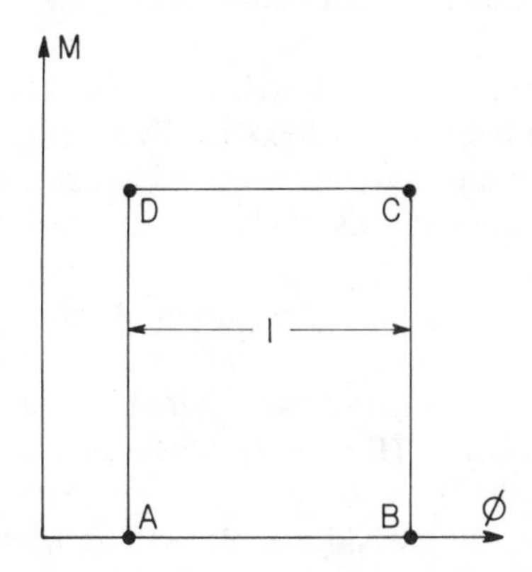

FIGURE 18. A rectangular path $ABCDA$ is constructed on the plane of stimulus descriptions in Figure 15.

heterogeneous populations be loosely rationalized. But this field entirely lacks any proper mathematical models that take into account the now widespread notion that an organism is a *clockshop* more nearly than a single clock. I think this will necessarily become a new domain of model-making in the 1980's. We will have to have models that enable graceful quantitative prediction of the distribution of phase in a population, following various treatments, assuming either a hypothetical mechanism for the individual clock and a mode of coupling to others, or (which would probably suffice) assuming only the known resetting behavior of the single clock or of a well-synchronized population.

XXIII. HOW CAN THE FIRST AND SIMPLEST SUCH MODEL BE FORMULATED, IN WHICH PHASE IS NOT A NUMBER BUT A DISTRIBUTION FUNCTION, AND THE OBSERVABLE PHASE IS SOMETHING LIKE THE MEAN OF THAT DISTRIBUTION?

The urgency of acquiring such a distribution model is seen in starkest form in the case of the simple clocks. I personally doubt that such fixed amplitude models are very realistic. But distributional aspects of phase play an equally essential role in populations of limit-cycle oscillators because they endow the population with a virtual amplitude, which reflects the width of the distribution of the phases on the attracting cycle. The second of the two problems alluded to above concerns not so much the phase of a collective rhythm as its ostensible amplitude.

3. *Amplitude resetting in a population of independent clocks*. Returning to Figure 16 we see that the amplitude of the fundamental harmonic $\Sigma_i X(\phi_i)$ is necessarily less than the single oscillator's amplitude. It is also as stable as is the distribution of phase. Thus a population of limit cycle oscillators can dissemble quite convincingly as a harmonic oscillator if we are allowed to measure only the first harmonic of its collective output. I have extensive and well-prepared data which measure this Hamiltonian-like behavior in the fruit fly's circadian clock. But no one has yet published an analytic model which rationalizes these observations in terms of the collective behavior of a population of independent limit-cycle oscillators.

4. *Amplitude resetting in a population of mutually coupled clocks*. As noted above, a population of mutually entrained oscillators can behave similarly by virtue of having two alternative stable phases of entrainment of each oscillator (Figure 7). Once an oscillator is pushed into one "potential well" by a stimulus, it stays there. The amplitude of the collective rhythm is thus adjustable and is perfectly stable but only within limits. It would be nice to have not just my computer simulations but analytic models by which to recognize the characteristic features of each kind of mechanism.

XXIV. WHAT QUALITATIVE FEATURES OF PHASE AND AMPLITUDE RESPONSE DISTINGUISH A POPULATION OF INDEPENDENT CLOCKS FROM A POPULATION IN WHICH MUTUAL COUPLING RESTRICTS THE PHASE DISTRIBUTION TO A MORE LIMITED CONTINUUM?

H. Exit. I think that in posing these 24 questions, I am serving the community of "clocks" physiologists as spokesman for widely recognized needs. Accordingly I imagine that looking back from the 1980's, the questions I pose here will seem primitive or silly in the light of a considerable mass of intervening experimental and mathematical work. This selection of questions is of course biased by my own experiences and prejudices and that is what you gamble on if you take these questions seriously. But if I were looking for mathematical problems in this area right now these are the ones I would choose among for a focused attack.

BIBLIOGRAPHY

The following constitutes a sampling (only) of excellent papers pertinent to the themes explored above. Fuller discussion, examples, and a fairly complete bibliography can be found in Winfree [**1979**] cited below.

Aizawa, Y. 1976. *Synergetic approach to the phenomena of mode-locking in nonlinear systems*, Prog. Theoret. Phys. **56**, 703–716.

Best, E. 1976. *Null space and phase resetting curves for the Hodgkin-Huxley equations*, Ph. D. Thesis, Purdue Univ.

Bourret, A., R. G. Lincoln and B. H. Carpenter. 1969. *Fungal endogenous rhythms expressed by spiral figures*, Science **166**, 763–764.

Brown, B. H., H. L. Duthie, A. R. Horn and R. H. Smallwood. 1975. *A linked oscillator model of electrical activity of human small intestine*, Amer. J. Physiology **229**, 384–388.

Coddington, E. A. and N. Levinson. 1955. *Theory of ordinary differential equations*, Maple Press, York, Pa. Theorem 2.2, p. 323.

Cohen, D. S., J. C. Neu and R. R. Rosales. 1978. *Rotating spiral wave solutions of reaction-diffusion equations*, SIAM J. Appl. Math. **35**, 536–547.

Davenport, H. W. 1977. *The digestive tract*, Motility of the Small Intestine, Yearbook Medical Publishers, Chicago, Ill., Chapter 4, pp. 58–71.

Engelmann, W., H. G. Karlsson and A. Johnsson. 1973. *Phase shifts in the Kalanchoe petal rhythm caused by light pulses of different durations*, Internat. J. Chronobiol. **1**, 147–156.

Clay, J. R. and R. L. DeHaan. 1979. *Fluctuations in interbeat interval of spontaneously beating embryonic heart cells arise from membrane voltage noise*, Biophys. J. (to appear).

Erneux, T. and M. Herschkowitz-Kaufman. 1977. *Rotating waves as asymptotic solutions of a model chemical reaction*, J. Chem. Phys. **66**, 248–250.

FitzHugh, R. 1961. *Impulses and physiological states in theoretical models of nerve membrane*, Biophys. J. **1**, 445–466.

Grattarola, M. and V. Torre. 1977. *Necessary and sufficient conditions for synchronization of nonlinear oscillators with a given class of coupling*, IEEE Trans. Circuits and Systems **CAS-24**, 209–215.

Greenberg, J. M. 1976. *Periodic solutions to reaction-diffusion equations*, J. Appl. Math. **30**, 199–205.

______. 1978. *Axi-symmetric, time-periodic solutions of reaction-diffusion equations*, SIAM J. Appl. Math. **34**, 391—397.

Guckenheimer, J. 1975. *Isochrons and phaseless sets*. J. Math. Biol. **1**, 259–273.

Guckenheimer, J., G. Oster and A. Ipaktchi. 1977. *The dynamics of density dependent population models*, J. Math. Biol. **4**, 101–147.

Jongsma, H. J, M. Masson-Pevet, C. C. Hollander and J. de Bruyne. 1975. *Synchronization of the beating frequency of cultured rat heart cells*, Developmental and Physiological Correlates of Cardiac Muscle, M. Leiberman and T. Sano, eds., Raven Press, New York, pp. 185–196.

Gul'ko, F. B. and A. A. Petrov. 1972. *Mechanism of formation of closed pathways of conduction in excitable media*, Biofizika **17**, 261–270.

Hale, J. K. 1963. *Oscillations in nonlinear systems*, McGraw-Hill, New York, Theorem10-1, p. 94.

______. 1969. *Ordinary differential equations*, Wiley-Interscience, New York, Theorem 2.1, p. 217.

______. 1977. *Theory of functional differential equations*, Applied Math. Sciences, vol. 3, Springer-Verlag, New York and Heildelberg.

Hayashi, C. 1964. *Nonlinear oscillations in physical systems*, Electrical and Electronic Engineering Series, McGraw-Hill, New York.

Kalmus, H. and L. A. Wigglesworth. 1960. *Shock excited systems as models for biological rhythms*, Cold Spring Harbor Sympos. Quantitative Biol., vol. 25, A. Chovnick ed., pp. 211–216.

Kaus, P. 1976. *Possible adaptive values of the resonant slave in biological clocks*, J. Theoret. Biol. **61**, 249–265.

Kauffman, S. A. 1969. *Metabolic stability and epigenesis in randomly constructed genetic nets*, J. Theoret. Biol. **22**, 437–467.

Kawato, M. and R. Suzuki. 1978. *Biological oscillators can be stopped—topological study of a phase response curve*, Biol. Cybernet. **30**, 241–248.

Kopell, N. and L. N. Howard. 1973. *Plane wave solutions to reaction-diffusion equations*, Studies in Appl. Math. **52**, 291–328.

Kuramoto, Y. and T. Yamada. 1976. *Pattern formation in oscillatory chemical reactions*, Progr. Theoret. Phys. **56**, 724–740.

Kuramoto, Y. 1975. *Self-entrainment of a population of coupled nonlinear oscillators*, Internat. Sympos. Math. Problems in Theoret. Physics, H. Araki ed., Lecture Notes in Physics, vol. 39, Springer, Berlin, pp. 420–422.

Linkens, D. A. 1974. *Analytical solution of large numbers of mutually coupled nearly sinusoidal oscillators*, IEEE Trans. Circuits and Systems **CAS-21**, 294–300.

May, R. M. 1976. *Simple mathematical models with very complicated dynamics*, Nature **261**, 459–467.

Neu, J. C. 1979. *Large populations of coupled chemical oscillators*, SIAM (to appear).

Ortoleva, P. and J. Ross. 1974. *On a variety of wave phenomena in chemical reactions*, J. Chem. Phys. **60**, 5090–5107.

Peskin, C. S. 1975. *Mathematical aspects of heart physiology*, Sci. Publ. Courant Inst. of Math., New York.

Pittendrigh, C. S. 1965. *On the mechanism of the entrainment of a circadian rhythm by light cycles*, Circadian Clocks, J. Aschoff, ed., North-Holland, Amsterdam, pp. 277–297.

Rössler, O. E. 1978. *Chemical turbulence—A synopsis*, Synergetics, H. Haken, ed., Lecture Notes in Physics, Springer-Verlag, Berlin.

Stratonovich, R. L. 1967. *Topics in the theory of random noise*, Vol. II, Gordon and Breach, New York, Chapter 9.

Winfree, A. T. 1967. *Biological rhythms and the behavior of populations of coupled oscillators*, J. Theoret. Biol. **16**, 15–42.

______. 1970. *Oscillatory control of cell differentiation in Nectria*? Proc. IEEE Sympos. Adaptive Process **9**, 23.4.1–23.4.7.

______. 1973. *Polymorphic pattern formation in the fungus Nectria*, J. Theoret. Biol. **38**, 363–382.

______. 1974. *Patterns of phase compromise in biological cycles*, J. Math. Biol. **1**, 73–95.

______. 1976. *On phase resetting in multicellular clockshops*, The Molecular Basis of Circadian Rhythms, J. W. Hastings, H. G. Schweiger, eds., Dahlem, Berlin, pp. 109–129.

______. 1978. *Chemical clocks: a clue to biological rhythms*, New Scientist **80**, 10–13.

______. 1979. *The geometry of biological time*, Springer-Verlag, Berlin.

Yamada, T. and Y. Kuramoto. 1976. *Spiral waves in a nonlinear dissipative system*, Progr. Theoret. Phys. **55**, 2035–2036.

Zwanzig, R. 1976. *Interactions of limit-cycle oscillators*, Topics in Statistical Mechanics & Biophysics, R. A. Piccirelli, ed., Amer. Inst. of Physics, New York.

Department of Biological Sciences, Purdue University, Lafayette, Indiana 47907

Lectures in Applied Mathematics
Volume 17, 1979

Stochastic Modelling and Nonlinear Oscilliations

Donald Ludwig

The lectures gave an introduction to stochastic modelling, followed by some examples of nonlinear oscillations where stochastic effects are important. A naive approach to stochastic modelling would be to take a deterministic model and replace one or more of the parameters by randomly fluctuating quantities. Shortcomings of this approach are exemplified in the problem of population growth and extinction in a random environment. There the choice of stochastic calculus can determine whether the population grows exponentially or goes extinct, although the corresponding deterministic model predicts exponential growth. Further details are given in Keiding (1976), Turelli (1977) and Ludwig (1976).

The Ornstein-Uhlenbeck process was introduced. It illustrates how the long-term behavior of a system can be determined jointly by deterministic and stochastic effects. The corresponding deterministic model has an exponentially stable equilibrium point. Random fluctuations then result in a Gaussian distribution centered at that point. Such results can be generalized, using perturbation techniques for small random fluctuations. See, for example, Ludwig (1974, 1975). Similar methods were employed by White (1977) to study the effects of random fluctuations on a system with an asymptotically stable limit cycle. The variance of the approximating Gaussian can be obtained by solving a Riccati equation along the limit cycle.

The phenomenon of recurrent epidemics illustrates how stochastic effects can be important even for populations of 6×10^5–10^6. Bartlett (1957) proposed a model for measles epidemics which predicts that the period between outbreaks should depend upon the community size. Such a relationship is borne out approximately by data from the United Kingdom. These data also indicate that in larger cities, the epidemic

AMS (*MOS*) *subject classifications* (1970). Primary 60-02, 92-02.

may never fade out. This point is confirmed by a study of measles in North American cities. The critical community size for fade outs is approximately 5×10^5, as shown in Bartlett (1960).

The firing of neurons is an analogous case where stochastic effects are important. According to a model due to Stein (1965), there is a threshold level of current input, below which the neuron will not fire. A stochastic modification of this model predicts that the neuron will fire repetitively, although the average frequency may be very low. The importance of such effects for estimation of physiological parameters is indicated by Tuckwell and Richter (1978).

Other important stochastic effects are studied in Schuss (1977), Matkowsky and Schuss (1977), and Schuss and Matkowsky (1977). This work has applications in the determination of chemical reaction rates.

A Selected Bibliography on Stochastic Modelling

General

M. Iosifescu and P. Tautu, 1973. *Stochastic processes and applications in biology and medicine*, vols. I, II, Springer-Verlag, Berlin.

L. Arnold, 1974. *Stochastic differential equations: theory and applications*, Wiley-Interscience, New York.

E. Wong, 1971. *Stochastic processes in information and dynamical systems*, McGraw-Hill, New York.

N. S. Goel and N. Richter-Dyn, 1974. *Stochastic models in biology*, Academic Press, New York.

D. Ludwig, 1974. *Stochastic population theories*, Lecture Notes in Biomathematics, vol. 3, Springer-Verlag, Berlin.

Z. Schuss, 1977. *Notes on stochastic differential equations*, Univ. of Delaware, Dover.

Problems of extinction, and the stochastic calculus

M. Turelli, 1977. *Random environments and stochastic calculus*, Theoret. Population Biology **12**, 140–178.

N. Keiding, 1975. *Extinction and exponential growth in random environments*, Theoret. Population Biology **8**, 49–63.

_____, 1976. *Population growth and branching processes in random environments*, Inst. of Math. Statist., Univ. of Copenhagen, Copenhagen.

D. Ludwig, 1976. *A singular perturbation problem in the theory of population extinction*, Proc. AMS-SIAM Sympos. on Asymptotic Methods and Singular Perturbations (New York, 1976), Amer. Math. Soc., Providence, R. I., pp. 87–104.

R. M. May, 1973. *Stability and complexity in model ecosystems*, Princeton Univ. Press, Princeton, N. J.

A. H. Gray and T. K. Caughey, 1965. *A controversy in problems involving random parametric excitation*, J. Math. and Phys. **44**, 288–296.

Stability and fluctuation problems

D. Ludwig, 1975. *Persistence of dynamical systems under random perturbations*, SIAM Rev. **17**, 605–640.

B. J. Matkowsky and Z. Schuss, 1977. *The exit problem for randomly perturbed dynamical systems*, SIAM J. Appl. Math. **33**, 365–382.

Z. Schuss and B. J. Matkowsky, 1977. *The exit problem: a new approach to diffusion across potential barriers*, Northwestern U. Applied Math. Tech. Rep. #7706.

B. S. White, 1977. *The effects of a rapidly-fluctuating random environment on systems of interacting species*, SIAM J. Appl. Math. **32**, 666–693.

Epidemics

N. T. J. Bailey, 1975. *The mathematical theory of infectious diseases and its applications*, 2nd ed., Hafner, New York.

M. S. Bartlett, 1957. *Measles periodicity and community size*, J. Roy. Statist. Soc. Ser. A **120**, 48–70.

_____, 1960. *The critical communtiy size for measles in the United States*, J. Roy. Statist. Soc. Ser. A **123**, 37–44.

W. London and J. Yorke, 1973. *Recurrent outbreaks of measles, chicken pox and mumps*, Amer. J. Epidemiology **92**, 453–482.

Neurons

A. V. Holden, 1976. *Models of the stochastic activity of neurons*, Springer-Verlag, New York, Heidelberg, Berlin.

J. Kielson and H. Ross, 1975. *Passage time distributions for Gaussian Markov (Ornstein-Uhlenbeck) statistical processes*, Selected Tables in Mathematical Statistics, Vol. III, Amer. Math. Soc., Providence, R. I.

B. K. Roy and D. R. Smith, 1969. *Analysis of the exponential decay model of the neuron showing frequency threshold effects*, Bull. Math. Biophys. **31**, 341.

R. B. Stein, 1965. *A theoretical analysing neuronal variability*, Biophys. J. **5**, 1973.

H. C. Tuckwell and W. Richter, 1978. *Neuronal interspike time distributions and the estimation of neurophysiological and neuroanatomical parameters*, J. Theoret. Biol. **71**, 167.

G. L. Yang and T. C. Chen, 1978. *On statistical methods in neuronal spiketrain analysis*, Math. Biosci. **38**, 1.

DEPARTMENT OF MATHEMATICS, UNIVERSITY OF BRITISH COLUMBIA, VANCOUVER, BRITISH COLUMBIA, CANADA

Lectures in Applied Mathematics
Volume 17, 1979

Computer Studies of Nonlinear Oscillators

Frank C. Hoppensteadt

Analytic studies of nonlinear oscillation problems for ordinary differential equations have been quite powerful in establishing the existence of various kinds of oscillatory solutions. Methods used in such studies are taken from many areas of mathematics. Standard methods based on geometric arguments in the phase plane or analytic arguments based on perturbation schemes like the methods of averaging or matching can resolve many questions of the behavior of oscillatory states.

On the other hand, determination of the stability of oscillatory solutions to nonlinear problems is a different matter. Here there are only a few techniques which are practically useful, even for straightforward problems. For example, the linearization of second order oscillator models about periodic solutions leads to Hill's or Mathieu's equations, which have quite rich structure in themselves. The stability of more complicated systems of oscillators can in principle, but rarely in practice, be determined by the methods of Floquet theory or generalizations of them. Sometimes averaging methods can be used to obtain approximations to the original system that are autonomous (i.e., time invariant), and these in turn can be studied to describe the stability of oscillatory states. These methods, however, have been successfully applied only in a (relatively) small number of cases.

Numerical calculations can be used to study stable responses of nonlinear oscillators. The sample of nonlinear oscillation problems presented here illustrates several numerical methods which have been successfully applied to the study of oscillatory phenomena. First, studies of Mathieu's equation suggest computer experiments which can be performed to study the response of a pendulum to oscillation of its support point. In particular, it will be shown that a pendulum having a vertically oscillating support point can have three (at least) stable

---

AMS (MOS) subject classifications (1970). Primary 65Lxx.

responses coexisting for the same parameter values of forcing and tuning: the straight up and straight down positions as well as a continual rotation are all stable solutions of this problem. Computer experiments can be used in these cases to determine the domains of attraction of the various modes of response. This example also demonstrates the interesting fact that otherwise unstable static states can be stabilized by external oscillatory forcing.

Next, stable oscillatory responses of the van der Pol equation to periodic forcing are described. It is shown how the concept of rotation number can be used to summarize numerical experiments on subharmonic solutions (i.e., solutions having periods that are integer multiples of the forcing period). Then it is shown how computation of the power spectrum can be used to study the presence of higher harmonics (i.e., solutions having periods which are some fraction of the forcing function).

Finally, computer studies of certain chaotic dynamical systems are discussed. Numerical simulations are used to describe density functions characterizing random behavior of deterministic oscillators.

1. Static states of a pendulum with a vertically moving support. Consider a mass hanging on a massless rod from a support point. The support point will be allowed to oscillate in the vertical direction, and we consider a geometry which moves with it. In this, we let x denote the angle in the positive (counterclockwise) direction which the pendulum makes with the straight down position. Then x satisfies the equation

$$b^2x'' + rx' + (1 + a \sin wt)\sin x = 0$$

where b denotes the ratio of the forcing frequency to the free frequency of the pendulum, w is typically taken to be 2π and a is the amplitude of the acceleration applied to the support point. Obviously, $x = 0, y = 0$, and $x = \pi, y = 0$, are static states of this problem. Linearization of the equation about these states leads to the (damped) Mathieu equation

$$b^2x'' + rx' + (1 + a \sin wt)x = 0.$$

The stability properties of this equation have been extensively studied, and the results are summarized in graphical form in **[1]**.

The stability diagram shows that there is an overlap region where both the up and down positions are stable. This region can be studied by introducing the scalings of the parameters

$$\mu = 1/b^2, \qquad E = a/b^2,$$

and studying the system for small values of μ. With this scaling, the problem becomes

$$dx/dt = \mu y,$$
$$dy/dt = -\mu r y - \mu \sin x - c \sin wt \sin x,$$

where $c = a/b$.

The method of averaging (or the two-time scale method) can be applied to this system. This results in a system of averaged equations for the evolution of the solutions on the slow time scale $s = \mu t$:

$$dx/ds = y - (c/w) \sin x,$$
$$dy/ds = \left[(r\, c/w) - 1 - (3\, c^2/w^2)\cos x \right] \sin x + y\left[(c/w)\cos x - r \right]$$

(see [5]).

As long as $r > 0$, the two rest points of the averaged system, $x = 0$, $y = 0$, and $x = \pi$, $y = 0$, are asymptotically stable. All points in the phase plane are attracted to one of these points (modulo 2π). Phase portraits of this system for various choices of the parameters can easily be constructed on a computer.

The behavior of solutions at the upper end of the overlap region can be studied by rescaling the problem. This leads to the zero-order problem in the averaging procedure equivalent to the unforced pendulum. The appropriate coordinates for performing the averaging of this system are therefore given in terms of elliptic functions. (See Chester [2].) It is useful to construct the Poincaré mapping for this system numerically, and thereby study subharmonic responses. Since this procedure is described in detail for the van der Pol equation in the next section, it is not carried out here.

Note that if the oscillation is removed from the support point, then only the down position is stable. Thus, the oscillating environment of the pendulum acts to stabilize a static state which, although present for the original system, is unstable. E. Sel'kov [3] makes an analogous observation in the context of cell metabolism. There the point is that *in vivo* a cell's environment may oscillate rapidly (relative to cell function) in many ways. This environment is not reproduced *in vitro*, and so there is the possibility of stable states *in vivo* which are not observable *in vitro*.

There are solutions near the edge of the overlap region which are stable, but not static. In particular, there is a running periodic solution which has the form

$$x(t) = vt + p(t)$$

where $p(t)$ is periodic. Such solutions describe the "clocking" of the pendulum, whose velocity is a periodic function. There are also subharmonic solutions which are stable. The Poincaré mapping is quite complicated for this problem; however, it is similar to other problems which have been studied in some detail (see Chester [2], Moser [4], and Levi, Hoppensteadt and Miranker [5]). Incidentally, the mean velocity of the pendulum in the case of the running periodic solutions (v above) can be calculated by accelerated averaging procedures derived in [6] for studying almost periodic systems.

2. Subharmonic and higher harmonic responses.

A. *Subharmonics of van der Pol's equation.* The van der Pol equation arises in many different contexts. It is in some sense the canonical description of systems into which energy is put and taken. It is discussed here in the form

$$u'' + k(u^2 - 1)u' + u = kBw \cos wt$$

where k is a tuning parameter, B is the amplitude of the external forcing, and w is 2π. We first describe some results for the case where $k \gg 1$. The work described here was reported in [7] where further references relating to this problem can be found.

It is known that for large values of k, the free problem ($B = 0$) has a unique periodic solution which is globally stable, except for an unstable rest point. The period of this solution is given by the asymptotic expansion

$$T = (3 - 2 \log 2)k + 7.014\, k^{1/3} - 1.325 + O(\log k/k).$$

As k increases, the period of the free oscillation passes through various multiples of the period of the forcing function. Therefore, we might expect that at such values of k, the forced problem might have a stable periodic solution which has this period. This expectation is analyzed by the computations presented in Figure 1.

Solutions of the van der Pol equation for various initial data $(x(0), x'(0))$ define a transformation of the (x, x')-plane into itself; namely, the Poincaré mapping which is defined by

$$P\colon (x(0), x'(0)) \to (x(1), x'(1)).$$

Since the components in the equation have period 1 in t, the Poincaré mapping conveniently describes solutions. Using this transformation, we see that subharmonic solutions (i.e., solutions having periods which are integers) form finite invariant sets for the Poincaré mapping. For example, a solution having period three will define a three point invariant set for P. If this subharmonic is stable, then a neighborhood of any one of these points will approach the point under iterations of P^3.

A given oscillatory solution can further be described by its rotation number. This is defined by

$$\rho = \lim_{n\to\infty} \frac{[\arg(P^n(\xi))]}{2\pi n}.$$

A subharmonic of period three defines a three point invariant set under P, and this might have rotation number $\frac{1}{3}$ or $\frac{2}{3}$ depending on whether a radius vector emanating from the origin to the trajectory winds around the origin once or twice over one period. The response of the van der Pol equation to external periodic forcing can be described in terms of the rotation number. The procedure used in [7] is the following.

Values of k and B are chosen, the solution is calculated until it has equilibrated, and then its rotation number is calculated. Some results presented in [7] are given in Figure 1. In this figure, the rotation number is given as a function of the two parameters B and $\varepsilon = 1/k$. Further structure of the rotation number surface suggested by this figure is also derived in [7].

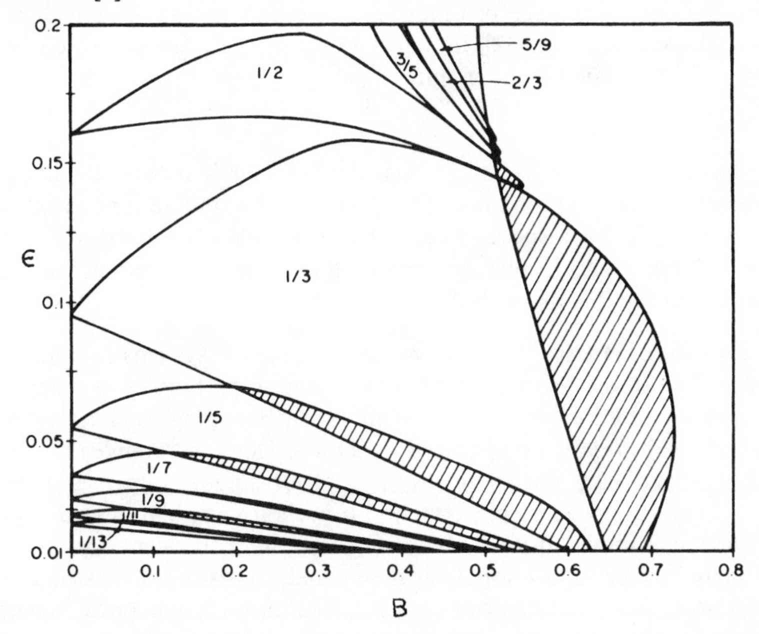

FIGURE 1. Contour map of the rotation number of stable responses (ρ). The shaded regions indicate parameter values for which ρ is double valued, and the labeled unshaded regions describe the regions of phase locking. Values in the unlabeled regions have been left out.

B. *Higher harmonics*. Higher harmonics are also observed as stable responses of the van der Pol equation. These are discussed to some extent in the book by Hayashi [**8**], mainly through analog computations. The Poincaré mapping is not much use in the study of higher harmonics since a higher harmonic will simply be a fixed point of P. Therefore, another method must be used. One that works efficiently in many cases is a method based on the computation of the power function of the calculated solution. In this, the solution, say $x(t)$, is computed, as well the two-point correlation function,

$$g(s) = \lim_{t\to\infty}\left(\frac{1}{t}\right)\int_0^t x(t'+s)x^*(t')\,dt',$$

is calculated. The Fourier transform of this function gives the modulus of the amplitudes of the various modes in $x(t)$. There are many texts written on signal processing, and most of these contain various algorithms for computing the power function as well as examples. (See Oppenheim-Schafer **[9]**.)

Another technique for numerical investigations of higher harmonics involves determination of a half-plane in the (t, x, y)-space containing the t-axis and being transverse to solutions of the equation. A solution then defines a sequence of times $\{t_n\}$, at which the transverse plane is crossed. As shown in **[12]**, in some problems the sequence of crossing times can be related by a mapping

$$t_{n+1} = \phi(t_n),$$

which can be viewed as being a mapping of a circle into itself when the sequence is reduced modulo 1 (the period of the forcing function). This is done in **[12]** for a special equation which is closely related to the van der Pol equation. Higher harmonics then correspond to a finite invariant set for this mapping.

3. Chaotic behavior. As we have seen, dynamical systems can often be reduced to studies of iterates of certain mappings. Studies of mappings of an interval into itself, or more complicated mappings such as those of annuli, lead to quite complicated dynamics. However, simpler transformations such as a quadratic mapping of an interval into itself, which have been proposed as reproduction models for populations, also lead to similar complications. (Many of these are equivalent, or at least strongly similar to the topological solenoid consisting of nested torii.) Interesting computations can be performed on such dynamical systems, giving insights into their behavior and structure. The ones presented here are based on the concept of invariant measures.

Consider a mapping of the unit interval into itself given by $x \to f(x)$. A fixed point of this mapping can be thought of as a static state, a fixed point of the iterate $f(f(x))$ corresponds to a solution of period two, etc. It is known for some mappings that most points in the unit interval migrate over an entire subinterval under iterates of f, never exactly returning. This kind of behavior is closely related to various stochastic processes, and can be characterized in terms of the probability of finding an iterate in a given set. This in turn is given by a measure which is defined on the interval covered by the iterates.

Measures corresponding to fixed points to various iterates of f are

Dirac measures having support at the points. For example, a fixed point gives rise to a measure having unit mass assigned to it, a fixed point of order two (i.e., a point x such that $f(f(x)) = x$) gives rise to a measure consisting of the sum of two Dirac measures assigning mass $\frac{1}{2}$ to each of x and $f(x)$. And, a mapping whose iterates fill up an interval can give rise to a measure which is absolutely continuous with respect to Lebesgue measure.

Simulations of the iteration process can be used in these cases to construct graphs of density functions for "stable" invariant measures. A numerical procedure based on an ergodic theorem was proposed and implemented in [**10**]. The following is a description of those results.

To fix ideas, we consider the mapping

$$f(x) = rx(1 - x)$$

where r is a parameter. If $0 \leqslant r \leqslant 4$, this is a mapping of the unit interval into itself. In particular, for $r = 4$, this maps the interval onto itself. It has been shown that, for $r = 4$, there is a measure defined on the unit interval which is invariant and absolutely continuous with respect to Lebesgue measure. Its density function is proportional to the function d given by the formula

$$d(x) = 1/[x(1 - x)]^{1/2}.$$

Simulation of iteration of this mapping yields a graph of this function: First take a point in the unit interval and iterate it 500 times under f. Record the number of times iterates enter each of 1000 cells on the interval, but delete the first 50 iterates in this to reduce extraneous results caused by transients. Then repeat the computation for 100 other initial points selected at random from the interval. The result, suitably normalized, then gives a description of the invariant measures. This fact follows from the ergodic theorem which states (roughly; see [**11**] for precise statements) that if P is a mapping which has an absolutely continuous invariant measure μ and if $g(x)$ is a simple function, then

$$\int_0^1 g(x)d\mu = \lim_{n\to\infty} \frac{1}{n} \sum_{k=0}^{n-1} g(P^k(x)).$$

If $g(x)$ is taken to be equal to one on a small interval, say $(x, x + \delta)$, and zero otherwise, then, for large N,

$$\frac{1}{N} \sum_{k=0}^{N-1} \frac{g(P^N(x))}{\delta} = \frac{1}{\delta} \int_x^{x+\delta} d\mu(x) \doteq \frac{d\mu}{dx} = d(x).$$

The result of such an experiment is described in Figure 2. It is interesting to perform this experiment for many values of r in the range $3 < r < 4$, although it is difficult to interpret correctly the results in that range.

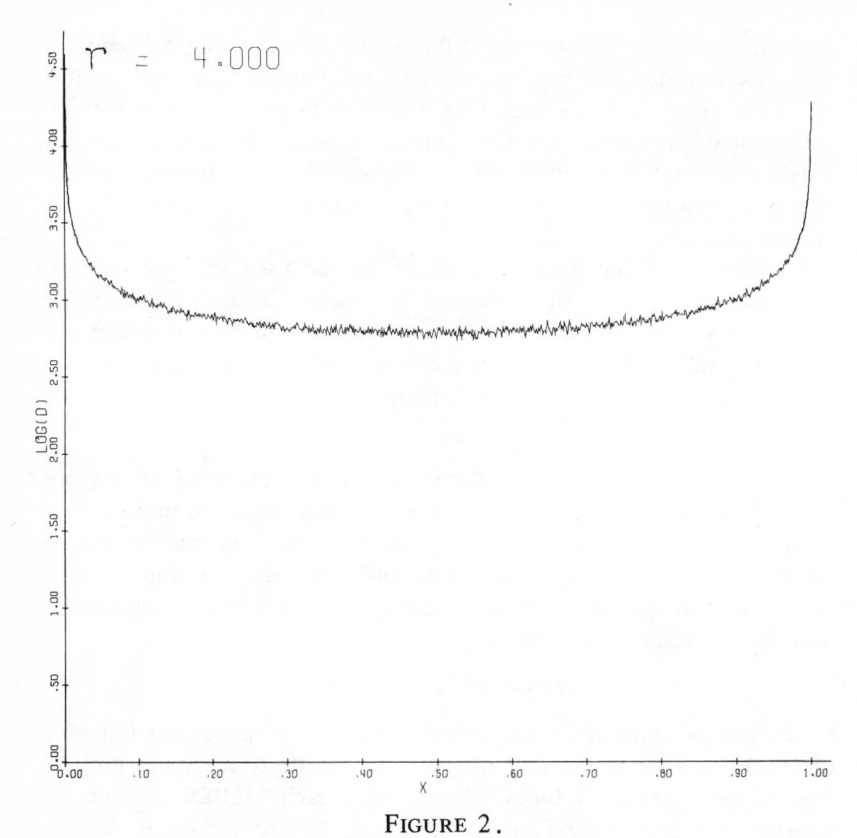

FIGURE 2.

Simulation methods like these have been used to study aperiodic (chaotic or ergodic) behavior in more complicated dynamical systems; for example, integrate-and-fire models of the response to nerves to oscillatory stimulation can lead to studies of discontinuous mappings of a circle. Analytical and numerical studies of such models, as well as references to further reading on these models and methods, can be found in [**12**].

BIBLIOGRAPHY

1. J. J. Stoker, *Nonlinear vibrations*, Wiley-Interscience, New York, 1950.

2. W. Chester, *The forced oscillations of a simple pendulum*, J. Inst. Math. Appl. **15** (1975), 289–306.

3. E. Sel'kov, *Stabilization of energy charge, generation of oscillations and multiple steady states in energy metabolism as a result of purely stoichiometric regulation*, European J. Biochem. **59** (1975), 151–157.

4. J. Moser, *Stable and random motions in dynamical systems*, Ann. of Math. Studies, vol. 77, Princeton Univ. Press, Princeton, N. J., 1973.

5. M. Levi, F. Hoppensteadt and W. Miranker, *Dynamics of the Josephson junction*, Quart. J. Appl. Math. **36** (1978/79), no. 2, 167–198.

6. F. Hoppensteadt and W. Miranker, *An extrapolation method for the numerical solution of stiff differential equations* (in preparation).

7. J. Flaherty and F. Hoppensteadt, *Frequency entrainment of a forced van der Pol oscillator*, Studies in Appl. Math. **58** (1978), 5–15.

8. C. Hayashi, *Nonlinear oscillations in physical system*, McGraw-Hill, New York, 1964.

9. A. V. Oppenheim and R. W. Schafer, *Digital signal processing*, Prentice-Hall, Inc., Englewood Cliffs, N. J., 1975.

10. F. Hoppensteadt and J. Hyman, *Periodic solutions of a logistic difference equation*, SIAM J. Appl. Math. **32** (1977), 73–81.

11. M. E. Munroe, *Introduction to measure and integration*, Addison-Wiley, Reading, Mass., 1953.

12. F. C. Hoppensteadt, J. P. Keener and J. Rinzel, *Integrate-and-fire models of nerve membrane response to oscillatory input* (in press).

DEPARTMENT OF MATHEMATICS, UNIVERSITY OF UTAH, SALT LAKE CITY, UTAH 84112

Lectures in Applied Mathematics
Volume 17, 1979

Chaotic Oscillations: An Example of Hyperchaos

O. E. Rössler

Introduction. Chaos, or dynamical behavior with a nontrivial basic set in the sense of Smale [**1**] (that is, with a Cantor set of singular solutions, which includes an infinite number of periodic solutions [**2**]), occurs readily in many natural and artificial systems [**3**]. One intuitive example is the irregularly dripping faucet. Here two suboscillations–the relaxation oscillation of droplet formation and the damped oscillation of the budding next drop–continually get into each other's way.

The reason why chaos occurs so easily lies in the structure of three-space. The 'lateral ordering' of trajectories characteristic of two dimensions is lost with three. Part of the flow can now be 'elevated' into the third dimension and then be 'put back', with the result of entanglement. This *principle of displaced re-injection*, as it may be called, is operational in the sense that it allows an effortless design of many different chaotic systems which each have prescribed behavior in the limit of some singular perturbation parameter tending to zero [**4**]. Sometimes the Poincaré map can be written down explicitly (see [**5**] for an appropriate, overlappingly piecewise linear system).

Thereafter the simpler relatives to those asymptotic examples can be sought; astonishingly, they all exist. Of course, developing the right approximation techniques by which to prove that the nonidealized *flintstone models* (as compared to the former *paper models*) still have cross-sections that are folded over (with or without 'cutting') to a sufficient degree to warrant formation of a Cantor set under iteration, is not an easy task. The computer is helpful for guidance here. It is an ironic instrument when simulating chaos: it proves itself constantly wrong in a quantitative sense, but at the same time underscores the robustness of the essential gross features (e.g., that one trajectorial region is placed on top of another). This suggests theorems that depend

AMS (*MOS*) *subject classifications* (1970). Primary 34-XX.

on the gross features alone. Sometimes, a very crude abstract computer (with error estimates much above those admissible for integrating devices useful in real computers) can be used to show that a given nonidealized equation indeed yields chaos.

Generality of the principle of displaced re-injection. Is the third dimension special? Chaos arises with the third dimension and stays for all higher dimensions. Oscillations in the same way arise with the second, however. Critical points emerge with the first. How about the fourth dimension?

The principle of displaced re-injection allows a unified view on the problem. Starting from a **one**-dimensional system, *nondisplaced* re-injection (right back to the starting point) turns an unstable steady state into a homoclinic saddle. *Displaced* re-injection, under the same conditions, leads to the new structure of the limit cycle as characteristic for two-dimensional flows. Similarly, starting with an unstable point in **two** dimensions, re-injection through the third dimension yields, if perfectly rotation-symmetrical (*nondisplaced*), a homoclinic saddle-focus or -node (cf. [**6**]). *Displacement* allows for the new structure of the torus, the latter equipped with a cross-section much like the phase portrait of the above limit cycle. However, to obtain this, only one degree of freedom of the displacement needed to be used, so to speak: rotation symmetry has been preserved. *Full displacement* allows for more complicated kinds of re-injection, each yielding a flow that can be said to lie on a distorted (folded over) torus. If the overlap is sufficiently strong, chaos emerges (in two basic types, with the re-injection hitting a spiral or a screw, respectively). Simple paper models and simulations may be found in [**4**]. More complicated re-injections, starting from more than one singularity, can be derived in a similar way.

'And so on.' These three words mean that the next levels are too difficult to consider in detail. However, the four-dimensional level is still partially accessible: Starting with an unstable (nonsaddle) point in **three** dimensions, nondisplaced re-injection yields a homoclinic hypersaddle (again with a single stable eigenvector). Relaxing one of the two rotation symmetries to a sufficient degree yields chaos on a torus, relaxing the other also yields chaos on a torus, but with different constituents. Full displacement then destroys both symmetrical regimes and replaces them by a folded hypertorus, the latter being folded over independently in two directions. If the overlap is sufficiently strong, a new type of flow emerges which may be called hyperchaos [**5**].

Hyperchaos. Hyperchaos at first sight defies imagination. Fortunately, however, it has a three-dimensional cross-section. The latter is completely open to inspection. Everything known from the two-dimensional cross-sections (the generalized-horseshoe or, synonymously, walking-stick map [3]) should carry over, but generalized. This generalization is the problem. Still, as the two-dimensional problem already proved easiest to describe in terms borrowed from three dimensions (*folding*, *sausage-in-the-sausage* [3]), the step may be easy. A picture of the anticipated cross-section is shown in Figure 1. There is a torus (solid, hollow) which is folded over part of itself in two different ways. Its

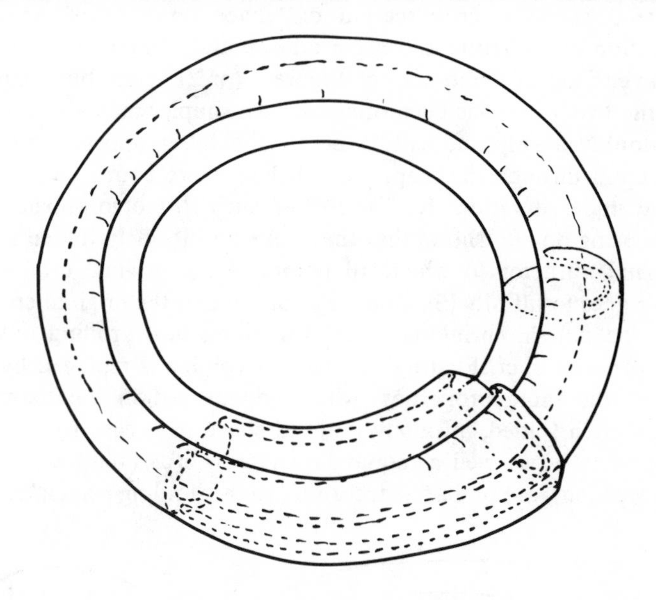

FIGURE 1. 'Doubly folded torus map'. The map has to be pictured to lie inside the original cross-section which is thicker and lacks foldings.

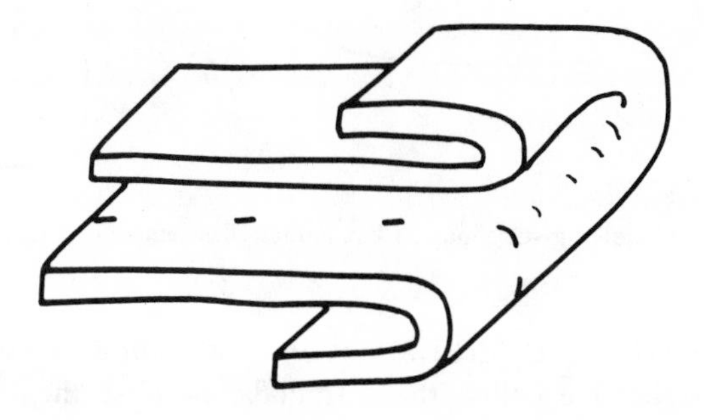

FIGURE 2. 'Folded towel map'. Again only the first iterate is shown; the original is a rectangular block.

essential ingredient, acting as an attracting submap at some iterate, is the double-laid *folded towel map* of Figure 2.

What are the properties of this hyperhorseshoe or hyperwalking-stick, respectively? The ordinary horseshoe contained a doubly asymptotic saddle, with formation of a homoclinic point between the saddle's stable and unstable manifold. Is there, now, something like a triply asymptotic hypersaddle present? There is a triply asymptotic set, but it is not necessarily a point any more.

Figure 3 shows a cross-section of 'chaos on a torus', that is, a combination of a chaotic and a periodic motion. The map is drawn in two conventions. It is the map of Figure 1 (or 2) again, but simplified: one of the two 'ruffs' has been omitted. This simpler map also no longer has a doubly asymptotic saddle point: the latter appears only on a cross-section through the map. Nonetheless, there is one very intuitive property under iteration. Realization of such diffeomorphisms in the kitchen, using dough, shows that they have an infinitely folded sheet as an internal attractor to which all points of the original cross-section tend as t goes to infinity [5]. One may conceive of the original cross-section as red dough, shrinking under the elongation ('rolling out') and folding procedure, cf. Figure 4. If the lost volume is replaced by white dough of the same properties, what remains red in the limit is an infinitely often folded, infinitely thin red sheet, as is easy to see. Fluffy pastry and taffy, as well as some Japanese swords (Winfree, personal communication), are prepared according to this folding principle.

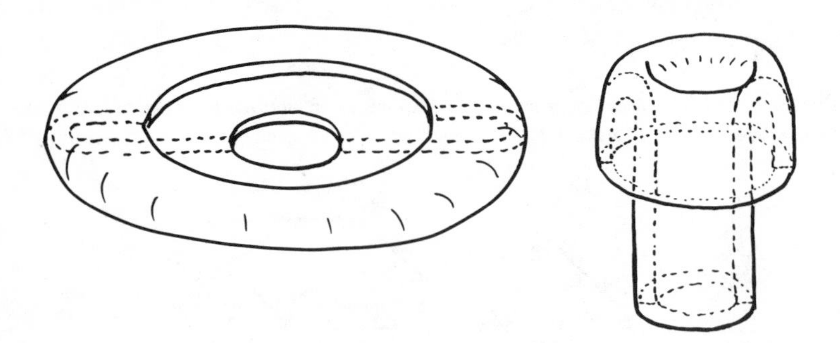

FIGURE 3. 'Rotated walking-stick map'. First iterate. Two equivalent representations are given.

Mandelbrot [7] recently drew attention to structures like those formed in folded doughs. He called them fractals, because their (Hausdorff) dimensionality is not an integer in general. Imagining how the attracting surface formed inside the map of Figure 2 looks in the limit of an infinite number of foldings in one direction and, on top of these (in fact,

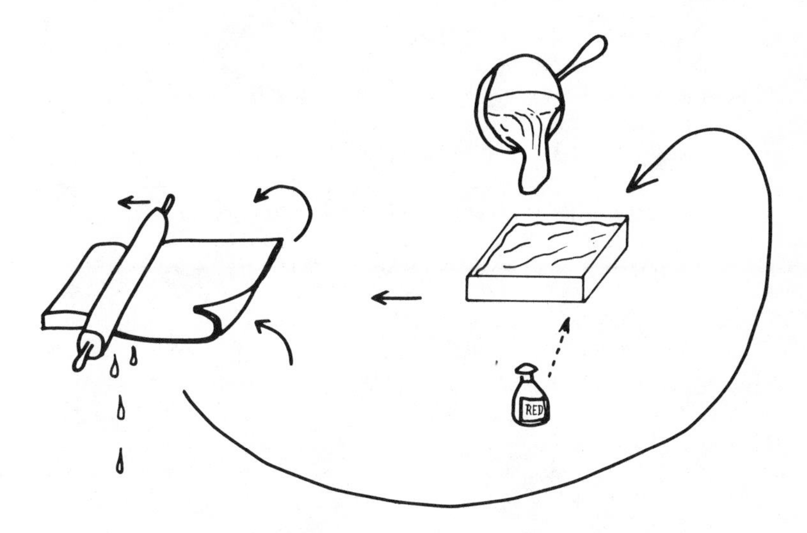

FIGURE 4. One possible way how to realize the map of Figure 2 and its higher iterates.

interspersed), another infinite number in the other direction, shows that not just a fractal sheet is being formed, as was the case in Figure 3, but something new: a 'hyperfractal,' if the term is allowed.

A hyperfractal of the kind encountered here is locally the product of a two-dimensional Cantor set and a one-dimensional arc. This definition is formed in analogy to Smale's definition of a solenoid which is an ordinary fractal [7]. There the definition involved an ordinary Cantor set [1]. A two-dimensional Cantor set as postulated here is the product of two Cantor sets, each one-dimensional in its own induced topology. Thus, there seem to be ways to define a hyperfractal as being different from a fractal, in fact as different as the latter is from a one-dimensional line. However, a dimensionality criterion consisting of a single (complex) number has yet to be found.

An example. The following ordinary four-variable differential equation yields hyperchaos in the limit of certain parameters ($\mu, \nu/\mu$) approaching zero.

$$\dot{x} = \omega y + \frac{x}{r}\left(-z + F(s) + \frac{x}{r}G(z) - wH(z)J(s)\right),$$

$$\dot{y} = -\omega x + \frac{y}{r}\left(-z + F(s) + \frac{x}{r}G(z) - wH(z)J(s)\right),$$

$$\dot{z} = \mu(s + K(z)),$$

$$\dot{w} = \nu(s^2 + L(w)), \tag{1}$$

where $r = \sqrt{x^2 + y^2}$, $s = r - 3$, and

$$F(s) = s + p(s+1) - \sqrt{p^2(s+1)^2 + \varepsilon} + p(s-1)$$
$$+ \sqrt{p^2(s-1)^2 + \varepsilon} - \delta p(s+2) + \sqrt{\delta^2 p^2(s+2)^2 + \varepsilon}$$
$$- \delta p(s-2) - \sqrt{\delta^2 p^2(s-2)^2 + \varepsilon} - \sqrt{(\delta p(s+1) + p)^2 + \varepsilon}$$
$$+ \sqrt{\delta^2 p^2(s+1)^2 + \varepsilon} + \sqrt{(\delta p(s-1) - p)^2 + \varepsilon}$$
$$- \sqrt{\delta^2 p^2(s-1)^2 + \varepsilon},$$

$$G(z) = 1.5z\left(-0.1 + \sqrt{(z+0.1)^2 + \varepsilon} - \sqrt{z^2 + \varepsilon}\right),$$

$$H(z) = 2.5\left(0.1 - \sqrt{(z-0.1)^2 + \varepsilon} + \sqrt{z^2 + \varepsilon}\right),$$

$$J(s) = s + \sqrt{s^2 + \varepsilon},$$

$$K(z) = 0.375(z - \sqrt{z^2 + \varepsilon}),$$

$$L(w) = -3 + w + \sqrt{w^3 + \varepsilon}.$$

Ideally, $\delta \to \infty$, $\varepsilon \to 0$, $p = 1.25$, and $\mu \to 0$, $\nu/\mu \to 0$.

The equation looks a bit 'messy', but this is only because this way it is simpler, so to speak. This becomes clear from Figure 5, which shows that $F(s)$ is just a 'double-z'. When taken as the basis of a two-dimensional system, F produces a relaxation oscillator with two stable limit cycles, each regularly oscillating between plus one and minus one, or plus two and minus two respectively, at the same frequency: consider the system

$$\dot{s} = -z + F(s),$$
$$\dot{z} = \mu s. \tag{2}$$

Thus, $F(s)$ is equivalent to a seventh-degree polynomial, but has the asset that in the limit of $\varepsilon \to 0$ there are straight line segments between the 'knees'. According to LaSalle **[8]**, a seventh-order polynomial is convenient for obtaining a double-limit cycle relaxation oscillator. The dashed arrows in Figure 5 mark the limiting positions of the fast transients when $\mu \to 0$ in equation (2). The inner and the outer loops are attracting, the intermediate loop (dotted arrows) corresponds to an unstable limit cycle.

The functions G, H, J are for convenience. They make sure that the shifting effects of x and w on the nullcline $\dot{s} = 0$ in the subsystem s, z occur only in a certain quadrant of its state space, so that the different events that occur in equation (1) are neatly kept apart. K and L similarly separate velocities in different orthants of state space.

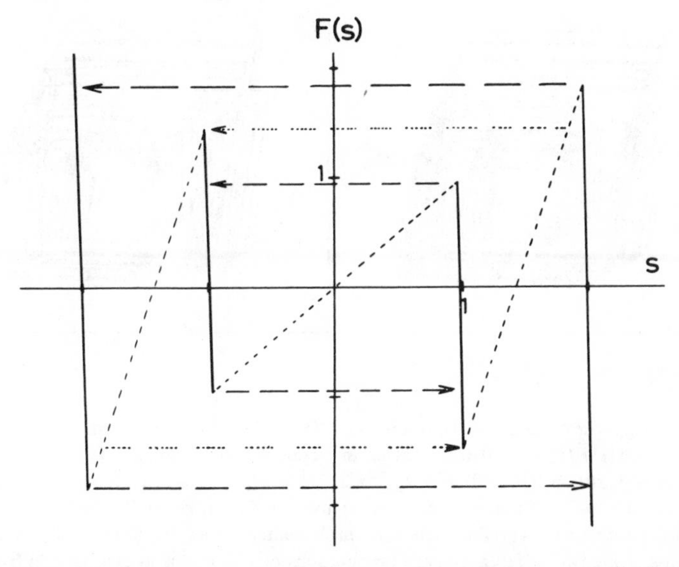

FIGURE 5. The function $F(s)$ of equation (1). $p = 1.25$, $\delta = 100$, $\varepsilon = 10^{-6}$. The dashed arrows correspond to fast transitions if F is used as a nullcline in a relaxation oscillator like equation (2). For $\varepsilon \to 0$, the 'knees' become arbitrarily sharp; for $\delta \to \infty$, the slopes of the vertical segments approach infinity; p can be used to re-adjust thresholds if δ is small.

There are two chaotic subregimes possible in (1). The first is performed by the subsystem

$$\begin{aligned} \dot{s} &= -z + F(s) - 0.5sw \cdot H(z)H(s), \\ \dot{z} &= \mu s, \\ \dot{w} &= \nu\left(s^2 - 3 + 2w \cdot H(w)\right), \qquad 0 < \nu \ll \mu \ll 1, \end{aligned} \tag{3}$$

which produces a relaxation oscillation occurring between the two limit cycles of equation (2), cf. Figure 5. Here, $H(\xi)$ is the Heaviside function $H(\xi) = 1$ if $\xi \geqslant 0$ and $H(\xi) = 0$ if $\xi < 0$. The quantitative conditions have been so adjusted that the flow, when falling over the wedge-shaped cliff at positive values of w (there is a positive threshold value of w beyond which the two inner limit cycles of s, z disappear), comes to cover appreciably more than one revolution when being injected onto the flow on the outer cylinder. At the other end, where the two outer limit cycles disappear (beyond the negative threshold value of w), there is no such phase amplification (although if there were it would not matter). The two switches ($H(z)H(s)$, $H(w)$) in (3) could be replaced by the constant 1 each, but then the involved map would be somewhat more complicated. The explanation for chaos lies (a) in the phase-dependent threshold, and (b) in the phase-amplification at one transition. This is discussed further below. A computer simulation is presented in Figure 6.

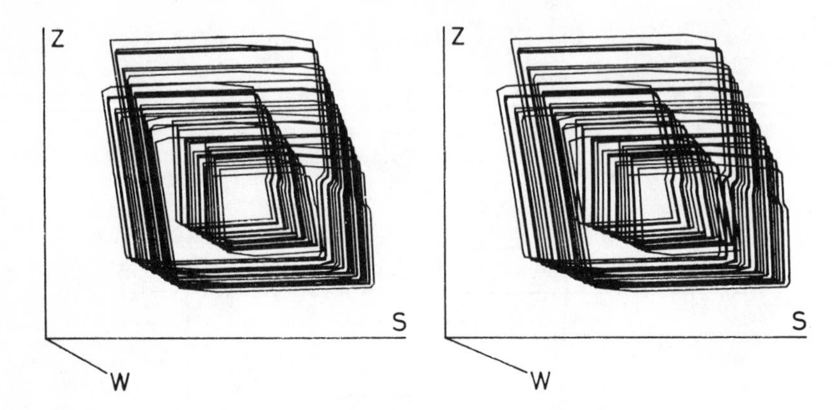

FIGURE 6. Upper-level chaos in equation (1). $G(z) = 0$ so that the system is 3-variable in s, z, w (see text). $K(z) = 0$. Parameters as in Figure 8, except for $\mu = 0.1$. Initial conditions: $s(0) = 4$, $z(0) = 0.8$, $w(0) = 0$. $t_{\text{end}} = 2171$. Axes: $-1, \ldots, 7$ for s, $-4, \ldots, 4$ for z, $-8, \ldots, 0.8$ for w. Numerical simulation using a fourth order Runge-Kutta-Merson integration routine with variable step size, implemented on an HP 9845 with peripherals. Stereoplots, using two different linear combinations of s, z, and w as can be seen from the axes. Not all computed points have been plotted.

The second chaotic ingredient of equation (1) is

$$\begin{aligned} \dot{x} &= \omega y + \frac{x}{r}\left(-z + F(s) - \frac{x}{r}\cdot 0.3z\cdot(1 - H(z))\right), \\ \dot{y} &= -\omega x + \frac{y}{r}\left(-z + F(s) - \frac{x}{r}\cdot 0.3z\cdot(1 - H(z))\right), \\ \dot{z} &= \mu(s + 0.75z\cdot(1 - H(z))), \end{aligned} \tag{4}$$

or, equivalently,

$$\begin{aligned} \dot{s} &= -z + F(s) - (\cos\omega t)\cdot 0.3z(1 - H(z)), \\ \dot{z} &= \mu(s + 0.75z\cdot(1 - H(z))), \end{aligned} \tag{5}$$

where H is as in (3) and s as in (1). The equivalence of both equations is based on the following identity: the two-dimensional system

$$\dot{x} = \omega y + \frac{x}{r}\cdot f(s), \qquad \dot{y} = -\omega x + \frac{y}{r}\cdot f(s), \tag{$*$}$$

where $r = \sqrt{x^2 + y^2} > 0$, $s = r - C$, $C =$ a constant > 0, and $y(0) = 0$, mimics for $s > -C$ the one-dimensional system

$$\dot{s} = f(s), \qquad s(0) = x(0) - C,$$

plus forcing function $x/r \equiv \cos\omega t$.

Of course, both (4) and (1) are unnecessarily complicated because of their implementing the last-mentioned identity up to the point. But this is again in the interest of clarity. Simpler equations will be discussed later. As to the switching nonlinearities $H(z)$ in (4) and (5), they can again be replaced by unity if a little more complication in understanding the flow is put up with.

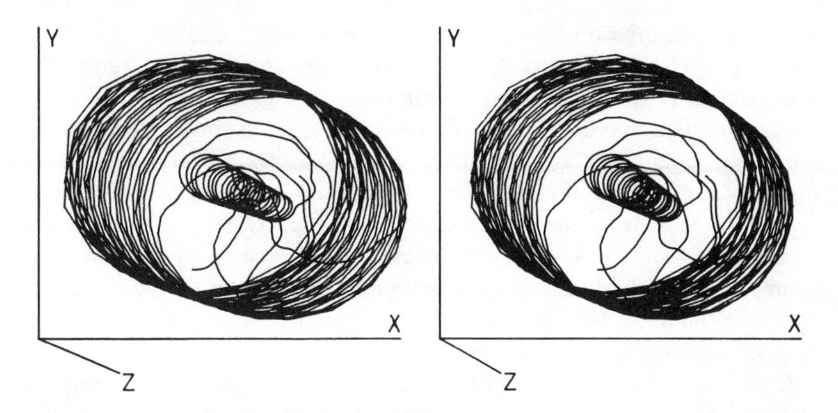

FIGURE 7. Lower-level chaos in equation (1). $\nu = 0$ so that the system is 3-variable in x, y, z. Stereoplots as in Figure 6, parameters as in Figure 8. Initial conditions: $x(0) = 3.793$, $y(0) = w(0) = 0$, $z(0) = -1.81$. $t_{\text{end}} = 347.5$. Axes: $-4, \ldots, 10$ for x and y, $-1, \ldots, 3$ for z. There is a second such chaotic regime (for example, for $x(0) = 2$, $y(0) = 0$, $z(0) = 1$).

As shown in Figure 7, the flow determined by (4) is just a rotation-symmetrical analogue to that of Figure 6. Again, one trajectorial revolution on one of the two jackets of the cylinder is mapped onto considerably more than one revolution on the other. The result is chaos. Either flow is related to Smale's solenoid (see [**1**]), but contains on its annular cross-section not only pieces whose lengths are continually enlarged, but also two 'compression zones' where the foldings-over occur. Thus, the effective cross-section is again the walking-stick map (see above).

There are, of course, two such chaotic regimes present in (5), both attracting, with a third repelling one (acting as a separatrix) present in between. Such a structure has apparently not been described yet. Another remark concerns the relationship of (5) to the forced van der Pol oscillator [**9**]–[**13**]. Equation (5), with H replaced by 1, is a van der Pol type oscillator, but with a nonlinear forcing term. The mechanism of chaos-generation is therefore possibly different. For example, a periodic transition between an oscillating and a mere excitable state seems to be essential in all cited cases. A comparative three-dimensional study of both mechanisms may be rewarding.

After having set up the two ingredients, it remains to be demonstrated that the subsystems (3) and (4) are still functioning in (1) without losing their character.

The whole system was devised in such a way that the slowest oscillatory component made chaos with the next faster one (subsystem

(3)), and the latter again with the fastest one (subsystem (4)). What was folded over in each interaction was the phase of the respective slower component. A chaotic phase shift in the respective faster component has but little influence on the next-higher folded map. Conversely, a slow upper-level chaotic motion should leave a fast chaotic submotion relatively undisturbed over relatively long periods of time.

The simulations of Figures 8 and 9 are in accordance with the preceding qualitative arguments. Figure 8 shows three three-dimensional projections of the four-dimensional flow. Figures 9(c)–9(d) give the time behavior of the four components. Figures 9(a)–9(b) show the time behavior of the two chaotic subsystems for comparison. The slow component clearly dominates the fast one. Nonetheless the fast one also survives: the ninth full impulse of s in Figure 9(d) is about 15% broader than its counterpart in Figure 9(a).

A next step has to be investigating the cross-section actually present in (1). Both numerical and analytical methods can be applied. After letting the parameters p, ε, and δ approach their indicated limits, everything takes place in a two-dimensional endomorphism which, moreover, can be computed analytically. The map will show one more folding-over (in its 'slow' direction) than indicated in Figure 2. The formed attractor will look 'space-filling'. Nonetheless, not every space-filling attractor in a two-dimensional endomorphism is hyperchaotic. The map of Figure 3 produces, when flattened into an endomorphism, also space-filling behavior. Specific criteria for hyperchaos are needed. Further experimentation with hyperchaotic systems may prove helpful.

Hyperchaos in simpler systems. At least a cubic (or two quadratic) nonlinearity seems to be required in order to obtain hyperchaos in four-variable systems. (See, however, [**20**].) The simplest example that can be envisioned so far is a three-variable *double-screw type chaos* producing equation [**4**] (see also [**14**] for a similar equation) combined with a fourth linear variable:

$$\begin{aligned} \dot{x} &= -0.1y(1 - x^2) - 0.003x, \\ \dot{y} &= -2z + 0.3x + y, \\ \dot{z} &= x + 2y - 0.5z + w, \\ \dot{w} &= az - bw, \end{aligned} \tag{6}$$

where w is the added fourth variable. If it is possible to say that in Lorenzian chaos [**15**] a relaxation oscillation goes on between two planar limit cycles separated by a saddle [**3**], then here a relaxation oscillation goes on between two three-dimensional chaotic regimes separated by a saddle limit cycle. Appropriate parameters a, b have yet to be found. The map that will describe the system of equation (6) is,

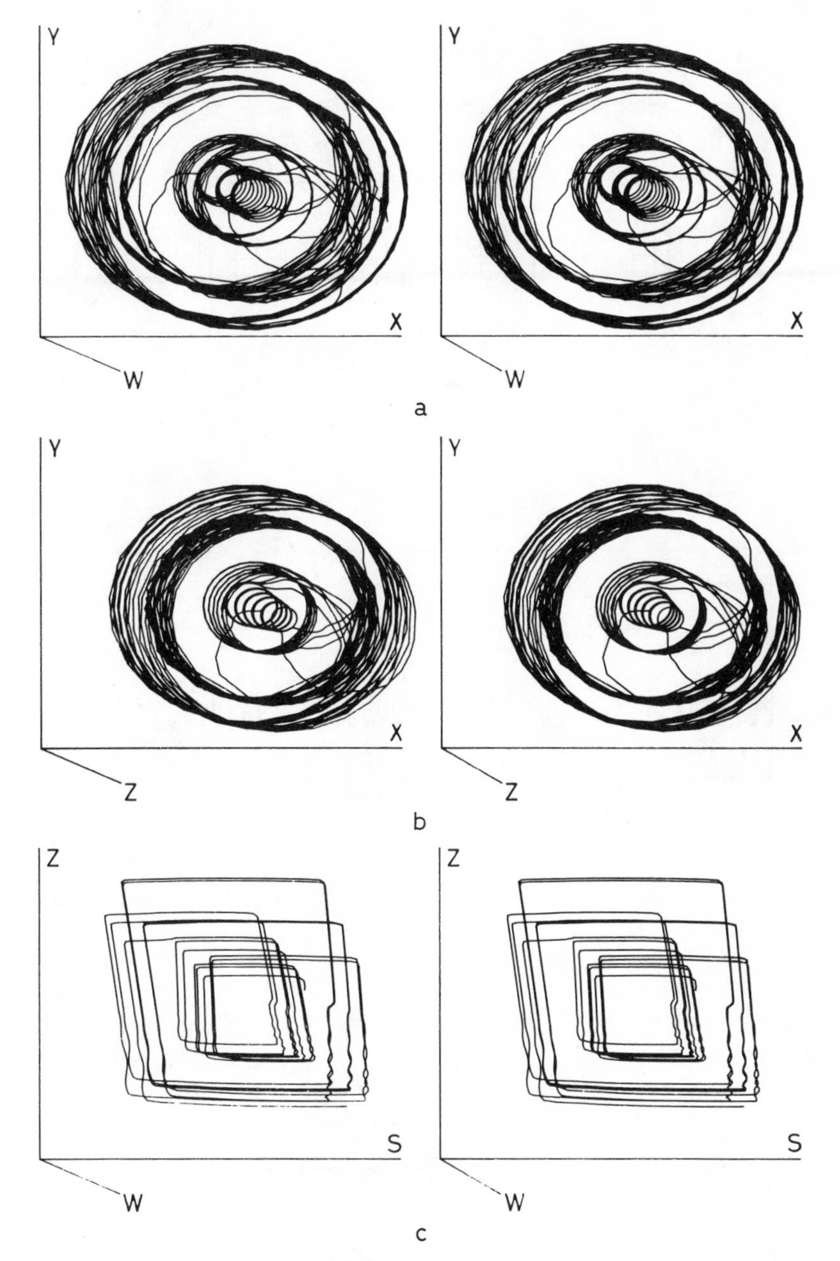

FIGURE 8. Hyperchaos in equation (1). Three partial projections (stereoplots of three variables) are shown. (a): x, y, w, (b): x, y, z, (c): s, z, w. Parameters: $\varepsilon = 10^{-3}$, $\delta = 4$, $p = 2$, $\omega = 0.85$, $\nu/\mu = 0.05$, $\mu = 0.05$. Initial conditions: $x(0) = 4$, $y(0) = w(0) = 0$, $z(0) = 0.8$, $t_{\text{end}}(a, b) = 459$, $t_{\text{end}}(c) = 1360$. Axes (a): $-3, \ldots, 9$ for x and y, $-1, \ldots, 1$ for w; (b): $-4, \ldots, 10$ for x and y, $-5, \ldots, 5$ for z; (c): $-1, \ldots, 7$ for s and z, $-1, \ldots, 1$ for w. The wriggles seen in the lower portion of (c) disappear for large δ.

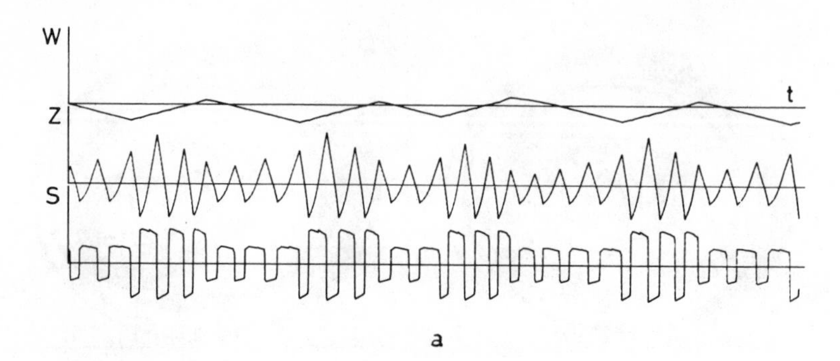

a

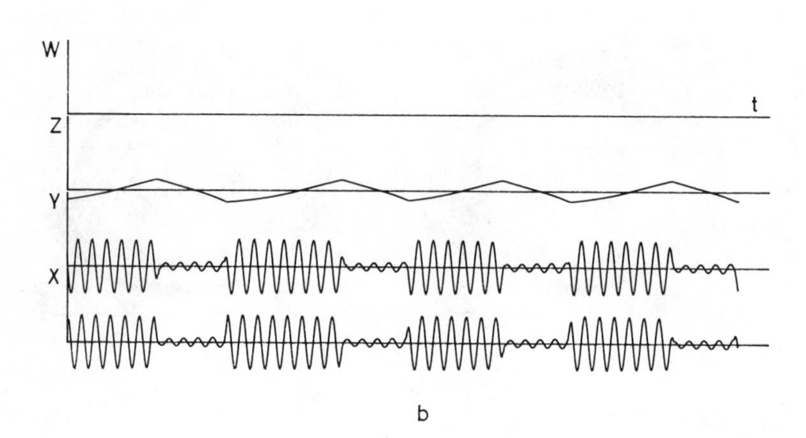

b

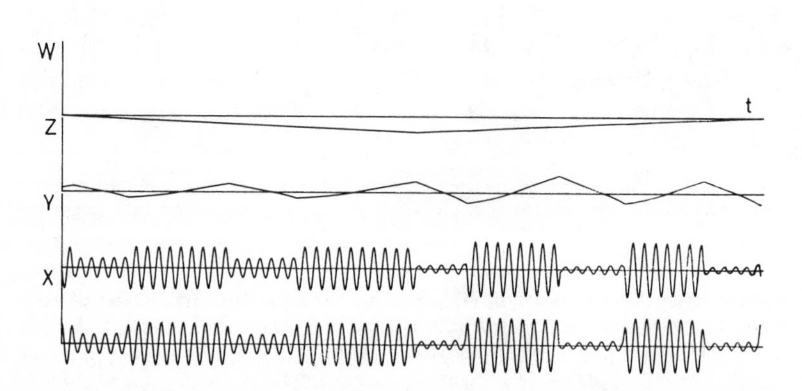

c

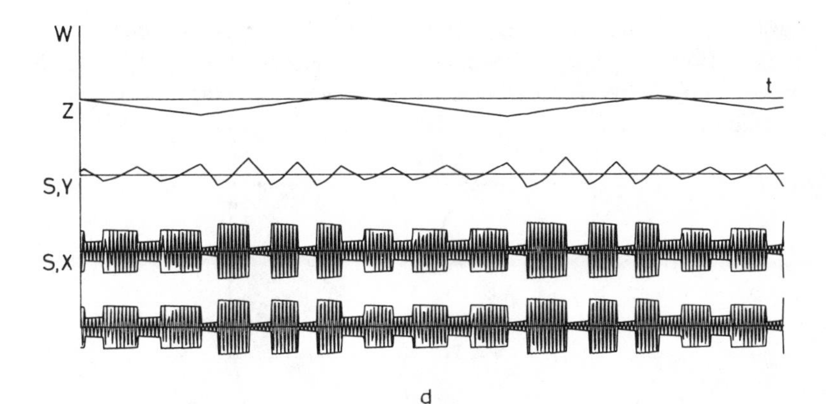

FIGURE 9. Time-plots for equation (1). (a) Upper-level chaos (see Figure 6), (b) lower-level chaos (see Figure 7), (c) hyperchaos, short-term (see Figure 8(a), (b)), (d) hyperchaos, long-term (see Figure 8(c)). Axes: (a) 3, . . . , 8 for s, 0, . . . , 5 for z, w, 0, . . . , 2829 for t. (b) 0, . . . , 15 for x, y, z, w, 0, . . . , 347.5 for t. (c) 0, . . . , 5 for x, y, z, w, 0, . . . , 459 for t. (d) 0, . . . , 15 for s, x, y, z, 0, . . . , 5 for w, 0, . . . , 1360 for t. In (d), the 'hull curve' of x and y, s, has also been entered.

incidentally, related to the 'big map' of **[5]** (which is a folded towel that in addition is cut at the folding edges), but so only in one direction. In the other direction, it will have an infinite number of foldings.

A conceptually simpler type of hyperchaos is obtained when a single-quadratic-nonlinearity chaotic system like that of **[16]** is periodically forced:

$$\begin{aligned}\dot{x} &= -y - z + a\cos\omega t,\\ \dot{y} &= x + 0.2y,\\ \dot{z} &= 0.2 + z(x - 5.7).\end{aligned} \tag{7}$$

If $a = 0$ (decoupled case), the corresponding four-dimensional flow (to be obtained by implementing the forcing term in the manner of (*)) has the cross-section of Figure 3. It is highly likely that by making the coupling not only finite but also sufficiently strong, again a folding over in an independent direction can be obtained, leading to a map like that of Figure 1. This prediction is based on the fact (exploited above) that a two-dimensional relaxation type system, when periodically forced, also is capable of developing a strong distortion of the originally toroidal (annular) map.

Conclusions. A procedure that allows one to identify hyperchaotic systems from their observed behavior has yet to be indicated. Genuine hyperchaos may have to be distinguished from 'apparent hyperchaos' as can be produced already by one-dimensional maps. An example is the map of Figure 10. Here a system switches irregularly between two chaotic subregimes which happen to possess a little 'door' open between them. A flow like this was found in the Lorenz equation [**17**]. Nonetheless, the possibility for detecting specific spectral features of genuine hyperchaos is still open.

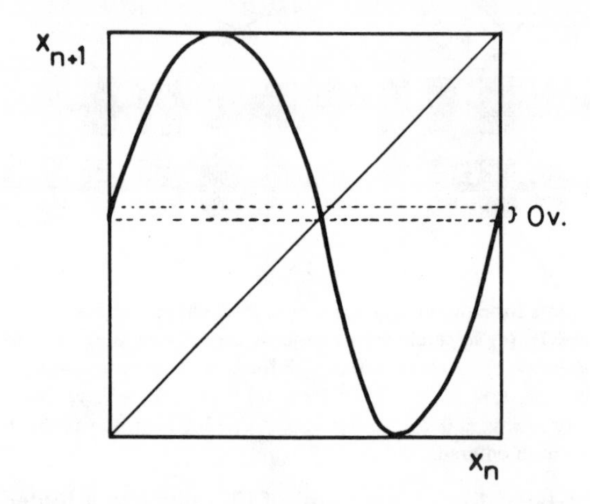

FIGURE 10. A map for pseudohyperchaos (see text). Ov = 'region of overlap' between two chaotic substates.

One physical and mathematical area where the notion of hyperchaos may have concrete implications is that of turbulence in distributed dynamical systems. In two coupled cellular oscillators of Rashevsky-Turing type, not only a *chaotically modulated morphogenesis* was found with two symmetrically positioned differentiated chaotic states, but also a higher chaotic state with a nonperiodic shuttle between the two chaotic substates [**18**]. It thus may be related to the 'higher-order Lorenzian flow' expected from (6). Kuramoto recently showed that cut (Lorenzian) chaos is typical for originally symmetric media [**19**].

In a discussion with Kuramoto, the hypothesis was reached that higher types of Lorenzian chaos, like that expected from (6) and the above mentioned Turing system, are characteristic of turbulent media. If this hypothesis can be corroborated, a picture of turbulence would emerge that is less different from the old Landau-Hopf picture (which assumed an infinite number of periodic components combined in an almost periodic motion) than the first discovery of chaotic behavior in simple models of turbulence suggested [**15**]. Once more, an infinite hierarchy of coupled subregimes would be the hallmark of 'full-fledged turbulence'.

Summary. A 4-variable ordinary differential equation is indicated which produces a hypertoroidal attracting flow. The parameters can be so adjusted that there are two different chaotic regimes possible simultaneously, each in a different subtorus, and coupled. A computer simulation is presented. The equation has the asset that its properties are retained in the limit of certain parameters approaching zero (in fact this is how the equation was constructed). Then a two-dimensional endomorphism applies as a cross-section to the degenerate flow. This map can be calculated in principle since the underlying differential equation is overlappingly linear. Two much simpler equations (one containing but a single cubic nonlinearity) likely to show similar behavior are also presented. Hyperchaotic flows are characterized by the presence in their cross-sections of an attracting invariant set which is a hyperfractal; that is, a fractal based on two Cantor sets. The formation of fluffy pastry involves a similar map. Full-fledged turbulence may be hyper^n-chaos, with n very large. Diagnostic criteria for the presence of hyperchaos have yet to be developed.

References

1. Smale, S. *Differentiable dynamical systems*, Bull. Amer. Math. Soc. **73** (1967), 747–817.

2. Li, T. Y. and J. A. Yorke. *Period three implies chaos*, Amer. Math. Monthly **82** (1975), 985–992.

3. Rössler, O. E. *Different types of chaos in two simple differential equations*, Z. Naturforsch. **31a** (1976), 1664–1670.

4. ______. *Continuous chaos*, in Synergetics–A Workshop, H. Haken, ed., Springer-Verlag, Heidelberg, New York, 1977, pp. 184–197.

5. ______. *Continuous chaos–Four prototype equations*, Ann. New York Acad. Sci. **316** (1979), 376–394.

6. Šil'nikov, L. G. *On the generation of a periodic motion from trajectories doubly asymptotic to an equilibrium state of saddle type*, Math. USSR-Sb. **6** (1968), 428–438.

7. Mandelbrot, B. B. *Fractals: Form, chance and dimension*, Freeman, San Francisco, 1977.

8. LaSalle, J. *Relaxation oscillations*, Quart. Appl. Math. **7** (1949), 1–19.

9. Cartwright, M. L. and J. E. Littlewood. *On nonlinear differential equations of the second order*. I. *The equation* $\ddot{y} - k(1 - y^2)\dot{y} + y = b\lambda k \cos(\lambda t + \alpha)$, *k large*, J. London Math. Soc. **20** (1945), 180–189.

10. Levinson, N. *A second order differential equation with singular solutions*, Ann. of Math. (2) **50** (1949), 127–153.

11. Flaherty, J. E. and F. C. Hoppensteadt. *Frequency entrainment of a forced van der Pol oscillator*, Studies in Appl. Math. **58** (1978), 5–15.

12. Lloyd, N. G. *On the non-autonomous van der Pol equation with large parameter*, Proc. Camb. Phil. Soc. **72** (1972), 213–227.

13. Rössler, O. E., Rössler, R. and H. D. Landahl. *Arrhythmia in a periodically forced excitable system: chaos in FitzHugh's equation with* $a = 0.467$, $b = 0.8$, $c = 3$ *and* $z = 0.045 \sin 0.2t$, Sixth Int'l. Biophysics Congr., Kyoto, Sept. 1978.

14. Rabinovich, M. I. *Stochastic self-oscillations and turbulence*. Uspehi Fiz. Nauk **125** (1978), 123–168.

15. Lorenz, E. N. *Deterministic nonperiodic flow*, J. Atmospheric Sci. **20** (1963), 130–141.

16. Rössler, O. E. *An equation for continuous chaos*, Phys. Lett. A **57** (1976), 397–398.

17. ______. *Horseshoe map chaos in the Lorenz equations*, Phys. Lett. A **60** (1977), 392–393.

18. ______. *Chemical turbulence*: *Chaos in a simple reaction-diffusion system*, Z. Naturforsch. **31a** (1976), 1168–1172.

19. Kuramoto, Y. *Diffusion-induced chaos in reaction systems*, Progr. Theoret. Phys. **64** (1978), Suppl. (in press).

20. Rössler, O. E. *An equation for hyperchaos*, Phys. Lett. A **71** (1979).

INSTITUTE FOR PHYSICAL AND THEORETICAL CHEMISTRY, UNIVERSITY OF TUBINGEN, 7400 TUBINGEN, W. GERMANY

Lectures in Applied Mathematics
Volume 17, 1979

Nonlinear Oscillations in Equations with Delays[1]

Jack K. Hale

1. Introduction. These lectures are concerned only with some aspects of bifurcation theory in the local theory of nonlinear oscillations in equations with delays; that is, the behavior of solutions near an equilibrium. In particular, we study how the qualitative behavior of solutions changes as parameters vary. A detailed study of the local theory is important in order to know the types of solutions to expect in a global problem. Of course, there is no reason to study only local theory near an equilibrium. One should study how the qualitative behavior changes near any invariant set; for example, behavior near a periodic orbit, behavior near an orbit which connects a saddle point to itself, etc. More complicated behavior is expected near these large invariant sets. One can obtain invariant torii, homoclinic points which exhibit a chaotic behavior, etc. We restrict ourselves in these lectures to behavior near equilibrium.

The simplest type of smooth bifurcation is from an equilibrium to a periodic orbit, the so-called Hopf bifurcation. In §2, we discuss the Hopf bifurcation in equations with finite delays permitting the bifurcation parameters to be the delays themselves. At first glance, such a result does not seem possible because the vector field in the equation is not differentiable in the parameters. The theorem does require some new ideas and, for this reason, the proof is given in some detail. Several examples are given in §3. In §4, similar results are presented for equations with infinite delays. In §5, we give an example in two dimensions for which stable Hopf bifurcation occurs with decreasing

AMS (MOS) subject classifications (1970). Primary 34K15, 34C15.

[1]This research was supported in part by the Air Force Office of Scientific Research under AF-AFOSR 76-3092A, in part by the National Science Foundation under NSF-MCS 76-07247 and in part by the United States Army under ARO-D-31-124-73-G130.

delay. In §6, we give an introduction to some of the methods available for nonautonomous equations.

2. The Hopf bifurcation theorem. One of the simplest ways in which nonconstant periodic solutions of autonomous equations can arise is when an equilibrium point changes from being stable to being unstable as parameters vary in the equation. A periodic orbit can bifurcate from the equilibrium, the process being generally referred to as Hopf bifurcation. For ordinary differential equations, several proofs of the existence of a Hopf bifurcation have been given. In some way, they all involve ultimately the implicit function theorem and the technicalities of the proofs are minimal.

In the statement of the Hopf bifurcation theorem, one must always impose differentiability conditions on the vector field in order to obtain smooth bifurcation curves. For ordinary differential equations, these conditions are not very restrictive. For functional differential equations, the obvious differentiability conditions can eliminate the discussion of variations in important parameters. For example, consider the equation

$$\begin{aligned}\dot{x}(t) = A(\delta)x(t) + B(\delta)x(t-r) + C(\delta)x(t-s) \\ + f(\delta, x(t), x(t-r), x(t-s)) \qquad (2.1)\end{aligned}$$

where $\delta \in \mathbf{R}$, $r > 0$, $s > 0$, are considered as parameters, $A(\delta)$, $B(\delta)$, $C(\delta)$ are $n \times n$ constant matrices, $f(\delta, x, y, z)$ as well as the first derivatives with respect to x, y, z vanish at $x = y = z = 0$. Suppose the characteristic equation of the linear part of equation (2.1),

$$\det \Delta(\lambda, \delta, r, s) = 0, \qquad (2.2)$$

$$\Delta(\lambda, \delta, r, s) = \lambda I - A(\delta) - B(\delta)e^{-\lambda r} - C(\delta)e^{-\lambda s}$$

for $(\delta, r, s) = (\delta_0, r_0, s_0)$, has a pair of purely imaginary roots $i\nu_0$, $-i\nu_0$, $\nu_0 > 0$, and all other roots have negative real parts.

The fundamental problem is to discuss the existence and stability of small nonconstant periodic solutions of (2.1) for (δ, r, s) near (δ_0, r_0, s_0) which vary in a smooth way in (δ, r, s). It is not very restrictive to assume that the right-hand side of equation (2.1) is continuous and has a continuous first derivative in δ. One can then prove that the solution of equation (2.1) is also continuously differentiable in δ. However, for most of the spaces of initial data for equation (2.1), the solution map will not be differentiable in (r, s). At first sight, this makes it unclear how to solve the above problem. The important feature that makes the problem tractable is that every periodic solution of equation (2.1) must have one more derivative in t than $f(\delta, x, y, z)$ has in x, y, z.

The Hopf bifurcation theorem is not an elementary exercise for functional differential equations. The verification of the hypotheses necessary to apply the implicit function theorem uses a number of special identities for linear systems with constant coefficients. Since we are going to state the theorem so that it will be applicable to variations in the delays, a proof of the theorem will be indicated and is based on the proof of a less general result in [7, p. 246].

Suppose $r \geqslant 0$ is a given real number, $\mathbf{R} = (-\infty, \infty)$, $\mathbf{R}^n$ is an n-dimensional linear vector space over the reals with norm $|\cdot|$, $C([a, b], \mathbf{R}^n)$ is the Banach space of continuous functions mapping the interval $[a, b]$ into $\mathbf{R}^n$ with the topology of uniform convergence. If $[a, b] = [-r, 0]$, we let $C = C([-r, 0], \mathbf{R}^n)$ and designate the norm of ϕ by $|\phi| = \sup_{-r \leqslant \theta \leqslant 0} |\phi(\theta)|$. Suppose $\Omega \subset \mathbf{R}^k$ is an open set (the parameter space), $f\colon \Omega \times C \to \mathbf{R}^n$ and $L\colon \Omega \times C \to \mathbf{R}^n$ are continuous, $L(\alpha)\phi$ is linear in ϕ, $f(\alpha, \phi)$ has continuous first and second derivatives in ϕ, $f(\alpha, 0) = 0$, $\partial f(\alpha, 0)/\partial\phi = 0$ and consider the equation

$$\dot{x}(t) = L(\alpha)x_t + f(\alpha, x_t), \tag{2.3}$$

where $x_t(\theta) = x(t + \theta)$, $-r \leqslant \theta \leqslant 0$.

Our first hypothesis is the following:

(H_1) *The characteristic matrix*

$$\Delta(\alpha, \lambda) = \lambda I - L(\alpha)e^{\lambda \cdot}I, \tag{2.4}$$

where I is the $n \times n$ identity matrix, is continuously differentiable in α, there is a purely imaginary characteristic root $\lambda_0 = i\nu_0$, $\nu_0 > 0$, for $\alpha = \alpha_0$ and no other characteristic root $\lambda_j \neq \lambda_0, \bar{\lambda}_0$ of the characteristic equation

$$\det \Delta(\alpha, \lambda) = 0 \tag{2.5}$$

for $\alpha = \alpha_0$ satisfies $\lambda_j = m\lambda_0$, m an integer.

Since $L(\alpha, \phi)$ is continuous and linear in ϕ, there is an $n \times n$ matrix function $\eta(\alpha, \theta)$ of bounded variation in θ, $-r \leqslant \theta \leqslant 0$, such that

$$L(\alpha)\phi = \int_{-r}^{0} [d\eta(\alpha, \theta)]\phi(\theta). \tag{2.6}$$

Along with the linear equation

$$\dot{x}(t) = L(\alpha)x_t \tag{2.7}$$

consider the formal adjoint equation

$$\dot{y}(\tau) = -\int_{-r}^{0} y(\tau - \theta)\, d\eta(\alpha, \theta) \tag{2.8}$$

and the bilinear form

$$(\psi, \phi) = \psi(0)\phi(0) - \int_{-r}^{0}\int_{0}^{\theta} \psi(\xi - \theta)[d\eta(\alpha, \theta)]\phi(\xi)\, d\xi \tag{2.9}$$

defined for $\phi \in C$ and $\psi \in C([0, r], \mathbf{R}^{n*})$, $\mathbf{R}^{n*}$ is the space of n-dimensional row vectors.

For the characteristic values $(i\nu_0, -i\nu_0)$ of the characteristic equation $\det \Delta(\alpha_0, \lambda) = 0$ corresponding to the linear equation (2.7) for $\alpha = \alpha_0$, there are two linearly independent solutions $b \cos \nu_0 t$, $b \sin \nu_0 t$ for some n-dimensional column vector b. In the same way, there are two linearly independent solutions $c \cos \nu_0 \tau$, $c \sin \nu_0 \tau$ of the formal adjoint equation (2.8) for $\alpha = \alpha_0$. In Chapter 7 of [7], the following remarks are proved. If

$$\Phi_{\alpha_0} = (\phi_1, \phi_2),$$

$$\phi_1(\theta) = b \cos \nu_0 \theta, \qquad \phi_2(\theta) = b \sin \nu_0 \theta, \qquad -r \leqslant \theta \leqslant 0,$$

$$\Psi^*_{\alpha_0} = \begin{pmatrix} \psi_1^* \\ \psi_2^* \end{pmatrix},$$

$$\psi_1^*(s) = c \cos \nu_0 s, \qquad \psi_2^*(s) = c \sin \nu_0 s, \qquad 0 \leqslant s \leqslant r;$$

$$(\Psi^*_{\alpha_0}, \Phi_{\alpha_0}) = \begin{bmatrix} (\psi_1, \phi_1) & (\psi_1, \phi_2) \\ (\psi_2, \phi_1) & (\psi_2, \phi_2) \end{bmatrix},$$

where $(\ ,\ )$ is the bilinear form in (2.9), then the 2×2 matrix $(\Psi^*_{\alpha_0}, \Phi_{\alpha_0})$ is nonsingular. If $\Psi_{\alpha_0} = (\Psi^*_{\alpha_0}, \Phi_{\alpha_0})^{-1}\Psi^*_{\alpha_0}$, then $(\Psi_{\alpha_0}, \Phi_{\alpha_0}) = I$ and one can decompose C as

$$C = P_{\alpha_0} \oplus Q_{\alpha_0},$$

$$P_{\alpha_0} = \{\phi \in C: \phi = \Phi_{\alpha_0} a,\ a \in \mathbf{R}^2\},$$

$$Q_{\alpha_0} = \{\phi \in C: (\Psi_{\alpha_0}, \phi) = 0\}. \tag{2.10}$$

This decomposition of C defines a projection π_{α_0} with $\pi_{\alpha_0} C = P_{\alpha_0}$, $(I - \pi_{\alpha_0})C = Q_{\alpha_0}$. When a decomposition is made in this way, we say C *is decomposed by the set of characteristic values* $\{i\nu_0, -i\nu_0\}$.

The following result is proved in exactly the same way as Lemma 2.2 in [7, p. 171].

LEMMA 2.1. *If* (H_1) *is satisfied, then there is a* $\delta > 0$ *and a simple characteristic root* $\lambda(\alpha)$ *of* (2.5) *which is continuous together with its first derivative*, $\operatorname{Im} \lambda(\alpha) > 0$, $|\alpha - \alpha_0| < \delta$, $\lambda(\alpha_0) = i\nu_0$. *Furthermore, C can be decomposed by* $\{\lambda(\alpha), \bar{\lambda}(\alpha)\}$ *as* $C = P_\alpha \oplus Q_\alpha$, $\dim P_\alpha = 2$, *and the corresponding projection operator* π_α *is continuous together with its first derivative in* α.

With this lemma and $\Phi_\alpha = (\phi_1, \phi_2)$ a basis for P_α, it follows that there is a 2×2 matrix $B(\alpha)$, continuous and continuously differentiable in α, such that

$$\Phi_\alpha(\theta) = \Phi_\alpha(0) \exp B(\alpha)\theta, \qquad -r \leqslant \theta \leqslant 0.$$

Also, the eigenvalues of $B(\alpha)$ are $\lambda(\alpha)$, $\bar{\lambda}(\alpha)$. Furthermore, we may assume by a change of coordinates that

$$B(\alpha) = \nu_0 B_0 + B_1(\alpha),$$

$$B_0 = \begin{bmatrix} 0 & 1 \\ -1 & 0 \end{bmatrix}, \quad B_1 = \begin{bmatrix} (\alpha - \alpha_0)\cdot\zeta(\alpha) & (\alpha - \alpha_0)\cdot\gamma(\alpha) \\ -(\alpha - \alpha_0)\cdot\gamma(\alpha) & (\alpha - \alpha_1)\cdot\zeta(\alpha) \end{bmatrix} \tag{2.11}$$

where $\zeta(\alpha), \gamma(\alpha) \in \mathbf{R}^k$ are continuous and continuously differentiable in α for $|\alpha - \alpha_0| < \delta$.

We may now state a generalization of the Hopf bifurcation theorem.

THEOREM 2.1. *Suppose $L(\alpha)\phi, f(\alpha, \phi)$ satisfy* (H_1),

(H_2) *for any $K > 0$, $\phi \in C$ with $d\phi/d\theta \in C$, $|d\phi/d\theta| \leqslant K$, the functions $L(\alpha)\phi$, $f(\alpha, \phi)$ have a first derivative in α which is continuous in α.*

(H_3) $\zeta(\alpha_0) \neq 0$ *where $\zeta(\alpha)$ is defined in equation* (2.11).

Then there is an $\varepsilon > 0$ such that for $a \in \mathbf{R}$, $|a| < \varepsilon$, there is a C^1-manifold $\Gamma_a \in \mathbf{R}_k$ of codimension 1, *Γ_a is continuous and continuously differentiable in a,*

$$\Gamma_0 = \{\alpha \in \mathbf{R}^k\colon \operatorname{Re}\lambda(\alpha) = 0, |\alpha - \alpha_0| < \varepsilon\}$$

where $\lambda(\alpha)$ is given in Lemma 2.1, such that for every $\alpha \in \Gamma_a$, there is a function $\omega(\alpha, a)$, an $\omega(\alpha, a)$-periodic function $x^*(\alpha, a)$ continuous together with their first derivatives in a and α, $\omega(\alpha_0, 0) = \omega_0 = 2\pi/\nu_0$, $x^*(\alpha, 0) =$ 0,

$$x_0^*(\alpha, a) = \Phi_\alpha \operatorname{col}(a, 0) + o(|a|) \quad \text{as } |a| \to 0$$

and $x^*(\alpha, a)$ is a solution of equation (2.3). Furthermore, for $|\alpha - \alpha_0| < \varepsilon$, $|\omega - \omega_0| < \varepsilon$ every ω-periodic solution of equation (2.3) with $|x_t| < \varepsilon$ must be of the above type except for a translation in phase.

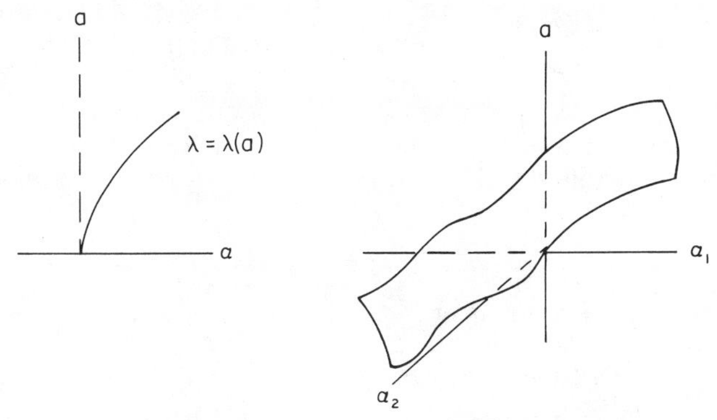

FIGURE 1. FIGURE 2.

Before proving the theorem, let us discuss some implications.

If $k = 1$, that is, $\alpha \in \mathbf{R}$, and $\lambda(\alpha) = \mu(\alpha) + i\nu(\alpha)$ is the characteristic root given by Lemma 2.1, then (H_3) is equivalent to $d\lambda(\alpha_0)/d\alpha \neq 0$. Thus, Theorem 2.1 is the usual Hopf bifurcation theorem. In fact, for $k = 1$, the sets Γ_a can be described by a smooth curve $\alpha(a)$. Thus, for $|a| < \varepsilon$, there is a function $\alpha(a)$ continuous together with its first derivative such that $\alpha(0) = \alpha_0$ and $x^*(a) = x^*(\alpha(a), a)$ is an $\omega(a) = \omega(\alpha(a), a)$ periodic solution of equation (2.3). The bifurcation curve $\omega(a)$ is shown schematically in Figure 1.

If $k = 2$, then the bifurcation surface in the (a, α)-space is shown schematically in Figure 2. For each a, there is a smooth curve of α values for which there exist nonconstant periodic solutions of equation (2.3). If $k > 2$, the picture is similar except Γ_a is a surface of codimension 1 in $\mathbf{R}^k$.

To prove Theorem 2.1, we first obtain the bifurcation equations for the small amplitude periodic solutions of the equation (2.3) for $|\alpha - \alpha_0| < \delta$ which have a period close to $2\pi/\nu_0$. As in ordinary differential equations, we normalize the period to 2π. Let $\beta \in (-1, 1)$, $\omega_0 = 2\pi/\nu_0$, $t = (1 + \beta)\tau$,

$$x(t + \theta) = u(\tau + \theta/(1 + \beta)), \qquad -r \leqslant \theta \leqslant 0,$$

and define $u_{\tau,\beta}$ as an element of the space $C([-r, 0], \mathbf{R}^n)$ given by

$$u_{\tau,\beta}(\theta) = u(\tau + \theta/(1 + \beta)), \qquad -r \leqslant \theta \leqslant 0.$$

Equation (2.3) is then equivalent to

$$\frac{du(\tau)}{d\tau} = (1 + \beta)\left[L(\alpha)u_{\tau,\beta} + f(\alpha, u_{\tau,\beta})\right]. \tag{2.12}$$

If this equation has an ω_0-periodic solution, then equation (2.3) has a $(1 + \beta)\omega_0$-periodic solution, and conversely.

Equation (2.12) may be rewritten as

$$\frac{du(\tau)}{d\tau} = L(\alpha_0)u_\tau + N(\beta, \alpha, u_\tau, u_{\tau,\beta}), \tag{2.13}$$

$$N(\beta, \alpha, u_\tau, u_{\tau,\beta}) = (1 + \beta)L(\alpha)u_{\tau,\beta} - L(\alpha_0)u_\tau + (1 + \beta)f(\alpha, u_{\tau,\beta}).$$

We consider equation (2.13) as a perturbation of the autonomous linear equation

$$du(\tau)/d\tau = L(\alpha_0)u_\tau. \tag{2.14}$$

We know that the columns of

$$U(\tau) \overset{\text{def}}{=} \Phi_{\alpha_0}(0)\exp B(\alpha_0)\tau, \qquad \tau \in \mathbf{R},$$

form a basis for the ω_0-periodic solutions of equation (2.14) and the rows of

$$V(\tau) \overset{\text{def}}{=} \exp[-B(\alpha_0)\tau]\Psi_{\alpha_0}(0), \qquad \tau \in \mathbf{R},$$

form a basis for the ω_0-periodic solutions of the formal adjoint equation (2.8).

We need the following lemma from [7, p. 209].

LEMMA 2.2. *Let* $\mathcal{P}_{\omega_0}$ *be the Banach space of continuous* ω_0*-periodic functions with values in* $\mathbf{R}^n$ *and the topology of uniform convergence. For any* $f \in \mathcal{P}_{\omega_0}$, *the equation*

$$\frac{du(\tau)}{d\tau} = L(\alpha_0)u_\tau + f(\tau) \tag{2.15}$$

has a solution in $\mathcal{P}_{\omega_0}$ *if and only if*

$$\int_0^{\omega_0} y(t)f(t)\,dt = 0 \tag{2.16}$$

for all ω_0*-periodic solutions of the formal adjoint equation* (2.8). *Furthermore, there is a projection* J: $\mathcal{P}_{\omega_0} \to \mathcal{P}_{\omega_0}$ *such that the set of* f *in* $\mathcal{P}_{\omega_0}$ *satisfying* (2.16) *is* $(I - J)\mathcal{P}_{\omega_0}$ *and there is a continuous linear operator* K: $(I - J)\mathcal{P}_{\omega_0} \to (I - \tilde{\pi})\mathcal{P}_{\omega}$ *such that* Kf *is a solution of* (2.15) *for every* $f \in (I - J)\mathcal{P}_{\omega_0}$. *The operator* $\tilde{\pi}$ *is any projection onto the* ω_0*-periodic solutions of* (2.14).

The operator J *is given by*

$$Jf = V'\left[\int_0^{\omega_0} V(s)V'(s)\,ds\right]^{-1}\int_0^{\omega_0} V(s)f(s)\,ds \tag{2.17}$$

where V' *is the transpose of* V.

This lemma simply states necessary and sufficient conditions for the existence of periodic solutions of nonhomogeneous linear equations. Knowing these conditions, one can always subtract off part of any $f \in \mathcal{P}_{\omega_0}$, namely Jf, so that the orthogonality conditions (2.16) are satisfied for $(I - J)f$. Thus, a solution will exist. We can always choose a particular solution to be orthogonal to the solutions of the homogeneous equation. This is $K(I - J)f$.

Using Lemma 2.2, we know that every ω_0-periodic solution of equation (2.13) except for a translation in phase is a solution of the equations

$$u(\cdot) = U(\cdot)\mathrm{col}(a, 0) + K(I - J)N(\beta, \alpha, u_\cdot, u_{\cdot,\beta}), \tag{2.18a}$$

$$JN(\beta, \alpha, u_\cdot, u_{\cdot,\beta}) = 0, \tag{2.18b}$$

and conversely.

One can now apply the implicit function theorem to solve equation (2.18a) for $u = u^*(a, \beta, \alpha)$ for a, β, α in a sufficiently small neighborhood of zero, $u^*(a, 0, 0) - U(\cdot)\mathrm{col}(a, 0) = o(|a|)$ as $|a| \to 0$. The function u^* is continuous together with its derivatives up through order two from an application of the implicit function theorem. However, without some additional information, we cannot prove it is differentiable in β, α.

Since $u^*(a, \beta, \alpha)(t)$ satisfies (2.18a), it is automatically continuously differentiable in t with bounded derivative for $|\alpha - \alpha_0| < \delta$. If we use this fact and the form of the implicit function theorem given in Lemma 2.2 in [7, p. 236], one obtains the following important result.

LEMMA 2.3. *If* (H_1), (H_2) *are satisfied, then the function* $u^*(a, \beta, \alpha)$ *is continuous together with its first derivatives in* a, β, α *and second derivatives in* a.

From the above lemma, all ω_0-periodic solutions of equation (2.13) are obtained by finding the solutions a, β, α of the bifurcation equations

$$JN(\beta, \alpha, u^*_\cdot(a, \beta, \alpha), u^*_{\cdot,\beta}(a, \beta, \alpha)) = 0. \tag{2.19}$$

Using the definition of J in equation (2.17) equation (2.19) is equivalent to the equation

$$G(a, \beta, \alpha) \overset{\text{def}}{=} \int_0^{\omega_0} e^{-B(\alpha_0)s}\Psi_{\alpha_0}(0) N(\beta, \alpha, u^*_s(a, \beta, \alpha), u^*_{s,\beta}(a, \beta, \alpha))\, ds$$
$$= 0. \tag{2.20}$$

From the above discussion, it follows that it remains to solve the equation $G(a, \beta, \alpha) = 0$, which represents two equations in the parameters a, β, α.

Since $G(0, \beta, \alpha) = 0$ for all β, α, define

$$H(a, \beta, \alpha) = G(a, \beta, \alpha)/a. \tag{2.21}$$

This function is continuous together with its first derivatives in a, β, α.

The nonzero small amplitude periodic solutions of equation (2.3) are given as $x(t) = u^*(a, \beta, \alpha)(t/(1 + \beta))$ where a, β, α satisfy the equation

$$H(a, \beta, \alpha) = 0. \tag{2.22}$$

Our next objective is to compute the Jacobian of $H(a, \beta, \alpha)$ with respect to β, α evaluated at $(a, \beta, \alpha) = (0, 0, \alpha_0)$. From the definition of G in (2.20), N in (2.13) and u^* one easily observes that, for $e_1 = \mathrm{col}(1, 0)$,

$$H(0, 0, \alpha) = \int_0^{\omega_0} e^{-B(\alpha_0)s}\Psi_{\alpha_0}(0)\{L(\alpha)U_s e_1 - L(\alpha_0)U_s e_1\}\, ds$$
$$= \int_0^{\omega_0} e^{-B(\alpha_0)s}\Psi_{\alpha_0}(0)[L(\alpha) - L(\alpha_0)]\Phi_{\alpha_0} e^{B(\alpha_0)s} e_1\, ds,$$

since $U_s = \Phi_{\alpha_0} \exp B(\alpha_0)s$.

We need the following Lemma 3.9 in [7, p. 179].

LEMMA 2.4. *Suppose the conditions of Lemma* 2.1 *are satisfied. Then*

$$\frac{dB(\alpha_0)}{d\alpha} = -\Psi_{\alpha_0}(0)\frac{dL(\alpha_0)}{d\alpha}\Phi_{\alpha_0}.$$

Using Lemma 2.4,

$$\frac{\partial H(0, 0, \alpha_0)}{\partial \alpha}\nu = \int_0^{\omega_0} e^{-B(\alpha_0)s}\frac{dB(\alpha_0)}{d\alpha}e^{B(\alpha_0)s}e_1 ds$$

$$= \omega_0 \frac{dB(\alpha_0)}{d\alpha}e_1 = \omega_0\begin{bmatrix} \nu \cdot \zeta(\alpha_0) \\ -\nu \cdot \gamma(\alpha_0) \end{bmatrix} \tag{2.23}$$

where $\zeta(\alpha_0)$, $\gamma(\alpha_0)$ are defined in (2.11).

To compute the derivative with respect to β, we note first that

$$H(0, \beta, \alpha_0) = \int_0^{\omega_0} e^{-B(\alpha_0)s}\Psi_{\alpha_0}(0)\{(1+\beta)L(\alpha_0)U_{s,\beta}e_1 - L(\alpha_0)U_s e_1\}\, ds.$$

For the evaluation of the derivative of this function, it is convenient to use the fact that Ψ_{α_0}, Φ_{α_0} are the real and imaginary parts of a complex function. Choose $w \neq 0$, $v \neq 0$ satisfying

$$w\Delta(\alpha_0, i\nu_0) = 0, \qquad \Delta(\alpha_0, i\nu_0)v = 0.$$

For an appropriate choice of w, v, we have

$$\Psi_{\alpha_0}(\theta) = \mathrm{col}(\mathrm{Re}\, we^{-i\nu_0\theta}, \mathrm{Im}\, we^{-i\nu_0\theta}),$$

$$\Phi_{\alpha}(\theta) = (\mathrm{Re}\, ve^{i\nu_0\theta}, \mathrm{Im}\, ve^{i\nu_0\theta}),$$

$$(we^{-i\nu_0 \cdot}, ve^{i\nu_0 \cdot}) = w\left[\frac{\partial \Delta(\alpha_0, i\nu_0)}{\partial \alpha}\right]\nu = 2.$$

Furthermore, if $H = (H_1, H_2)$, $Z = H_1 + iH_2$, then

$$Z(0, \beta, \alpha_0) = \omega_0 w\Delta\left(\alpha_0, \frac{i\nu_0}{1+\beta}\right) - \omega_0 \beta w \int_{-r}^{0} d\eta(\theta)e^{i\nu_0\theta/(1+\beta)}v,$$

$$\frac{\partial Z(0, 0, \alpha_0)}{\partial \beta} = 2\pi i w\Delta'(\alpha_0, i)v = 4\pi i.$$

Thus,

$$\frac{\partial H(0, \beta, \alpha_0)}{\partial \beta} = \begin{pmatrix} 0 \\ -4\pi \end{pmatrix}. \tag{2.24}$$

Relation (2.24) implies

$$H_2(0, 0, \alpha_0) = 0, \qquad \frac{\partial H_2(0, 0, \alpha_0)}{\partial \beta} = -4\pi.$$

Consequently, the implicit function theorem implies there is an $\varepsilon > 0$ and a function $\beta^*(\alpha, a)$ continuous together with its first derivatives in α, a, $|\alpha - \alpha_0| < \varepsilon$, $|a| < \varepsilon$, $\beta^*(\alpha_0, 0) = 0$, such that $\beta^*(\alpha, a)$ satisfies the equation

$$H_2(a, \beta^*(\alpha, a), \alpha) = 0 \tag{2.25}$$

and is the only solution for $|\beta| < \varepsilon$, $|\alpha - \alpha_0| < \varepsilon$, $|a| < \varepsilon$.

Up to this point, we have the following lemma.

LEMMA 2.5. *If* (H_1), (H_2) *are satisfied, then there is an* $\varepsilon > 0$, *functions* $\beta^*(\alpha, a)$, $u^*(a, \beta, \alpha)$, *continuous together with their first derivatives,* $u^*(a, \beta, \alpha)$ *is* ω_0*-periodic,* $|a| < \varepsilon$, $|\alpha - \alpha_0| < \varepsilon$, $\beta^*(\alpha_0, 0) = 0$, $u^*(0, 0, \alpha_0) = 0$, *such that for* $|\omega - \omega_0| < \varepsilon$, $|\alpha - \alpha_0| < \varepsilon_0$, *equation* (2.3) *has an* ω*-periodic solution* x *with* $|x| < \varepsilon$ *if and only if*

$$x(t) = u^*(a, \beta^*(\alpha, a), \alpha)(t/(1 + \beta^*)\alpha, a), \qquad \omega = (1 + \beta^*(\alpha, a))\omega_0,$$

except for a translation in phase where α, a *satisfy the bifurcation equation*

$$h(\alpha, a) \stackrel{\text{def}}{=} H_1(a, \beta^*(\alpha, a), \alpha) = 0. \tag{2.26}$$

The function $u^*(a, \beta, \alpha)$ *satisfies* (2.18a) *and* $\beta^*(\alpha, a)$ *satisfies* (2.25).

REMARK 2.1. It is clear from the proof of the above results that if we assume $L(\alpha)\varphi + f(\alpha, \varphi)$ has k derivatives with respect to φ which are continuous and hypothesis (H_2) is satisfied for the kth derivatives with respect to α when φ has k continuous derivatives, then the function $h(\alpha, a)$ has continuous derivatives in α, a up through order k.

From the expressions for the partial derivatives of H with respect to α, β at $(0, 0, \alpha_0)$, one easily determines that

$$h(\alpha, a) = \xi_0(a) + (\alpha - \alpha_0)\xi_1(a) + o(|\alpha - \alpha_0|), \qquad \xi_1(0) = \zeta(\alpha_0), \tag{2.27}$$

where $\zeta(\alpha_0)$ is given in (2.11).

PROOF OF THEOREM 2.1. If $\Gamma_a = \{\alpha\colon h(\alpha, a) = 0, |a| < \varepsilon\}$, then the implicit function theorem implies the assertions about Γ_a if $k = 1$. For $k > 1$, it is the transversality theorem (see [**1**, p. 45]). Then

$$\omega(\alpha, a) = (1 + \beta^*(\alpha, a))\omega_0,$$

$$x^*(\alpha, a)(t) = u^*(a, \beta^*(\alpha, a), \alpha)(t/(1 + \beta^*(\alpha, a)))$$

satisfy all of the properties stated in the theorem. This proves Theorem 2.1.

Theorem 2.1 uses only knowledge of the linear operator $L(\alpha)$ in equation (2.3). We know nothing about the specific structure of the sets Γ_α. To obtain a more complete picture of the bifurcation, one must consider nonlinear terms. Let us suppose $L(\alpha)\varphi$, $f(\alpha, \varphi)$ satisfy all conditions above with the differentiability conditions up through order $k \geqslant 3$. Then the function $h(\alpha, a)$ in relation (2.27) satisfies

$$\begin{aligned} h(\alpha, a) &= h(\alpha, 0) + p(\alpha)a^2 + o(|a|^2), \\ h(\alpha, 0) &= (\alpha - \alpha_0)\cdot\zeta(\alpha_0) + o(|\alpha - \alpha_0|), \end{aligned} \tag{2.28}$$

as $|a| \to 0$, $\alpha \to \alpha_0$. Let us assume that

$$p(\alpha_0) \neq 0. \tag{2.29}$$

The function $h(\alpha, a)/p(\alpha)$ will then have a minimum at $a = 0$ given by $h(\alpha, 0)/p(\alpha)$. The bifurcation equation will have a solution if and only if this minimum is less than or equal to zero and will have no solution if this minimum is greater than zero. The minimum equal to zero is therefore a bifurcation surface in the α-space. This is given by

$$0 = \frac{h(\alpha, 0)}{p(\alpha)} = \frac{1}{p(\alpha_0)}(\alpha - \alpha_0) \cdot \zeta(\alpha_0) + o(|\alpha - \alpha_0|) \qquad (2.30)$$

as $\alpha \to \alpha_0$.

Under hypothesis (H_3), equation (2.30) defines a smooth hyperplane Q in a neighborhood of α_0 in $\mathbf{R}^k$. In a sufficiently small neighborhood of the point $(\alpha_0, 0)$ in $\Omega \times C$, on one side of this hyperplane there are no nonconstant periodic solutions and on the other side there is a unique nonconstant periodic solution. For $k = 2$, the bifurcation diagram is similar to the one shown in Figure 3. The curve is the intersection of the surface in Figure 3 with the plane $a = 0$.

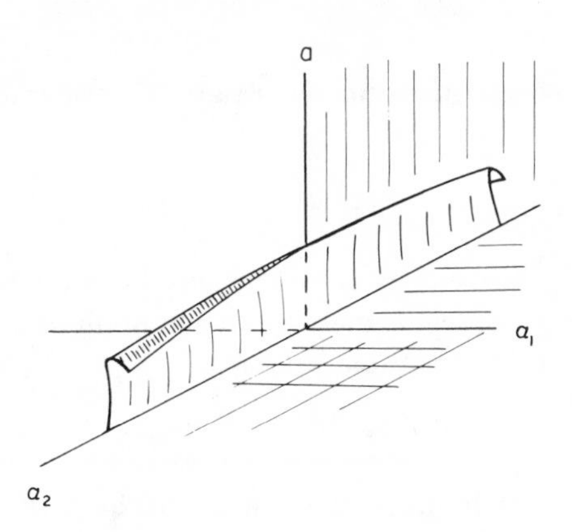

FIGURE 3.

The direction of bifurcation is determined by $p(\alpha_0)$.

These results are briefly summarized in the following result.

THEOREM 2.2. *Suppose* $L(\alpha)\phi$, $f(\alpha, \phi)$ *satisfy* (H_1), (H_2), *taking into account derivatives up through order three. If, in addition,* (H_3) *is satisfied and* $p(\alpha_0) \neq 0$ *where* $p(\alpha)$ *is given in* (2.28), *then there are neighborhoods* $V \subset \mathbf{R}^k$ *of* α_0, U *in* C *of zero and a smooth hyperplane* $\Gamma \subset \mathbf{R}^k$ *such that* $V \setminus \Gamma = A_1 \cup A_2$ *and*

(i) $\alpha \in A_1$ *implies no nonconstant periodic solution in* U,

(ii) $\alpha \in A_2$ *implies a unique nonconstant periodic solution in* U.

A less general version of the existence theorem 2.2 was obtained by Ruiz-Claeyssen [**22**, Theorem 6.2]. He considered equation (2.3) in the space $W^{1,\infty}$ in order to obtain the center manifold and then applied techniques similar to the ones employed by Chafee [**4**]. In addition, he was able to extend the analysis in Hale [**9**] and Hausrath [**11**] to obtain the stability of the bifurcation when (H_1) was strengthened to say that all characteristic values have negative real parts except the two on the imaginary axis. The stability properties are determined by the change in stability of the zero solution as Γ is crossed. One can actually improve the results in [**22**] as well as extend them to the case when equation (2.3) is considered in C. More specifically, one can prove the following result.

THEOREM 2.3. *If the conditions of Theorem* 2.2 *are satisfied and for* $\alpha = \alpha_0$, *all eigenvalues of equation* (2.7) *have negative real parts except two on the imaginary axis, then for* $\alpha \in V$ *all eigenvalues of equation* (2.7) *have negative real parts except two,* $\lambda(\alpha)$, $\bar{\lambda}(\alpha)$ *and either*

(i) Re $\lambda(\alpha) < 0$ *in* A_1, Re $\lambda(\alpha) > 0$ *in* A_2, *or*

(ii) Re $\lambda(\alpha) > 0$ *in* A_1, Re $\lambda(\alpha) < 0$ *in* A_2.

The bifurcating periodic orbit is stable if (i) *is satisfied and unstable if* (ii) *is satisfied.*

3. Examples of variations in the delays. Consider the linear scalar equation

$$\dot{x} = -\tfrac{1}{2}x(t - r_1) - \tfrac{1}{2}x(t - r_2) \tag{3.1}$$

where r_1, r_2 are positive constants. The specific coefficients in equation (3.1) are chosen in order to make the computations simple. The characteristic equation for equation (3.1) is

$$\lambda + \tfrac{1}{2}e^{-\lambda r_1} + \tfrac{1}{2}e^{-\lambda r_2} = 0. \tag{3.2}$$

To discuss the Hopf bifurcation for a nonlinear perturbation of equation (3.1), one must determine the curves in (r_1, r_2)-space at which the roots of equation (3.2) have zero real parts. Ruiz-Claeyssen [**22**] has analyzed this problem completely. There are infinitely many curves for which there are purely imaginary roots, but we are going to concentrate on the curve which has the property that if $(r_1, r_2) \in \Gamma$, then there is a pair of purely imaginary roots and all other roots of equation (3.2) have negative real parts.

For $r_1 = r_2 = r$, equation (3.2) is the familiar equation

$$\lambda + e^{-\lambda r} = 0. \tag{3.3}$$

There is a unique $r_0 = \pi/2$ such that equation (3.3) has all roots with negative real parts for $0 < r < r_0$ and a unique pair of purely imaginary roots $\lambda = \pm i$ for $r = r_0$. The curve Γ that interests us is the one in the (r_1, r_2)-plane which passes through the point $(\pi/2, \pi/2)$.

If $\lambda = i\gamma$ in equation (3.2), then we must have

$$0 = \cos \tfrac{1}{2}\gamma(r_1 + r_2)\cos \tfrac{1}{2}\gamma(r_1 - r_2),$$
$$\gamma = \sin \tfrac{1}{2}\gamma(r_1 + r_2)\cos \tfrac{1}{2}\gamma(r_1 - r_2). \tag{3.4}$$

Since $\gamma = 0$ is not a solution of equation (3.2), it follows that $\cos \frac{1}{2}\gamma(r_1 - r_2) \neq 0$. Thus, we must have $\gamma = \pi/(r_1 + r_2)$, and

$$\frac{\pi}{r_1 + r_2} = \cos \frac{\pi(r_1 - r_2)}{2(r_1 + r_2)}.$$

The curve Γ is, therefore, the solution of this equation and is shown schematically in Figure 4.

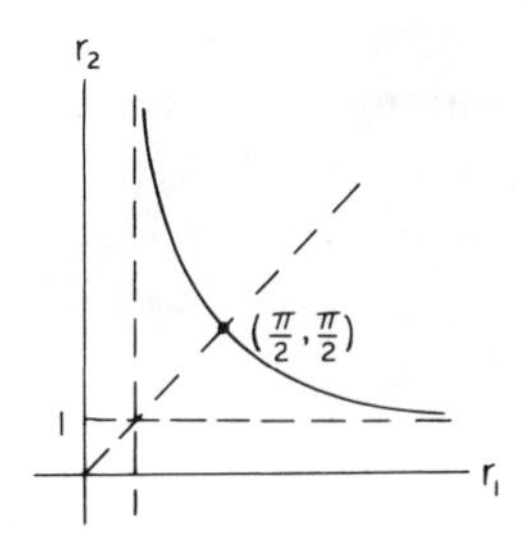

FIGURE 4.

Now consider a perturbation of equation (3.1):

$$\dot{x}(t) = -\tfrac{1}{2}x(t - r_1) - \tfrac{1}{2}x(t - r_2) + f(x(t), x(t - r_1), x(t - r_2))$$

where $f(x, y, z)$ is continuous together with derivatives up through order three, f and its first partial derivatives vanish at $x = y = z = 0$. Theorem 2.1 implies there is a Hopf bifurcation at every point in Γ.

To obtain more information, one must assume something about the nonlinearity f. In particular, consider the equation

$$\dot{x}(t) = -\tfrac{1}{2}x(t - r_1) - \tfrac{1}{2}x(t - r_2) + \beta x^3(t), \qquad \beta \neq 0. \tag{3.5}$$

If we let $\alpha_0 = (r_1^0, r_2^0) \in \Gamma$, then the characteristic roots $\lambda(\alpha_0)$, $\bar{\lambda}(\alpha_0)$ on the imaginary axis are given by $i\gamma_0$, $-i\gamma_0$, $\gamma_0 = \pi/(r_1^0 + r_2^0)$. Choose bases $\Phi(\alpha_0)$, $\Psi(\alpha_0)$ as in the previous section for the decomposition of C as $C = P_{\alpha_0} \oplus Q_\alpha$ for the characteristic roots $\{i\gamma_0, -i\gamma_0\}$. In **[22]**, it is shown that $p(a_0)$ in (2.28) is given by

$$p(\alpha_0) = -kc(\alpha_0)\beta \tag{3.6}$$

where k is a positive constant,

$$c(\alpha_0) = \left(1 - \gamma_0 \sin \tfrac{1}{2}\gamma_0 r_2^0\right). \tag{3.7}$$

If $c(\alpha_0) \neq 0$, Theorem 2.2 implies there is a neighborhood $V \subset \mathbf{R}^2$ of

$\alpha_0 \in \Gamma$ such that $V \setminus \Gamma = A_1 \cup A_2$ and

(i) there is no small nonconstant periodic solution of (3.4) for $(r_1, r_2) \in A_1$,

(ii) there is a unique small nonconstant periodic solution of (3.4) for $(r_1, r_2) \in A_2$.

Ruiz-Claeyssen [**22**] also showed that $c(\alpha_0)\beta < 0$ implies the periodic orbit is asymptotically stable and the bifurcation occurs as Γ is crossed from right to left. If $c(\alpha_0)\beta > 0$, the periodic orbit is unstable and the bifurcation occurs as Γ is crossed from left to right. There are some points α_0 on Γ where $c(\alpha_0) = 0$. Higher order approximations in the bifurcation equations will be needed to determine periodic orbits near these points. However, we do see that a two-parameter problem is very different from a one-parameter problem. The direction of bifurcation can change as we move along Γ. Stech [**24**] has also observed this phenomenon in problems with two delays.

An obvious generalization of equation (3.1) is

$$\dot{x}(t) = -(1-\alpha)x(t-r_1) - \alpha x(t-r_2), \qquad 0 \leqslant \alpha \leqslant 1, \quad (3.8)$$

whose characteristic equation is

$$\lambda + (1-\alpha)e^{-\lambda r_1} + \alpha e^{-\lambda r_2} = 0. \quad (3.9)$$

The analysis of equation (3.9) for all $\alpha \in [0, 1]$, $r_1 > 0$, $r_2 > 0$ is difficult. From our point of view, we would like to know for a fixed α the curve Γ in (r_1, r_2)-space which passes through the point $(\pi/2, \pi/2)$ and has the property that $(r_1, r_2) \in \Gamma$ implies that two roots are purely imaginary and the remaining roots have negative real parts. This curve changes in a complicated way as a function of α and it is not completely understood.

The equation for the purely imaginary roots $\lambda = i\gamma$ of equation (3.9) are

$$(1-\alpha)\cos \gamma r_1 + \alpha \cos \gamma r_2 = 0, \quad (3.10a)$$

$$(1-\alpha)\sin \gamma r_1 + \alpha \sin \gamma r_2 = \gamma. \quad (3.10b)$$

For $\alpha = 0$, the solutions of this equation are

$$\gamma r_1 = (2k+1)\pi/2, \qquad \gamma = (-1)^k, \qquad k = 0, 1, 2, \ldots,$$

and the special curve Γ is given by $r_1 = \pi/2$, r_2 arbitrary and corresponds to $k = 0$, $\gamma = 1$.

For $\alpha = \frac{1}{2}$, equation (3.10) can be written as equation (3.4) and has the solutions, $\gamma = (2k+1)\pi/(r_1 + r_2)$,

$$(-1)^k \frac{(2k+1)\pi}{r_1 + r_2} = \cos\left((2k+1)\frac{\pi}{2}\,\frac{r_2 - r_1}{r_2 + r_1}\right), \qquad k = 0, 1, 2, \ldots. \quad (3.11)$$

As remarked earlier, the special curve Γ corresponds to $k = 0$ and is shown schematically in Figure 4. To understand how Γ changes with α, it is necessary to investigate the solutions of equation (3.11) for other values of k. For $k = 1$, that is, $\gamma = 3\pi/(r_1 + r_2)$,

$$-\frac{3\pi}{r_1 + r_2} = \cos\left(\frac{3\pi}{2}\,\frac{r_2 - r_1}{r_2 + r_1}\right). \tag{3.12}$$

The solutions of equation (3.12) are depicted in Figure 5(a) for $r_2 \geqslant r_1$. In this figure, we have also included the original curve Γ.

The value of $\gamma = (2k + 1)\pi/(r_1 + r_2)$, $k = 0, 1$, approaches zero as r_1, r_2 approach infinity along either of these curves; that is, for large values of $r_1 + r_2$, there are many purely imaginary roots of equation (3.11) near zero. Consequently, for a small change in α from $\alpha = \frac{1}{2}$ in equation (3.10), it is conceivable that the curves representing the solutions of equation (3.10) for large values of $r_1 + r_2$ can have a different topological structure than for $\alpha = \frac{1}{2}$. This suspicion was verified on a computer by M. Michaud and T. Lewis for the values of α depicted in Figure 5 and corresponding to the two curves associated with $k = 0, 1$ for $\alpha = \frac{1}{2}$.

In Figure 5(b), the smooth parabolic-like curve corresponds to the curve $k = 1$, $\gamma = 3\pi/(r_1 + r_2)$ for $\alpha = \frac{1}{2}$ and the other curve to $k = 0$, $\gamma = \pi/(r_1 + r_2)$ for $\alpha = \frac{1}{2}$. At the cusp point, there are two distinct pairs of purely imaginary roots on the imaginary axis. The number of roots with positive real parts near this cusp is indicated in Figure 5(b). The Hopf bifurcation theorem can be applied along each smooth branch passing through this cusp. The end result is that it is possible to have the same number of periodic orbits as roots with positive real parts.

For $r_2 = 3r_1$, it is not difficult to show (see Stech [**24**]) that in a neighborhood of $\alpha = \frac{1}{4}$, equation (3.10) has one solution for $\alpha < \frac{1}{4}$ and three solutions for $\alpha > \frac{1}{4}$ with γr_1 near $\pi/2$. This is a different type of phenomenon from the one exhibited near the cusp where the purely imaginary parts were different depending on which branch was followed. For α near $\frac{1}{4}$, the curve is smooth but has a "kink" in it (see Figure 5(d)). Similar phenomena have been observed by Nussbaum [**21**].

For larger values of r_2 than shown in Figure 5, more "kinks" and cusps must appear because the functions in equation (3.10) are almost periodic functions in r_1, r_2. Therefore, similar bifurcation phenomena occur.

These results are important because they have the following implications. If the ratio of the delays is constant, say $r_2 = \beta r_1$, $\beta \geqslant 1$, then it is possible to have an α and numbers $1 < \beta_0 < \beta_1 < \beta_2$ such that along the line $r_2 = \beta r_1$:

(i) if $1 \leqslant \beta < \beta_0$, there is a unique intersection with Γ_α,

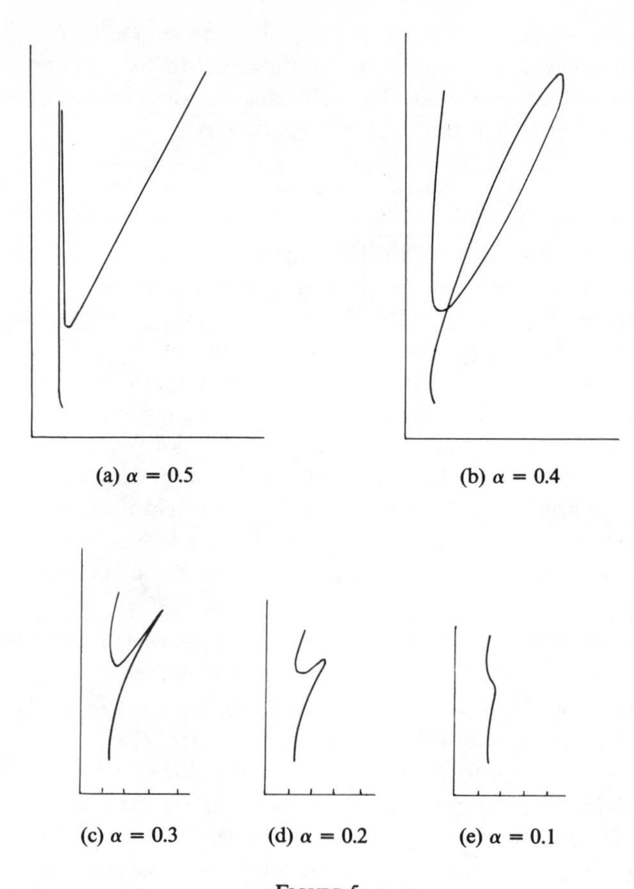

FIGURE 5.

(ii) if $\beta_0 < \beta < \beta_1$, there are several intersections with Γ_α,

(iii) if $\beta_1 < \beta < \beta_2$, there is a unique intersection with Γ_α.

Thus, there may be more than one bifurcation of a nonlinear perturbation of equation (3.8) as we move along a ray $r_2 = \beta r_1$.

For the particular problem

$$\dot{x}(t) = -\left[(1-\alpha)x(t-r_1) + \alpha x(t-r_2)\right]\left[1 + x(t)\right], \qquad 0 \leqslant \alpha \leqslant 1, \tag{3.13}$$

Nussbaum [**21**] has obtained some very interesting global results for r_1, r_2 not necessarily near the curve Γ_α but being in a wedge $r_2 = \beta r_1$, $1 \leqslant \beta \leqslant 3$. The technique is to use ejective fixed point theorems and he has proved the following results:

(i) if $1 \leqslant \beta \leqslant 2$, $r_2 = \beta r_1$ and r_1 is sufficiently large, there is a nonconstant periodic solution of equation (3.11).

(ii) if $2 < \beta \leqslant 3$, $r_2 = \beta r_1$, there are some ranges of α such that for all

r_1 sufficiently large there are at least two solutions of equation (3.13).

Our results on local bifurcation theory will apply equally as well to the equation

$$\dot{x}(t) = -\alpha\left[\int_{-1}^{-r} k(\theta)x(t+\theta)\,d\theta\right][1 + x(t)]. \tag{3.14}$$

One can study the bifurcation as a function of α, r. One of the main difficulties always consists of analyzing the characteristic equation

$$\lambda + \alpha\int_{-1}^{-r} k(\theta)e^{\lambda\theta}\,d\theta = 0.$$

The existence of a bifurcation can be then determined. The specific nature of the bifurcation depends on the nonlinear terms. As remarked earlier, the most efficient scheme for treating the nonlinear terms near $x = 0$ is the method of averaging (see **[5]**).

Some global results for equation (3.14) have been recently obtained by Walther **[27]** for $|r - 1|$ small.

4. Infinite delays. In this section, we summarize part of the thesis of Lima **[16]** on the Hopf bifurcation theorem with infinite delays. In the case of finite delays, the solution x_t of a retarded equation will belong to C for $t \geqslant r$ regardless of the space of initial data for which one defines solutions. This is the reason why the initial space C or $\mathbf{R}^n \times L^p([-r, 0], \mathbf{R}^n)$ or $W^{1,p}([-r, 0], \mathbf{R}^n)$ has little effect on the qualitative theory. For infinite delays, the space of initial functions becomes more crucial since all of the past history is contained in x_t. In the first part of this section, we consider a special case of the initial space in Lima **[16]** and later make some remarks about other spaces.

Suppose g, G: $(-\infty, 0] \to \mathbf{R}$ are nonnegative continuous functions satisfying

$$g(t+s) \leqslant G(t)g(s),$$

$$\int_{-\infty}^{0} g < \infty,$$

$$\text{there is a } \beta_0 \in (-\infty, 0] \text{ such that } G(\beta_0) < 1. \tag{4.1}$$

For example, if $g(\theta) = \exp\lambda\theta$, then $G(\theta) = \exp\lambda\theta$ and all hypotheses are satisfied if $\lambda > 0$. The function $g(\theta) = \theta^{-2}$ will not satisfy the hypotheses. Let

$$B_2 = \Big\{\text{equivalence classes of all measurable functions}$$

$$\phi\colon (-\infty, 0] \to \mathbf{R}^n \text{ such that}$$

$$|\phi|_2 \overset{\text{def}}{=} \left\{|\phi(0)|^2 + \textstyle\int_{-\infty}^{0} g(\theta)|\phi(\theta)|^2\,d\theta\right\}^{1/2} < \infty\Big\}. \tag{4.2}$$

Suppose $\Omega \subset \mathbf{R}^k$, $f, L: \Omega \times B_2 \to \mathbf{R}^n$ are continuous, $L(\alpha)\varphi$ is linear in φ, $f(\alpha, \varphi)$ has continuous derivatives in φ up through order two, $f(\alpha, 0) = 0$, and $\partial f(\alpha, 0)/\partial\varphi = 0$. Consider the linear equation

$$\dot{x}(t) = L(\alpha)x_t \tag{4.3}$$

as well as the perturbed equation

$$\dot{x}(t) = L(\alpha)x_t + f(\alpha, x_t). \tag{4.4}$$

Since $L(\alpha)\varphi$ is linear in φ, it can be written as

$$L(\alpha)\phi = A(\alpha)\phi(0) + \int_{-\infty}^{0} k(\alpha, \theta)\phi(\theta)g(\theta)\, d\theta \tag{4.5}$$

where $k(\alpha, \cdot) \in L^2[(-\infty, 0), \mathbf{R}^n]$ and $A(\alpha)$ is an $n \times n$ constant matrix. In the remainder of the section, we assume that $L(\alpha)\varphi$ is given by relation (4.5) and that $A(\alpha)$, $k(\alpha, \cdot)$ are continuous in α. The characteristic equation for equation (4.3) is

$$\det \Delta(\alpha, \lambda) = 0,$$

$$\Delta(\alpha, \lambda) = \lambda I - A(\alpha) - \int_{-\infty}^{0} k(\alpha, \theta)e^{\lambda\theta}g(\theta)\, d\theta. \tag{4.6}$$

The formal adjoint equation is

$$\dot{y}(t) = -y(0)A(\alpha) - \int_{-\infty}^{0} y(t-\theta)k(\alpha, \theta)g(\theta)\, d\theta \tag{4.7}$$

and the corresponding bilinear form is

$$\langle \psi, \phi \rangle = \psi(0)\phi(0) + \int_{-\infty}^{0}\left(\int_{\theta}^{0}\psi(\theta - u)k(\alpha, \theta)\phi(u)\, du\right)g(\theta)\, d\theta, \tag{4.8}$$

defined for $\varphi \in B_2$ and an n-dimensional row vector function ψ in some appropriate class of functions. The specific class is not important to us here and may be found in **[16]**. It is only necessary to remark that $\langle \psi, \varphi \rangle$ will be meaningful for all functions considered below.

Before stating the Hopf bifurcation theorem, let us point out the important implications of $G(\beta_0) < 1$ in relation (4.1). If $x(\phi)$ is the solution of equation (4.3) for $\phi \in B_2$ and $T(t) \overset{\text{def}}{=} x_t(\phi)$, $t \geqslant 0$, then $\{T(t), t \geqslant 0\}$ is a strongly continuous semigroup of linear operators on B_2. If $G(\beta_0) < 1$, then there is a $\mu > 0$ such that the essential spectrum of $T(t)$ belongs to the circle of radius $e^{-\mu t}$, $t \geqslant 0$ (see **[16]**, **[19]** or **[10]**). Therefore, only elements of the point spectrum of $T(t)$ lie outside this circle and these elements of the point spectrum are $e^{-\lambda t}$ where λ satisfies the characteristic equation (4.6). This makes the Hopf bifurcation theorem feasible since one can have two characteristic values on the imaginary axis and know that it is possible to have the other elements of the spectrum of $T(1)$ inside the unit circle.

The discussion below is a slight generalization of Lima [**16**] taking into account the fact that we know from §2 that one can prove bifurcation theorems without having $L(\alpha)\varphi, f(\alpha, \varphi)$ differentiable in α for all functions if we are very careful in the analysis.

The hypotheses imposed for a bifurcation theorem are the same as the ones in §2 for finite delay, stated slightly differently:

(H_1) *The characteristic matrix $\Delta(\alpha, \lambda)$ in relation (4.6) is continuously differentiable in α, there is a purely imaginary characteristic root $\lambda_0 = i\nu_0$, $\nu_0 > 0$, for $\alpha = \alpha_0$ and no other characteristic root $\lambda_j \neq \lambda_0, \bar{\lambda}_0$ for $\alpha = \alpha_0$ satisfies $\lambda_j = m\lambda_0$, m an integer;*

(H_2) *for any $K > 0$, ϕ: $(-\infty, 0] \to \mathbf{R}^n$ uniformly continuous together with its first derivative bounded by K, the function $L(\alpha)\phi + f(\alpha, \phi)$ has a derivative in α which is continuous in α,*

(H_3) $\zeta(\alpha_0) \neq 0$.

The number $\zeta(\alpha_0)$ is defined in the following manner. $\lambda_0 = i\nu_0$, $\bar{\lambda}_0 = -i\nu_0$ are characteristic roots of equation (4.3) for $\alpha = \alpha_0$. There are characteristic roots $\lambda(\alpha)$, $\bar{\lambda}(\alpha)$ of equation (4.3) which are continuously differentiable in α, $\lambda(\alpha_0) = i\nu_0$, and two linear independent solutions $b \exp \lambda(\alpha)t$, $\bar{b} \exp \bar{\lambda}(\alpha)t$ of equation (4.3). Defining

$$\Phi_\alpha = (\phi_1, \phi_2),$$

$$\phi_1(\theta) = \operatorname{Re} b \exp \lambda(\alpha)\theta, \qquad \phi_2(\theta) = \operatorname{Im} b \exp \lambda(\alpha)\theta,$$

then we may assume (by a change of basis if necessary) that

$$\Phi_\alpha(\theta) = \Phi_\alpha(0)e^{B(\alpha)\theta}, \qquad -\infty < \theta \leqslant 0,$$

$$B(\alpha) = \nu_0 B_0 + B_1(\alpha),$$

$$B_0 = \begin{bmatrix} 0 & 1 \\ -1 & 0 \end{bmatrix}, \qquad B_1(\alpha) = \begin{bmatrix} (\alpha - \alpha_0)\cdot\zeta(\alpha) & (\alpha - \alpha_0)\cdot\gamma(\alpha) \\ -(\alpha - \alpha_0)\cdot\gamma(\alpha) & (\alpha - \alpha_0)\cdot\zeta(\alpha) \end{bmatrix}.$$

Under hypotheses (H_1)–(H_3) the bifurcation theorem 2.1 holds. The proof follows the one given in [**16**] under less general hypotheses. The appropriate generalization of Theorem 2.2 is also valid.

In some applications, it may be more appropriate to have other spaces of initial data. For any $p \geqslant 1$, one can consider B_p defined in relation (4.2) with 2 replaced by p (see [**16**]). Another interesting space is the following. For any real number γ, let

$$C_\gamma = \{\phi\colon (-\infty, 0] \to \mathbf{R}^n \text{ continuous}, \\ e^{-\gamma\theta}\phi(\theta) \to \text{a limit as } \theta \to -\infty\} \tag{4.9}$$

with $|\phi|_\gamma = \sup_\theta\{|\phi(\theta)|\exp - \gamma\theta\}$. If $\gamma > 0$, the essential spectrum of the semigroup $T(t)$ has the same properties as in B_2 (see [**10**]). The Hopf bifurcation should, therefore, be valid for C_γ, $\gamma > 0$. A recent preprint of Naito [**20**] indicates it will be possible to supply a proof of this result in a manner similar to the one given in §2. Also, this preprint suggests that the same will be true for the more general spaces of initial data considered by Hale and Kato [**10**] (see also Schumacher [**28**]).

Consider the example in [**16**]; suppose k is nonnegative, continuous on $(-\infty, 0]$, $\int_{-\infty}^0 k = 1$ and consider the scalar equation

$$\dot{x}(t) = -\alpha \int_{-\infty}^{-r} k(\theta + r) f(x(t + \theta))\, d\theta \tag{4.10}$$

where $r \geqslant 0$, $\alpha \geqslant 0$ are constants, f is continuous together with derivatives up through order two, $f(0) = 0$, $f'(0) = 1$. We wish to study the Hopf bifurcation for equation (4.9) as a function of (α, r).

In [**12**], Israelson and Johnson and, in [**13**], Johnson and Karlsson proposed equation (4.10) as a mathematical model for studying the rhythmic movement of some plants. Experimental evidence indicates the kernel k should have the form shown in Figure 6. The particular form of k is not important below.

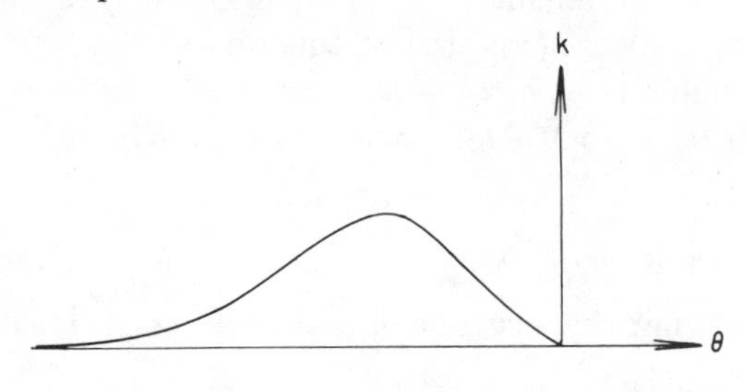

FIGURE 6.

The first problem is to choose the space of initial functions so that the previous theory applies. If there is a $\rho > 0$ and constant K such that $|k(\theta)| \leqslant K \exp \rho\theta$, for $-\infty < \theta \leqslant 0$, then we take $g(\theta) = \exp \mu\theta$, $0 < \mu < \rho$, and define

$$B_2 = \Big\{\phi\colon (-\infty, 0] \to \mathbf{R}\colon \phi \text{ is locally integrable,}$$
$$|\phi|^2 = |\phi(0)|^2 + \textstyle\int_{-\infty}^0 e^{\mu\theta} |\phi(\theta)|^2\, d\theta < \infty\Big\}.$$

In order for the right-hand side of equation (4.10),

$$F(r, \alpha, \phi) \overset{\text{def}}{=} -\alpha \int_{-\infty}^{-r} k(\theta + r) f(\phi(\theta))\, d\theta, \tag{4.11}$$

to be continuous in (r, α, ϕ) and continuously differentiable in α, ϕ, we suppose f satisfies

$$|f(x)| \leqslant a + bx^2, \qquad |f'(x)| \leqslant a_1 + b_1|x|$$

for all x and some constants a, b, a_1, b_1 (see [**16**] or [**26**]). The function

$F(r, \alpha, \phi)$ is not continuously differentiable in r but it does satisfy hypothesis (H_2) above.

The linearization of equation (4.9) around $x = 0$ is

$$\dot{x}(t) = -\alpha \int_{-\infty}^{-r} k(\theta + r) x(t + \theta)\, d\theta. \tag{4.12}$$

The characteristic equation is

$$\lambda + \alpha \int_{-\infty}^{-r} k(\theta + r) e^{\lambda\theta}\, d\theta = 0. \tag{4.13}$$

Stech [**23**] has shown that equation (4.13) has all roots with negative real parts if

$$\alpha \int_{-\infty}^{0} \theta k(\theta)\, d\theta > -1. \tag{4.14}$$

Therefore, no bifurcation can occur if relation (4.14) is satisfied.

If $r > 0$, $\tilde{\lambda} = r\lambda$, $\tilde{\alpha} = r\alpha$, then equation (4.12) is equivalent to

$$\tilde{\lambda} + \tilde{\alpha} e^{-\tilde{\lambda}} \int_{-\infty}^{-1} e^{\tilde{\lambda}\theta} k(r\theta)\, d(r\theta) = 0. \tag{4.15}$$

Stech [**23**] has shown that for $\tilde{\alpha} > \pi/2$ there is an $r_0 > 0$ such that equation (4.15) has roots with positive real parts. This certainly suggests there must be a Hopf bifurcation for equation (4.19) for some value of $\tilde{\alpha}, r$. To show that a smooth bifurcation occurs at some point, one must be able to say something about the derivatives of the roots as they cross the imaginary axis, generally a very difficult problem. Lima [**16**] has discussed some particular cases for the function k to show that smooth bifurcations do occur. In the class of all continuous kernels k one can say without any computations that generically there will always be smooth bifurcations.

Of course, once the linear equation (4.13) has ben analyzed, the existence of a Hopf bifurcation will also be true for the equation

$$\dot{x}(t) = -\alpha \left[\int_{-\infty}^{-r} k(\theta + r) x(t + \theta) \right] [1 + x(t)]$$

as well as more general equations.

It is possible also to consider that the linear equation has the form

$$\dot{x}(t) = -\alpha \int_{-\infty}^{-r} x(t + \theta)\, dK(\theta + r) \tag{4.16}$$

where $K(\theta)$ is a nondecreasing function on $(-\infty, 0]$. This allows discrete delays as well as integral dependence over the past history. If we suppose there is a $\mu > 0$ such that $\int_{-\infty}^{0} e^{-\mu\theta} dK(\theta) < \infty$ then the previous theory will be applicable by choosing the space B_2. Of course, there must be a few technical changes to take care of the fact that $\varphi(-1)$, for example, is not a function on B_2. A more convenient space would probably be C_μ defined in relation (4.9).

The results of Stech [23] are also valid for equation (4.16). Therefore, bifurcation problems can be considered for either

$$\dot{x}(t) = -\alpha \int_{-\infty}^{-r} f(x(t+\theta))\,dK(\theta + r),$$

$$\dot{x}(t) = -\alpha\left[\int_{-\infty}^{-r} x(t+\theta)\,dK(\theta + r)\right][1 + x(t)],$$

or more general equations.

5. Stabilizing effect of delays. In many applications of delay equations, an increase in the delay tends to have a destabilizing effect (see, for example, Cushing [**6**] and May [**18**]). In population dynamics, it has been noted that it can have a stabilizing effect. The rate of growth away from an unstable equilibrium can be decreased by an increase in the delay (see, for example, Cushing [**6**, pp. 35 et ff.]). For some population models representing two species, Cushing [**6**, pp. 80 et ff.] proves that an unstable equilibrium cannot be made stable by increasing the delay although the instability can be weakened in the above sense. He also gives an example [**6**, pp. 99 et ff.] in three dimensions for which an unstable equilibrium can be made stable by increasing the delays. This means *stable Hopf bifurcation can occur by decreasing the delays*.

It is possible to give an example in two dimensions where unstable equilibrium can be made stable by increasing the delays. The purpose of this section is to give an example of this type.

Consider the system

$$\begin{aligned} \dot{y}(t) &= -c_{11}y(t) + c_{12}z(t) + b_1 y(t - r_1), \\ \dot{z}(t) &= c_{21}y(t) - c_{22}z(t) + b_2 y(t - r_2), \end{aligned} \tag{5.1}$$

where $c_{ij} > 0$, $b_j > 0$, $r_j > 0$ are constants.

If $C = (c_{ij})$, we suppose

$$(c_{11} + c_{22})^2 - 4 \det C < 0. \tag{5.2}$$

This implies the eigenvalues of C are $-a_1 \pm ia_2$ where $a_1 > 0$, $a_2 > 0$; that is, the eigenvalues of C have negative real parts and are complex conjugate. To simplify computations, also suppose that $b_1 = b_2 = b$, $r_1 = r_2 = r$. By a change of coordinates

$$x = \begin{pmatrix} x_1 \\ x_2 \end{pmatrix} = D\begin{pmatrix} y \\ z \end{pmatrix},$$

we may assume system (5.1) has the form

$$\dot{x}(t) = Ax(t) + bx(t-r),$$
$$A = \begin{pmatrix} -a_1 & -a_2 \\ a_2 & -a_1 \end{pmatrix}. \tag{5.3}$$

If $-a_1 + b > 0$, $r = 0$, the ordinary differential (5.3) has eigenvalues with real parts positive and, thus, $x = 0$ is unstable. It has been shown by Tsen **[25]** that for $a_2 r = \pi/2$, $b = a_1^2 + 1/r^2$, there is an $r_0 > 0$ such that the zero solution of equation (5.3) is asymptotically stable for $r > r_0$. We reproduce some of these computations here.

By another change of coordinates, one can reduce equation (5.3) to a scalar equation in a single complex variable w,

$$\dot{w} = (-a_1 + ia_2)w + bw(t-r), \tag{5.4}$$

for which the characteristic equation is

$$\lambda = -a_1 + ia_2 + be^{-\lambda r}. \tag{5.5}$$

If $\lambda = \alpha + i\beta$, α, β real, then

$$\alpha = -a_1 + be^{-\alpha r}\cos\beta r,$$
$$\beta = a_2 - be^{-\alpha r}\sin\beta r, \tag{5.6}$$

or

$$\beta = a_2 \pm \left[b^2 e^{-2\alpha r} - (\alpha + a_1)^2\right]^{1/2},$$
$$\alpha = -a_1 + be^{-\alpha r}\cos\beta r. \tag{5.7}$$

Any real nonnegative solution of these equations must satisfy $0 \leqslant \alpha \leqslant \alpha_0(r)$ where $\alpha_0(r)$ is the unique positive zero of

$$f(\alpha, r) = b^2 e^{-2\alpha r} - (\alpha + a_1)^2. \tag{5.8}$$

For $0 \leqslant \alpha \leqslant \alpha_0(r)$, $f(\alpha, r) \geqslant 0$ and $f(\alpha, r)r \to 0$ as $r \to \infty$ uniformly in α since $b^2 = a_1^2 + 1/r^2$. Choose $\varepsilon > 0$ such that $|b|\cos(\pi/2 + \varepsilon) < a_1$ and choose $r_0 > 0$ such that $|f(\alpha, r)r| < \varepsilon$ for $0 \leqslant \alpha \leqslant \alpha_0(r)$, $r > r_0$. Define

$$\beta = a_2 + \left[b^2 e^{-2\alpha r} - (\alpha + a_1)^2\right]^{1/2}.$$

Since $a_2 r = \pi/2$, we have $|\beta r - \pi/2| < \varepsilon$ and

$$\alpha = -a_1 + be^{-\alpha r}\cos\beta r < -a_1(1 - e^{-\alpha})$$

for $r > r_0$. This latter inequality implies $\alpha < 0$ since $a_1 > 0$. This proves the assertion that increasing the delay stabilizes equation (5.1) for $r_1 = r_2 = r$, $a_2 r = \pi/2$, $b_1 = b_2$.

As remarked earlier, this implies that equation (5.1) under the above hypotheses and when subjected to nonlinear perturbations which vanish together with their first derivatives can have a stable Hopf bifurcation occur as the delay decreases. We start with $r > r_0$ where the zero solution is stable and decrease r.

6. Nonautonomous equations. The previous discussion is concerned only with the case when the vector field in the equations is independent of time–the autonomous case. When the equation is nonautonomous, the problem is much more difficult. The basic difficulty is not due to the delays, but is a consequence of the fact that it is difficult to determine the effect of nonlinearities in resonance phenomena. The best procedure available for discussing this problem locally is general transformation theory, averaging methods and the method of integral manifolds. For ordinary differential equations, the basic ideas in the transformation theory go back as far as Liapunov [**15**] (and perhaps further) when he was concerned with the problem of determining conditions for the stability of an equilibrium point for a nonlinear system when the coefficient matrix of the linear part has zero roots or purely imaginary roots. Certain aspects of the method of averaging were also encountered by Liapunov. The more general theory of averaging as well as the theory of integral manifolds was discovered by Krylov and Bogoliubov in the 1930's (see Bogoliubov and Mitropolskii [**2**] or Hale [**8**]). The transformation theory is extensive (see, for example, Malkin [**17**] and Brjuno [**3**]).

The generalization of these ideas to delay equations of retarded and neutral type began in the late 1950's. An idea of the contributions in this development can be found in [**17**]. At this time, we can safely say that most of the results that are known in this area for ordinary differential equations are known for delay equations or the necessary machinery is available to obtain them.

It is not feasible to present the elements of this theory in a few pages–a book would be more reasonable. Therefore, we are not going to discuss delay equations. Furthermore, we discuss only one problem for a very simple ordinary differential equation. To avoid any technicalities, we also assume all functions are very smooth in all variables.

Consider the second order system

$$\dot{x} = Bx + f(t, x, \varepsilon), \tag{6.1}$$

where $x = (x_1, x_2) \in \mathbf{R}^2, f = (f_1, f_2) \in \mathbf{R}^2$, B is a 2×2 constant matrix, the real parts of the eigenvalues of B are zero,

$$B = \begin{bmatrix} 0 & \sigma \\ -\sigma & 0 \end{bmatrix}, \qquad \sigma > 0, \tag{6.2}$$

the function $f(t, x, \varepsilon)$ is T-periodic in t,

$$\begin{aligned} &f(t, 0, \varepsilon) = 0, \\ &f(t, x, 0) = O(|x|^2) \quad \text{as } |x| \to 0. \end{aligned} \tag{6.3}$$

Hypothesis (6.3) implies $x = 0$ is a solution of equation (6.1). Our objective is to discuss the behavior of the solutions of equation (6.1) near $x = 0$. In the previous sections, we have considered this problem for $f(t, x, \varepsilon)$ independent of t and observed that it was possible for a Hopf bifurcation to occur at $\varepsilon = 0$. The bifurcating periodic orbit gave a closed curve Γ in $\mathbf{R}^2$ and a cylinder $\Gamma \times \mathbf{R}$ of solutions in (x, t)-space in $\mathbf{R}^2 \times \mathbf{R}$. The cylinder is obtained because to any nonconstant periodic solution of an autonomous equation, one obtains another solution by a phase shift (see Figure 7).

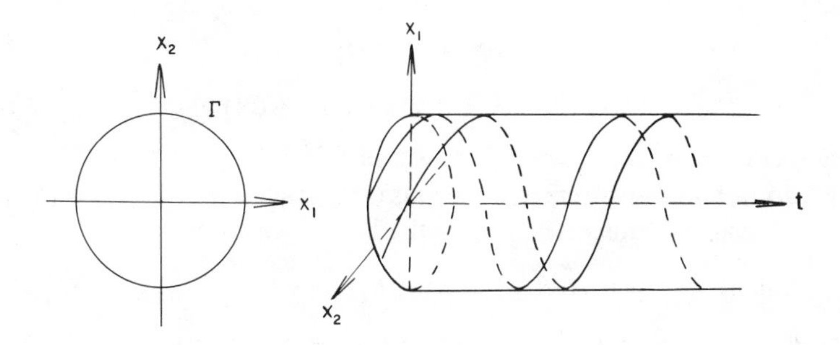

FIGURE 7.

If the bifurcating periodic orbit is asymptotically stable, then the cylinder $\Gamma \times \mathbf{R}$ is asymptotically stable in $\mathbf{R}^2 \times \mathbf{R}$. If the autonomous differential equation is subjected to small periodic disturbances, we do not expect to preserve the same kind of orbit structure as before (families of periodic solutions, for example), but we might expect to preserve a surface similar to the cylinder $\Gamma \times \mathbf{R}$ which would be asymptotically stable and completely filled up with solutions of the equation. A surface with this latter property is called an integral manifold.

With the above motivation, let us take as our objective the determination of integral manifolds in a neighborhood of the set $\{(x, t): x = 0, t \in \mathbf{R}\}$ which are like cylinders. More specifically, we wish to find a function $g: \mathbf{R} \times \mathbf{R} \to \mathbf{R}^2$, $g(t, \theta)$ periodic in θ of period $2\pi/\sigma$, periodic in t of period T such that the set

$$S = \{(x, t) \in \mathbf{R}^2 \times \mathbf{R}: x = g(t, \theta), (t, \theta) \in \mathbf{R}^2\}$$

is an integral manifold; that is if $(\xi, \sigma) \in S$, then the solution $x(t)$ of equation (6.1) through (ξ, σ) exists for all $t \in \mathbf{R}$ and $(x(t), t) \in S$ for $t \in \mathbf{R}$.

To obtain such integral manifolds, let

$$x_1 = \rho \sin \sigma\theta, \qquad x_2 = \rho \cos \sigma\theta, \tag{6.4}$$

which gives

$$\dot{\theta} = 1 + \Xi(t, \theta, \rho, \varepsilon),$$
$$\dot{\rho} = R(t, \theta, \rho, \varepsilon), \tag{6.5}$$

where

$$\Xi(t, \theta, \rho, \varepsilon) = (\rho\sigma)^{-1}[f_1 \cos \sigma\theta - f_2 \sin \sigma\theta],$$
$$R(t, \theta, \rho, \varepsilon) = f_1 \sin \sigma\theta + f_2 \cos \sigma\theta, \tag{6.6}$$

and the arguments of f_1, f_2 are (t, x, ε) with x given in relation (6.4). The functions Ξ, R satisfy

$$\Xi = O(|\rho| + |\varepsilon|),$$
$$R(t, 0, \varepsilon) = 0, \qquad R(t, \rho, 0) = O(|\rho|^2), \tag{6.7}$$

are T-periodic in t and $2\pi/\sigma$-periodic in θ.

We do not attempt to analyze equation (6.5) completely but only give a special result to illustrate some of the basic ideas. Let

$$\Xi(t, \theta, \rho, \varepsilon) = 1 + \alpha_1(t, \theta, \varepsilon) + \alpha_2(t, \theta, \varepsilon)\rho + \alpha_3(t, \theta, \varepsilon)\rho^2 + O(|\rho|^3),$$
$$R(t, \theta, \rho, \varepsilon) = \beta_1(t, \theta, \varepsilon)\rho + \beta_2(t, \theta, \varepsilon)\rho^2 + \beta_3(t, \theta, \varepsilon)\rho^3 + O(|\rho|^4). \tag{6.8}$$

The objective is to transform to new coordinates

$$\xi = \theta + u_0(t, \theta, \varepsilon) + u_1(t, \theta, \varepsilon)\rho + u_2(t, \theta, \varepsilon)\rho^2 + O(|\rho|^3),$$
$$r = \rho + v_1(t, \theta, \varepsilon)\rho + v_2(t, \theta, \varepsilon)\rho^2 + O(|\rho|^3), \tag{6.9}$$

where all functions are T-periodic in t and $2\pi/\sigma$-periodic in t in such a way that the new equations in ζ, r have the form

$$\dot{\zeta} = 1 + \mu_1(\varepsilon) + \mu_2(\varepsilon)r + \mu_3(\varepsilon)r^2 + O(|r|^3),$$
$$\dot{r} = \nu_1(\varepsilon)r + \nu_2(\varepsilon)r^2 + \nu_3(\varepsilon)r^3 + O(|r|^4). \tag{6.10}$$

If one formally substitutes relations (6.9) in equations (6.5) and equates coefficients of r, then one observes that each of the functions u_j, v_j satisfies a partial differential equation of the type

$$\frac{\partial u}{\partial t} + \frac{\partial u}{\partial \theta} = h(t, \theta) \tag{6.11}$$

where $h(t, \theta)$ is T-periodic in t and $2\pi/\sigma$-periodic in θ. To obtain a solution of equation (6.11) which is T-periodic in t and $2\pi/\sigma$-periodic in θ, one must impose some restrictions on ω, σ, where $\omega = 2\pi/T$. In fact, if

$$u(t, \theta) = \sum u_{kl} e^{i(k\omega t + l\sigma\theta)},$$
$$h(t, \theta) = \sum h_{kl} e^{i(k\omega t + l\sigma\theta)},$$

then we can uniquely determine u_{kl} as

$$u_{kl} = \left[i(k + l\sigma) \right]^{-1} h_{kl}$$

if $k\omega + l\sigma \neq 0$ for every h_{kl} that is not zero. If $h_{kl} = 0$ we take $u_{kl} = 0$. Therefore, if we assume $k\omega + l\sigma \neq 0$ for those $k \neq 0$, $l \neq 0$ which occur in the Fourier series for h, the function u will be uniquely determined if

$$\int_0^{2\pi/\sigma} \int_0^{2\pi/\omega} h(t, \theta)\, dt\, d\theta = 0. \tag{6.12}$$

In general, relation (6.12) will not be satisfied. The coefficients $\mu_j(\varepsilon)$, $\nu_j(\varepsilon)$ are chosen so that h will satisfy relation (6.12). Thus, $\mu_j(\varepsilon)$, $\nu_j(\varepsilon)$ are the mean values of some functions explicitly obtained in the process of computing the u_j and v_j. For more details on methods of computation, the reader is referred to **[2]**, **[3]**, **[17]**.

If we assume the nonresonance conditions are satisfied for ω, σ and equation (6.5) has been transformed to equation (6.10), then we can state the following result which is a consequence of the theory of integral manifolds.

THEOREM 6.1. *If*

$$\begin{aligned} &\nu_1'(0) \neq 0, \\ &\nu_2(0) = \nu_2'(0) = 0, \\ &\nu_3(0) \neq 0, \qquad \nu_3(0)\nu_1'(0) < 0, \end{aligned} \tag{6.13}$$

then there is an $\varepsilon_0 > 0$ *and a function* $g(t, \theta, \mu)$, $0 \leqslant \mu < \sqrt{\varepsilon_0}$, *T-periodic in* t, $2\pi/\sigma$*-periodic in* θ, $g(t, \theta, 0) = [-\nu_3(0)\nu_1'(0)]^{1/2}$ *such that the set*

$$\begin{aligned} S_\varepsilon = \big\{(x, t) \in \mathbf{R}^2 \times \mathbf{R};\ & x_1 = \sqrt{\varepsilon}\, g(t, \theta, \sqrt{\varepsilon})\sin \sigma\theta, \\ & x_2 = \sqrt{\varepsilon}\, g(t, \theta, \sqrt{\varepsilon})\cos \sigma\theta,\ \theta \in \mathbf{R},\ t \in \mathbf{R}\big\}, \end{aligned} \tag{6.14}$$

$0 \leqslant \varepsilon < \varepsilon_0$ *is an integral manifold of equation* (6.1). *Furthermore, this integral manifold is asymptotically stable if* $\nu_1'(0) > 0$ *and unstable if* $\nu_1'(0) < 0$.

If $\nu_3(0)\nu_1'(0) > 0$, *the same conclusions hold on existence if* $-\varepsilon_0 < \varepsilon \leqslant 0$ *and* S_ε *is unstable if* $\nu_1'(0) < 0$ *and asymptotically stable if* $\nu_1'(0) > 0$.

Let us make a few more remarks about second order ordinary differential equations. Consider the equation

$$\dot{y} = By + F(t, y, \varepsilon) \tag{6.15}$$

where B is the same as before, F is T-periodic in t,

$$\begin{aligned} F(t, 0, 0) &= 0, \\ F(t, y, 0) &= O\big(|y|^2\big). \end{aligned}$$

Assume the linear equation $\dot{y} = By + h(t)$ for a T-periodic function h has a unique T-periodic solution. Then the nonlinear equation (6.16) will have a unique T-periodic solution $y^*(t, \varepsilon)$ in a neighborhood of $x = 0$, $\varepsilon = 0$, $y^*(t, 0) = 0$ (see, for example, [**8**]). If $y = y^*(t, \varepsilon) + x$ then x satisfies an equation of the same form as equation (6.1). Thus, the above result applies to equation (6.16).

For equations with delays, results similar to Theorem 6.1 can be proved. However, much more machinery is needed. One must use the decomposition theory of linear autonomous equations (see Hale [**7**, Chapters 7, 9]), averaging methods (see Chow and Mallet-Paret [**5**] or Hausrath [**11**]) and the generalization of the methods of integral manifolds to delay equations (see, for example, Hale [**7**, Chapter 10, §10.4] for some references). If one is very familiar with these methods in ordinary differential equations, the ideas used in delay equations are not too surprising. Before tackling these problems for delay equations, I would suggest that the reader be able to supply all details in the previous example.

REFERENCES

1. R. Abraham and J. Robbin, *Transversal mappings and flows*, Benjamin, New York, 1967.

2. N. Bogoliubov and Yu. Mitropolskii, *Asymptotic methods in the theory of nonlinear mechanics*, Gordon and Breach, New York, 1961.

3. A. D. Brjuno, *Analytic form of differential equations*. I, Trans. Moscow Math. Soc. **25** (1971), 131–288. II, ibid. **26** (1972), 199–239.

4. N. Chafee, *A bifurcation problem for a functional differential equation of finitely retarded type*, J. Math. Anal. Appl. **35** (1971), 312–348.

5. S. Chow and J. Mallet-Paret, *Integral averaging and bifurcation*, J. Differential Equations **26** (1977), 112–159.

6. J. M. Cushing, *Integrodifferential equations and delay models in population dynamics*, Lecture Notes in Biomath., vol. 20, Springer-Verlag, Berlin and New York, 1977.

7. J. K. Hale, *Theory of functional differential equations*, Appl. Math. Sci., vol. 3, second ed., Springer-Verlag, Berlin and New York, 1977.

8. ______, *Ordinary differential equations*, Wiley-Interscience, New York, 1969.

9. ______, *Critical cases for neutral functional differential equations*, J. Differential Equations **10** (1971), 59–82.

10. J. K. Hale and J. Kato, *Phase space for retarded equations with infinite delays*, Funkcial. Ekvac. **21** (1978), 297–315.

11. A. Hausrath, *Stability in the critical case of purely imaginary roots for neutral functional differential equations*, J. Differential Equations **13** (1973), 329–357.

12. D. Israelson and A. Johnson, *A theory for circummutations in Helianthus Annus*, Physiol. Plant. **20** (1967), 957–976.

13. A. Johnson and H. G. Karlsson, *A feedback model for biological rhythms*. I, J. Theoret. Biol. **36** (1972), 153–174.

14. N. Keyfitz, *Introduction to the mathematics of populations*, Addison-Wesley, Reading, Mass., 1977.

15. A. M. Liapunov, *Problème général de la stabilité du mouvement*, Ann. of Math. Studies, no. 17, Princeton Univ. Press, Princeton, N. J., 1947.

16. P. Lima, *Hopf bifurcation in equations with infinite delays*, Ph.D. Thesis, Brown University, June, 1977.

17. J. G. Malkin, *Theory of stability of motion*, Tech. Report AEC-tr-3352, U. S. Atomic Energy Commission.

18. R. M. May, *Stability and complexity in model ecosystems*, second ed., Princeton Univ. Press, Princeton, N. J., 1974.

19. T. Naito, *On autonomous linear functional differential equations with infinite retardation*, J. Differential Equations **21** (1976), 297–315.

20. ______, *On linear autonomous retarded equations with an abstract phase space for infinite delays*, J. Differential Equations (to appear).

21. R. Nussbaum, *Differential delay equations with two time lags*, Mem. Amer. Math. Soc. (to appear).

22. J. Ruiz-Claeyssen, *Effect of delays on functional differential equations*, J. Differential Equations **20** (1976), 404–440.

23. H. W. Stech, *The effect of time lags on the stability of the equilibrium state of a population growth equation*, J. Math. Biol. (to appear).

24. ______, *The Hopf bifurcation: stability result and application*, J. Math. Anal. Appl. (to appear).

25. F.-S. Tsen, *Asymptotic stability for a class of functional differential equations of neutral type*, Ph.D. Thesis, Brown University, June, 1978.

26. M. M. Vainberg, *Variational method for the study of nonlinear operators*, Holden-Day, Inc., San Francisco, 1964.

27. H.-O. Walther, *Existence of a nonconstant periodic solution of a nonlinear autonomous functional differential equation representing the growth of a single species population*, J. Math. Biol. **1** (1975), 227–240.

28. K. Schumacher, *Existence and continuous dependence for functional differential equations with unbounded delay*, Arch. Rational Mech. Anal. **67** (1978), 315–335.

LEFSCHETZ CENTER FOR DYNAMICAL SYSTEMS, DIVISION OF APPLIED MATHEMATICS, BROWN UNIVERSITY, PROVIDENCE, RHODE ISLAND 02912

Lectures in Applied Mathematics
Volume 17, 1979

A Brief Introduction to Dynamical Systems[1]

John Guckenheimer

I. Introduction. The theory of dynamical systems is the study of the qualitative properties of the solutions of ordinary differential equations. It is a subject which has seen tremendous advances in the past twenty years. This is a brief and informal introduction to some of this work. The work we describe forms a coherent theory which describes a great deal about the behavior of solutions of a large class of differential equations. We limit ourselves to a description of central results from this body of material in an attempt to expose the reader to the nature of this theory as quickly (and, we hope, as painlessly) as possible. There are no proofs and much has been left out. I apologize to those whose work should have been included and was not and to the reader for having done so. The references are also chosen selectively rather than comprehensively. Our aim is to provide the reader with a maximally efficient introduction to this material which is intermediate in sophistication to the text of Hirsch and Smale [**14**] and the lecture notes of Bowen [**8**].

Before proceeding with a discussion of the material to be covered, let us try to place our subject in some perspective. Ordinary differential equations are often used as "models" of processes in the real world. We will not try to be precise about the relationship between the equations and the process being modeled. Solutions of the differential equations are expected to mirror the "dynamics" or "evolution" of the process. For example, the differential equations might describe the motion of a mechanical system (with a finite number of degrees of freedom), the evolution of a chemical reaction, or the dynamics of a population forming an ecosystem. We are *not* concerned here with how one writes down the appropriate system of equations. Rather, we assume that the

AMS (*MOS*) *subject classifications* (1970). Primary 58-XX.
[1]Partially supported by the National Science Foundation.

system of equations is given. Nor are we interested in how one solves the system of equations. The well-known existence and uniqueness theorems for ordinary differential equations do that for us. Rather we are interested in the properties of solutions. In particular, we are concerned about the asymptotic behavior of solutions for large time. Do solutions approach a limit with increasing time, do they approach regular oscillations, or do they do something complicated? What sorts of complicated asymptotic behavior can be described? How much information does one need to determine the asymptotic behavior of solutions? These are questions which are appropriate for our discussion.

Digital computers have made it possible (even easy!) to find approximate solutions to systems of ordinary differential equations for moderate amounts of time. This means that the short time behavior of solutions of a system of differential equations for particular initial conditions is accessible to us. From such information one would like to infer asymptotic properties of solutions (in time). For example, this information might include information about the existence of limit cycles and stability. Dynamical systems theory provides just this sort of information. Many readers will be familiar with the "classical" aspects of the theory which deal with systems of first order differential equations in two variables and originated with Poincaré near the end of the 19th century. This theory is often called "phase plane methods". With later work culminating with Peixoto's paper [**18**], these phase plane methods yield a good geometric description of "most" systems of equations in two variables.

Unfortunately the extension of these two dimensional methods to higher dimensions encounters serious obstacles. One finds systems of differential equations in three or more variables with solutions having complicated asymptotic behavior. We shall examine a number of examples. These equations are met in applications; for example, one of the first systems of this sort to be studied was the 3-body problem (by Poincaré who coined the term homoclinic orbit) in the context of celestial mechanics. Other more recent examples involving three variables come from the study of electrical circuits, turbulence of fluids, and population dynamics.

There are three strands of ideas which have been interwoven in attempting to analyze the complicated asymptotic behavior that emerges. These can be summarized by the terms: *genericity*, *structural stability*, and *symbolic dynamics*. The concept and tools of *hyperbolicity* tie these three strands together. Let us attempt to give a brief indication of the role of each of these ideas in turn.

The goal of dynamical systems theory is to describe and classify the geometry associated with solutions of ordinary differential and difference equations. It is an impossible task, just as a description of the zero set of an arbitrary smooth function of several variables is a hopeless task. The possible pathologies are too horrendous, so that one never is able to escape from a morass of exceptional cases. To overcome this problem one simply discards all the exceptional cases. The collection that remains consists of the *generic* cases. For example, Sard's theorem implies that in a suitable technical sense, the zero set of *almost all* smooth functions of several variables is a manifold (a set which is "locally" like Euclidean space of a particular dimension). The idea of genericity is not always an easy thing to specify. In the first place, it depends upon the mathematical universe which one is studying. What is generic in one universe may be exceptional in a larger universe. Moreover, once a universe is specified, the boundary between the generic and the exceptional may not be an easy thing to determine. It is also a matter of convention as we shall see later. A discussion of the concept of genericity in a number of mathematical universes can be found in Thom's book [**29**]. Later we shall examine the idea of genericity within the universe of systems of ordinary differential equations.

It may (and does) happen that some of the generic cases are still too complicated to allow a good description. Instead of trying to eliminate further exceptional cases, one can select desirable properties which allow one to describe objects with those properties. One such property is that the description of the object should not change if the object is varied a little bit. For example, consider the number of zeros of a function of a single variable. If the derivative of the function is nonzero at each zero of the function, then the number of zeros will not change when the function is varied slightly. (Again, the size of a perturbation is to be measured in an appropriate sense.) In the context of differential equations, one interpretation of this robustness property is called *structural stability*.

Our discussion will reverse the order in which we discuss these topics, beginning with the more constructive and concrete aspects first. The concept of *hyperbolicity*, a technical tool which frequently leads to stability results, appears early and reappears frequently.

It has been argued that the only physically relevant systems of differential equations are the structurally stable ones. The argument is based upon the presumption that one can never know the coefficients of a system of equations precisely. If the properties of the system change when the coefficients are altered slightly, then the system is not a good model because its qualitative features are sensitive to infinitesimal changes in measurements which determine the coefficients. For example, this is true of the Lotka-Volterra equations for interacting predator-prey species [**14**]. Despite this argument, there are systems of equations which are not structurally stable, but which do seem to be physically realistic models. My own view is that this philosophical standpoint should be modified to the hypothesis that the only physical features of a particular model which are relevant to physical measurements are those which are structurally stable in some sense. To say that the whole model should be structurally stable is too strong: it imposes an unreasonable restriction on the laws of nature.

After one has singled out a good class of describable systems, one is faced with the task of analyzing them. One of the most fruitful approaches to this task has been through *symbolic dynamics*. The idea of symbolic dynamics can be traced to the studies of geodesics on a surface of negative curvature by Hadamard **[12]** and Morse and Hedlund **[17]**. The technique involves partitioning an asymptotic limit set into pieces (think of them as states of a Markov chain) so that the limiting solutions are described by the sequence of transitions from one piece to another. The power of this technique comes from the fact that a sequence of transitions occurs for a solution if each of the transitions in the sequence is possible. The applications of symbolic dynamics have been impressive. They have allowed the construction of a statistical mechanics and ergodic theory for the asymptotic behavior of the solutions of a large class of differential equations.

I am grateful to a number of people, especially G. Oster, R. Palais, A. Winfree, and A. Kabani for their comments and corrections to this essay.

II. Foundations. In this section we shall review some basic theory of ordinary differential equations which will be used throughout the paper. We shall work with systems of n autonomous ordinary differential equations in n variables:

$$\begin{aligned} \frac{dx_1}{dt} &= f_1(x_1, \dots, x_n), \\ &\vdots \\ \frac{dx_n}{dt} &= f_n(x_1, \dots, x_n). \end{aligned} \tag{$*$}$$

Think of each x_i as being the population size of one of the populations in an ecological system or as the concentration of a chemical participating in a reaction. The system $(*)$ can be abbreviated to $dx/dt = F(x)$ where $x = (x_1, \dots, x_n)$ is an n vector and $F: \mathbf{R}^n \to \mathbf{R}^n$ is given by $F(x) = (f_1(x), \dots, f_n(x))$. We shall assume throughout that F is infinitely differentiable. In population and chemical models it is customary to consider only values of the variables for which $x_i \geqslant 0$. These restrictions are unimportant here. For convenience, we shall use coordinate systems in which these conditions need not be met (e.g., x_i

might denote the difference between the size of the ith population or concentration and some specified size).

The fundamental existence and uniqueness theorem for ordinary differential equations states that if $x(0) = (x_1(0), \ldots, x_n(0))$ is specified, then there is a unique solution $x(t)$ of $(*)$ with $x(0)$ being the specified *initial condition*. The solution may not be defined for all time unless F satisfies various hypotheses (e.g., $\|F(x)\|/\|x\|$ is bounded as $\|x\| \to \infty$).

In thinking about a solution $x(t)$ of the equations $(*)$, we shall consider $x(t) = (x_1(t), \ldots, x_n(t))$ as a curve in the multidimensional space $\mathbf{R}^n$ of population sizes or chemical concentrations. This multidimensional space is analogous to the state space of a classical mechanical system. The assumption of the model is that the population sizes or concentrations at any time determine the population sizes or concentrations at all later times. This is the meaning of the adjective deterministic applied to the model. The curve $x(t)$ represents a *trajectory* or an *orbit* moving through the space of population sizes or concentrations with time. A different perspective on the solutions can be obtained by introducing the *flow* of $(*)$. This is a map $\Phi: \mathbf{R}^n \times \mathbf{R} \to \mathbf{R}^n$ defined by $\Phi(x_1, \ldots, x_n, t) = (x_1(t), \ldots, x_n(t))$ = the value of the solution $x(t)$ at time t which has initial condition $x = (x_1, \ldots, x_n)$. The map Φ satisfies the basic flow property $\Phi(\Phi(x, t_1), t_2) = \Phi(x, t_1 + t_2)$. This expresses the fact that if one successively solves $(*)$ for times t_1 and t_2, using the first solution as initial condition for the second, then one obtains the same results as solving $(*)$ for time $(t_1 + t_2)$. All of the orbits together partition the space of population sizes. The map $F: \mathbf{R}^n \to \mathbf{R}^n$ is called a *vector field*. One can think of F as assigning a "free" vector or tangent vector to each point of $\mathbf{R}^n$. A solution to $(*)$ is then interpreted as a curve whose tangent vector at each point is the vector that F assigns to that point in $\mathbf{R}^n$. This is a useful point of view which lends itself well to geometric interpretations of the problems we consider.

It is often useful to study flows which have a discrete time rather than continuous time. These occur, for example, when one studies models for a population with nonoverlapping generations. The analog of $(*)$ for such a situation is a set of difference equations. We will express the set of difference equations by a map $f: \mathbf{R}^n \to \mathbf{R}^n$ where $f(x)$ is interpreted to be the sizes of the populations one unit of time later than the time at which the population sizes were x. The analog of the flow of a differential equation is determined by the sequence of iterates $f^n = f \circ \cdots \circ f$ of f. (The operation $\circ$ here is composition.) In time reversible situations, f is assumed to have an inverse f^{-1}. (Then f is called a *diffeomorphism*.) The trajectories or orbits in these models are given by the sequences $\{f^n(x)\}$ where n ranges over the integers. Mathematically, the theory which we consider here is much more fully developed when the map f does have an inverse. Biological models do not all have this property, but, for the sake of exposition, we will assume that discrete time difference equations are invertible unless we specify otherwise.

The basic problem upon which we focus in this paper is a description of all of the orbits of a system of differential equations (*) or of a system of difference equations. A description predicts the population sizes at all later (and former) times for any choice of current population sizes. Stated this broadly, we have assumed an impossible task. The best information we can hope for will be qualitative, in general. Our main goal will be to pursue information about the asymptotic nature of solutions for large times.

Certain types of solutions have special interest. An *equilibrium* or *stationary point* or *singular point* for (*) is a vector $x_0 \in \mathbf{R}^n$ such that $F(x_0) = 0$. The solution with initial condition x_0 is $x(t) = x_0$ for all t. If $\tau > 0$, a *periodic orbit* of (prime) period τ is a solution for which $x(t) \neq x(0)$ for $0 < t < \tau$ and $x(\tau) = x(0)$. Equilibria represent population sizes or concentrations which remain constant. Periodic orbits represent population sizes or concentrations which oscillate but return to precisely their initial sizes at some later time and then continually repeat this behavior. Typically, it is unreasonable to expect that initial conditions lie precisely at an equilibrium or on a periodic orbit. If they did, random influences (for example, times of births and deaths or thermodynamic fluctuations) would soon displace the population levels from such a point. Thus the stability of an equilibrium or periodic orbit is important to ascertain.

There are various interpretations of the meaning of stability in this context. Two of these warrant mention at this point. Precise statements require that we introduce geometric language to deal with the multidimensional space of population sizes.

An equilibrium point or periodic orbit of a system of differential equations is *asymptotically stable* if it has a neighborhood U with the property that if y is any point in U, the distances between the points $y(t)$ on the trajectory having initial condition y and the equilibrium point or periodic orbit tend to 0 uniformly as $t \to \infty$. An equilibrium point or periodic orbit is *neutrally stable* if it has the property that if U is any neighborhood of the equilibrium point or periodic orbit, then there is another neighborhood V such that a trajectory with initial condition in V remains in U for all future times.

There is a large literature on stability of equilibria for population models. The Lotka-Volterra equations have a neutrally stable equilibrium, for example. (See §XI.) A topic of recent interest in ecology centers around the stability of equilibria. There is an ecological maxim that "diversity breeds stability". Arguments against this principle have been based upon an analysis of the stability of equilibria. Roughly summarized, the thrust of the argument is that it is increasingly difficult for an equilibrium of a system of differential equations to be stable as one increases the size of the system of equations.

A good criterion exists for assessing the asymptotic stability of an equilibrium. The procedure is to compute the matrix of partial derivatives $\partial f_i / \partial x_j$ for the system of equations (*) at the equilibrium. If all of the eigenvalues of this matrix have negative real parts, then the equilibrium is asymptotically stable. The basis for the argument in the last paragraph is that it becomes less probable for a matrix to have no eigenvalue with nonnegative real part as the size of the matrix increases. Thus ecosystems with large numbers of species are less likely to be stable than ecosystems with a small number of species. This example illustrates how models can be used (correctly?) to make statements about population theory.

It is somewhat more difficult to establish criteria for stability of a periodic orbit. Suppose γ is a periodic orbit of period τ. Let Φ be the flow of the vector field and f be the map which moves points τ units along solutions: $f(x) = \Phi(x, \tau)$. The map f will send each point of γ to itself. If $p \in \gamma$, then $f(p) = p$. The matrix A of partial derivatives of f at p can be used to study the stability of the closed orbit γ. The matrix A must have an eigenvector tangent to γ with eigenvalue 1. If the remaining eigenvalues have absolute value smaller than 1, then the periodic orbit γ will be asymptotically stable. These eigenvalues are independent of the point p on the periodic orbit γ. If any eigenvalues of A have absolute value larger than 1, then γ is definitely not stable.

A large literature in a number of fields embodies discussion of the existence and stability of equilibria and periodic orbits of systems of differential equations. These two types of orbits have often been regarded (implicitly) as the only alternatives for the asymptotic behavior of solutions in deterministic models. This is a state of affairs brought about largely by the "state of the art" in interpreting mathematical models. Vector fields and flows are easier to visualize in two dimensions. Consequently, the theory of vector fields in the plane has been developed much more extensively than the theory of higher dimensional systems. Much of the two dimensional theory is "classical" in that it was created near the turn of the century and has been long included in differential equations texts. Thus it is this material which has been used most widely for the last fifty years. The theory for larger systems of differential equations is much more recent and less accessible. As a result, many applied mathematicians have been unable to deal with the asymptotic behavior of solutions of ordinary differential equations which do not tend to an asymptotically stable equilibrium or periodic orbit.

Indeed, there are two problems here. The first is understanding a model. The questions raised in trying to understand a model are the ones addressed in this review. What possibilities arise for the asymptotic behavior of integral curves of a vector field? If different orbits have very different behaviors, are there statistical statements that can be made about the "average" behavior of trajectories? What is the role of stability in these models? These are questions considered here. A second problem which confronts the theorist is the correlation of experimental data with the model. This is quite a different sort of problem. As far as I can gather, it is a problem which has been sorely neglected. It is largely ignored in this review. The usual criterion for a population model, for example, is whether its predictions "look pretty good" in a vague visual sense. There are few, if any, well-established criteria for judging when one has a good model. Part of the reason for this is that reliable data are often difficult to obtain. But the lack of standards for judging models makes for a situation in which it is difficult to test predictions or assess their value.

With these remarks, we turn now to the theory of dynamical systems.

III. An example [24]. Before proceeding with the development of general theory, it is important to understand something of the richness of behavior possible for a dynamical system. We begin with an example of a discrete time system.

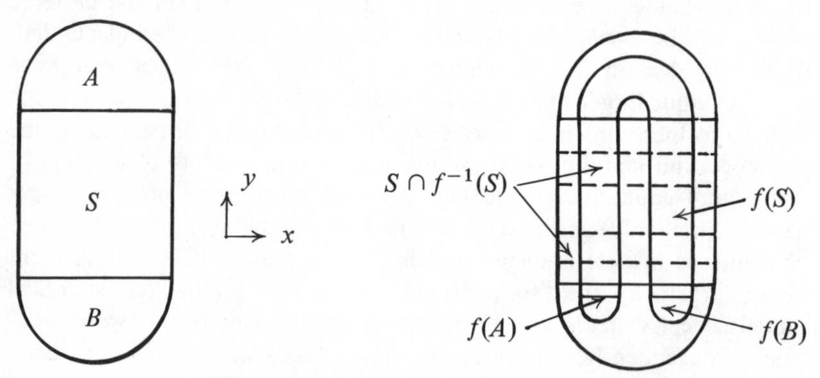

FIGURE 1. (a) The square with caps; (b) The image of the square with caps.

Smale's horseshoe. This map illustrates the richness of the dynamical behavior which can sometimes be deduced from a modest amount of information about a dynamical system. It will be important later to understand the essential features of the example, so we devote a good deal of attention to it.

Consider a unit square S in the plane with two semicircular caps A and B at top and bottom respectively. The map f we consider has its image shown in Figure 1(b). We are interested mainly in S and the behavior of f and its iterates on S. To make the analysis as simple as possible, assume that f is linear on the set $f^{-1}(S) \cap S$ of points in S whose images also are in S. The set $f^{-1}(S) \cap S$ consists of two horizontal bands cutting across S. On this set, assume f has the form $f(x, y) = (\pm ax + c_i, \pm by + d_i)$ where a, b, c_i, d_i are numbers with $0 < a < \frac{1}{2}$, $b > 2$, and $i = 1$ or 2. This establishes the linear "hyperbolic" nature of f on the set $f^{-1}(S) \cap S$.

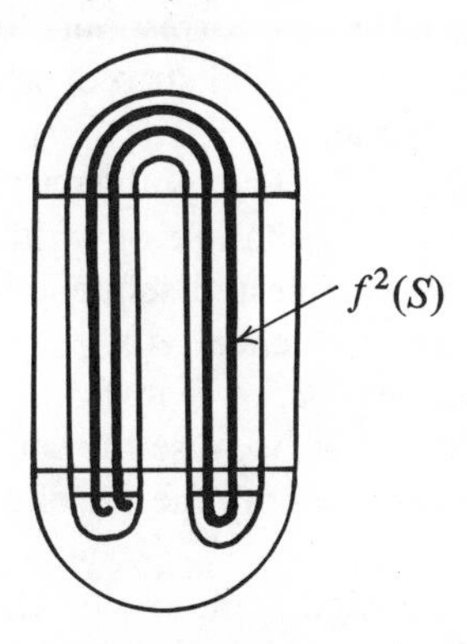

FIGURE 2. The second image of S under f.

Consider successively the sets S, $S \cap f(S)$, $S \cap f(S) \cap f^2(S)$, $\ldots, \bigcap_{i=0}^{n} f^i(S)$. It is not difficult to convince oneself that $\bigcap_{i=0}^{n} f^i(S)$ consists of 2^n vertical strips, each of width a^n as shown in Figure 2. It follows that $\bigcap_{i=0}^{\infty} f^i(S)$ is an uncountable collection of vertical segments. This is proved in the same way one constructs the "Cantor set" of elementary analysis.

Now consider where points come from which map into S under successive iterates of f. We have seen that $f^{-1}(S) \cap S$ consists of two horizontal bands across S whose height is b^{-1}. The set $f^{-2}(S) \cap f^{-1}(S) \cap S$ consists of four horizontal bands across S whose height is b^{-2}. Inductively, one finds that $\bigcap_{i=-n}^{0} f^i(S)$ consists of 2^n horizontal bands of height b^{-n}. From this one concludes that $\bigcap_{i=-\infty}^{0} f^i(S)$ is an uncountable collection of horizontal segments crossing S. Putting this together with our description of $\bigcap_{i=0}^{\infty} f^i(S)$, we find that the set $\Omega = \bigcap_{i=-\infty}^{\infty} f^i(S)$ is an uncountable collection of points. These are the points which always remain in S and which are the images of points of S under arbitrarily high iterates of f.

We want to go further by obtaining a description of the dynamics of f on the set Ω. To do so, we introduce a "symbolic" description of what happens. Choose two symbols, say 0 and 1, to label the two pieces of $S \cap f(S)$. If we follow a point of Ω under the iterates of f, at each time it lies in one of the two pieces of $S \cap f(S)$. Thus, corresponding to each point $p \in \Omega$ is a sequence $\{a_t\}_{t=-\infty}^{\infty}$ of 0's and 1's determined by the rule

$$a_t = \begin{cases} 0 & \text{if } f^t(p) \text{ is in rectangle 0 of } S \cap f(S), \\ 1 & \text{if } f^t(p) \text{ is in rectangle 1 of } S \cap f(S). \end{cases}$$

Smale [**24**] proved that this is a 1-to-1 correspondence between the points of Ω and the set of sequences of 0's and 1's.

The proof is not difficult. Those points to which we assign a particular finite sequence $\{a_0, \ldots, a_{n-1}\}$ correspond to one of the 2^n components of $\bigcap_{t=0}^{n} f^t(S)$. This is a vertical strip of width a^n. Those points to which we assign the sequence $\{a_{-n}, \ldots, a_{-1}, a_0\}$ correspond to a horizontal strip of height b^{-n}. The points assigned to the sequence $\{a_{-n}, \ldots, a_0, \ldots, a_n\}$ form a rectangle of width a^n and height b^{-n}. As $n \to \infty$, these rectangles form a nested sequence with exactly one point in all of them. Thus, to each sequence $\{a_t\}_{t=-\infty}^{\infty}$, there is a unique point p for which $f^t(p)$ is in rectangle a_t of $S \cap f(S)$. This shows us that the set of points in Ω is uncountable because the set of sequences of 0's and 1's is uncountable. We shall denote the set of sequences of 0's and 1's by Σ.

We gain more from this argument. If $p \in \Omega$, we want to follow its orbit under f. This can be done by using the 1-to-1 correspondence we have established between Ω and Σ. If $\{a_t\}$ is the sequence associated to $p \in \Omega$, what is the sequence associated to $f(p)$? If $f^t(p)$ lies in the rectangle a_t of $S \cap f(S)$, then $f^{t-1}(f(p))$ lies in the rectangle a_t of $S \cap f(S)$. This says that the sequence $\{b_t\}$ corresponding to $f(p)$ has the property $b_{t-1} = a_t$. Thus, in terms of the symbol sequences, the map f looks like a map which just shifts the indices of sequences. Formally, define the *shift* map σ: $\Sigma \to \Sigma$ by $\sigma(\{a_t\}) = \{b_t\}$ with $b_{t-1} = a_t$. To express the relationship between σ and f, let φ: $\Sigma \to \Omega$ be the 1-to-1 correspondence between sequences and points of Ω. One then has the equation $\varphi\sigma = f\varphi$, or $\sigma = \varphi^{-1}f\varphi$. This equation implies

$$\sigma^n = (\varphi^{-1}f\varphi)(\varphi^{-1}f\varphi) \cdots (\varphi^{-1}f\varphi) = \varphi^{-1}f^n\varphi.$$

Thus φ maps orbits of σ in Σ to orbits of f in Ω. We can use the map σ to study the map f on the set Ω. It is much easier to study σ.

As an illustration of how one can use σ to study f, consider the problem of finding periodic orbits of f. We assert that there are exactly 2^n sequences $\{a_t\}$ of Σ such that $\sigma^n(p) = p$. Any "block" $(a_1, \ldots, a_n)$ of length n repeated infinitely often in both directions gives such a sequence. Since there are 2^n blocks of length n, we conclude that there are exactly 2^n points p in Ω such that $f^n(p) = p$. This implies that there are an infinite number of periodic orbits of f in Ω with arbitrarily long periods.

As another illustration, we show that there is an orbit of f which is dense in Ω. This means that every point of Ω is the limit of a sequence of points selected from the orbit. Two points of Ω are very close together if there is a large $N > 0$ such that the corresponding sequences of Σ agree with one another for $|t| < N$. If the sequences are $\{a_t\}$ and $\{b_t\}$, this says $a_t = b_t$ when $|t| < N$. Now there are a countable number of finite sequences of 0's and 1's. These finite sequences can be listed by first listing the two sequences of length one, then the four sequences of length two, then the eight sequences of length three, etc. Thus, a sequence $\{a_t\}$ exists with the property that every finite sequence of 0's and 1's is contained somewhere in $\{a_t\}$. Applying the shift map to this sequence the proper number of times, we find that an iterate of $\{a_t\}$ contains any desired block in its central position. This implies that the point $p \in \Omega$ corresponding to $\{a_t\}$ has an f-orbit which comes as close as we like to any other point of Ω. Thus the orbit of this point p is dense in Ω.

As a final illustration of the power of the "symbolic" analysis of f on Ω, we discuss the problem of predicting the asymptotic behavior of orbits in Ω. If we know a point $p \in \Omega$ to a certain precision, to what extent can we predict where the orbit of p will go? Knowing p to a certain precision is equivalent to knowing the terms of the corresponding sequence $\{a_t\} \in \Sigma$ for t less than some integer N. For times t with $|t| \geqslant N$, we cannot predict whether $f^t(p)$ lies in rectangle 0 or in rectangle 1. In other words, after a certain time, one can say nothing about where in Ω the orbit of p goes. There is another perspective on this situation. Given a sequence $\{a_t\}$, inside Ω there is a point $p \in \Omega$ such that $f^t(p)$ is in rectangle a_t. Thus there is no correlation between a rectangle containing p and a rectangle containing $f^t(p)$ for any $t > 0$. Later, we shall see that a measure can be introduced on Ω for which the action of f looks like the Bernoulli process of flipping a coin an infinite number of times. In this case, the action of f on the set Ω is as random as it could possibly be.

There is an aspect of this example which detracts from its physical relevance. The complicated limit set it contains is not an "attractor". This means that most nearby initial conditions have trajectories which lead away from the complicated limit set. However, there are other examples which have complicated limit sets which are attractors. A geometric technique for constructing a large class of such examples has been described by Williams [**30**]. A simple example of this sort is pictured in Figure 3.

In $\mathbf{R}^3$, picture a solid torus T. The map $f\colon T \to T$ is described by first shrinking the torus in the directions around its central core, then stretching the torus along its central core so that the core is approximately twice as long, and finally placing the torus back inside itself so that the image approximately wraps around the central core twice. The limit set for this map is a complicated one dimensional set (called a

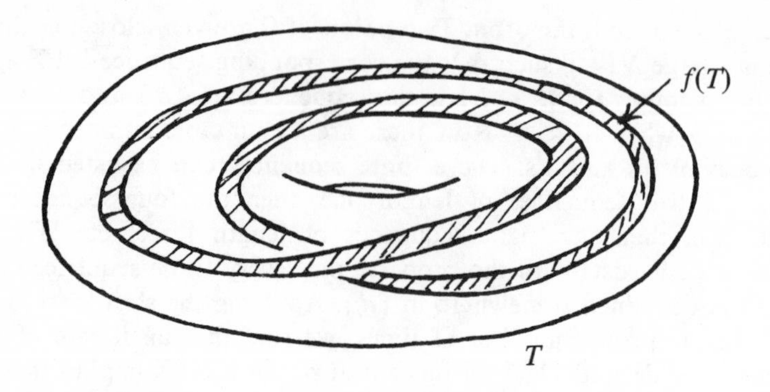

FIGURE 3.

solenoid) which intersects a radial disk of the torus in a Cantor set. It is possible to give an explicit "symbolic" description of this limit set similar to the one described for the example above. There is one additional complication here in that there will be some points which are represented by two symbol sequences (just as the two decimal expansions $0.999\cdots$ and $1.000\cdots$ represent the same number).

Our long-range goal is to derive a comprehensive theory which allows us to describe the dynamical behavior of as many systems of differential and difference equations as we can. We would like to include "almost all" systems into such a theory, but attempts to do so have been largely unsuccessful. The theory which we describe here does include a wide range of systems, but there are innocent looking equations which are outside the scope of our theory.

Such a system was discovered by Lorenz in his work on constructing finite dimensional models of fluid dynamics **[15]**. Lorenz considered the following system of differential equations in $\mathbf{R}^3$:

$$\dot{x} = -10x + 10y,$$

$$\dot{y} = xz + 28x - y,$$

$$\dot{z} = xy - \tfrac{8}{3}z.$$

This system of equations has a complicated attractor which contains sets of orbits which look like the analog of the horseshoe for continuous time systems **[11]**. This attractor also contains an unstable equilibrium point. We mention this example here for three reasons. First, it illustrates that systems of differential equations which might appear simple at first glance can produce *very* complicated behavior. Second, the example illustrates that this complicated dynamical behavior occurs in *attractors* as well as limit sets which are not attractors (as in the horseshoe). Third, the example indicates the incompleteness of the theory of dynamical systems and that the theoretical analysis of specific models still requires a good deal of ingenuity. Considerable progress has been made in the analysis of a geometric model of this system of equations but there remain numerical problems in verifying that the geometric model accurately depicts the behavior of Lorenz's system.

IV. Stable and unstable manifolds. Much of dynamical systems theory is based upon a study of which sets of points have the same asymptotic behavior in forward or backward time. In this section, we state a number of theorems of this sort. There is an extensive literature on this subject, most of it very technical. For our purposes, the ultimate reference is Hirsch-Pugh-Shub [**13**].

Consider a system of linear differential equations

$$\dot{x} = Ax, \qquad x \in \mathbf{R}^n,$$

and A an $n \times n$ matrix. The solution curves of this system of differential equations are given by $x(t) = e^{tA}x(0)$. If one chooses coordinates in $\mathbf{R}^n$ so that A is in Jordan canonical form, then it is easily seen that each solution $x(t)$ different from 0 has one of the following sorts of behavior:

(1) $\|x(t)\| \to \infty$ as $t \to \pm\infty$,

(2) $\|x(t)\| \to 0$ as $t \to +\infty$, $\|x(t)\| \to \infty$ as $t \to -\infty$,

(3) $\|x(t)\| \to 0$ as $t \to -\infty$, $\|x(t)\| \to \infty$ as $t \to +\infty$,

(4) $\|x(t)\|$ and $\|x(t)\|^{-1}$ are bounded.

Solutions with the behavior indicated by (4) occur if and only if the matrix A has an eigenvalue with zero real part. If 0 is an eigenvalue, then there is a linear subspace of equilibria. If there is a pair of nonzero, pure imaginary eigenvalues, then there are periodic orbits. If there is more than one pair of pure imaginary eigenvalues, then there may be almost periodic solutions. This depends on the ratios of these eigenvalues.

"Generically", a matrix A will not have pure imaginary eigenvalues. Consider this generic case. Each nonzero solution of the equations has asymptotic behavior described by properties (1), (2) or (3). Those solutions with asymptotic behavior (2) fill out a linear subspace of $\mathbf{R}^n$. This subspace of $\mathbf{R}^n$ is spanned by the (generalized) eigenvectors of the eigenvalues of A which have negative real parts. This linear subspace is called the *stable manifold* of the equilibrium point at 0. Similarly, the solutions with asymptotic behavior (3) fill a linear subspace called the *unstable manifold* of 0. Since A has no pure imaginary eigenvalues, the stable and unstable manifolds are complementary linear subspaces. The unstable manifold is spanned by (generalized) eigenvectors of A belonging to eigenvalues with positive real parts. When A has no pure imaginary eigenvalues, the equilibrium point at 0 is called *hyperbolic*. All of the solutions which are not on the stable or unstable manifolds follow trajectories which roughly look like hyperbolae. For example if A is the

2×2 matrix

$$A = \begin{pmatrix} \lambda_1 & 0 \\ 0 & \lambda_2 \end{pmatrix}$$

with $\lambda_1 < 0 < \lambda_2$, then the solutions lie along curves $x_1^{\lambda_2} x_2^{-\lambda_1} = \text{const}$.

Much of the behavior of these linear differential equations is displayed also by "generic" nonlinear differential equations. In its simplest form, there is the following:

STABLE MANIFOLD THEOREM. *Let X be a smooth vector field on $\mathbf{R}^n$ (or on an n-dimensional manifold M) with an equilibrium point p. Assume that the flow of X is defined for all time. Assume that the derivative of X at p has s eigenvalues with negative real parts and $n - s$ eigenvalues with positive real parts (counting multiplicities). Then there are submanifolds $W^s(p)$ and $W^u(p)$ with the following properties:*

(1) $p \in W^s(p) \cap W^u(p)$;

(2) *denote the flow of X by ϕ_t. Then*

$$W^s(p) = \{x \mid \phi_t(x) \to p \text{ as } t \to \infty\}$$

and

$$W^u(p) = \{x \mid \phi_t(x) \to p \text{ as } t \to -\infty\};$$

(3) *W^s is the image of a smooth* 1-*to*-1 *map of $\mathbf{R}^s$ into $\mathbf{R}^n$; W^u is the image of a smooth* 1-*to*-1 *map of $\mathbf{R}^{n-s}$ into $\mathbf{R}^n$;*

(4) *the tangent space to W^s at p is spanned by generalized eigenvectors of the derivative of X at p corresponding to eigenvalues with negative real parts. Similarly, the tangent space to W^u at p is spanned by the generalized eigenvectors of the derivative of X at p corresponding to eigenvalues with positive real parts.*

W^s is called the *stable manifold* of p; W^u is the *unstable manifold* of p. A similar theorem is true for fixed points of invertible maps.

STABLE MANIFOLD THEOREM (DISCRETE CASE). *Let f: $\mathbf{R}^n \to \mathbf{R}^n$ be a map such that f^{-1} exists. Assume f has a fixed point p whose derivative Df_p: $\mathbf{R}^n \to \mathbf{R}^n$ at p has no eigenvalues of absolute value one. Assume that s of the eigenvalues (counting multiplicities) have absolute value less than one. Then there are submanifolds $W^s(p)$ and $W^u(p)$ of $\mathbf{R}^n$ such that*

(1) $p \in W^s(p) \cap W^u(p)$;

(2) $W^s(p) = \{x \mid f^n(x) \to p \text{ as } n \to \infty\}$, $W^u(p) = \{x \mid f^n(x) \to p \text{ as } n \to -\infty\}$;

(3) *W^s [W^u] is the image of a smooth* 1-*to*-1 *map of $\mathbf{R}^s$ [$\mathbf{R}^{n-s}$] into $\mathbf{R}^n$.*

(4) *The tangent space of W^s [W^u] at p is spanned by the generalized eigenspace of Df_p corresponding to eigenvalues of absolute value less than [greater than] one.*

Once again, W^s and W^u are called the *stable* and *unstable manifolds* of p, and p is called a *hyperbolic* fixed point. Note that the eigenvalues in the discrete case correspond to the exponentials of the eigenvalues of a vector field.

A number of comments about the stable manifold theorem are in order. W^s and W^u are the images of maps defined on Euclidean spaces. It often happens that these maps are folded sufficiently that the topology their images inherit as subsets of $\mathbf{R}^n$ is not the same as their original topology as Euclidean spaces. For example, in the horseshoe example, the stable manifold of each fixed point $p \in \Omega$ contains a dense subset of Ω. Indeed, the closure of each stable manifold contains the Cantor set of horizontal segments in S which contains Ω.

Much simpler examples of changed topologies occur for vector fields in $\mathbf{R}^2$. Consider a vector field on $\mathbf{R}^2$ which has a *saddle*, i.e., an equilibrium point having one dimensional stable and unstable manifolds. In this case the stable and unstable manifolds each have two branches. A branch of the stable manifold and a branch of the unstable manifold may coincide. In this case, the stable manifold contains a loop. A specific example is given by the system of equations

$$\dot{x} = y,$$
$$\dot{y} = x - 2x^3.$$

Here the origin is a singular point at which the derivative of the vector field has eigenvalues ± 1. The trajectories of the system lie on the curves $y^2 + x^4 - x^2 = \text{const}$. The curve $y^2 = x^2 - x^4$ is a "figure-eight" which consists of the origin together with its stable and unstable manifolds. In this example, W^s is the "figure-eight", but nonetheless it is the 1-to-1 image of a map of the line into the plane as indicated in Figure 4(b). As a subset of the plane, the neighborhoods of the origin are different from the neighborhoods of the origin on the line.

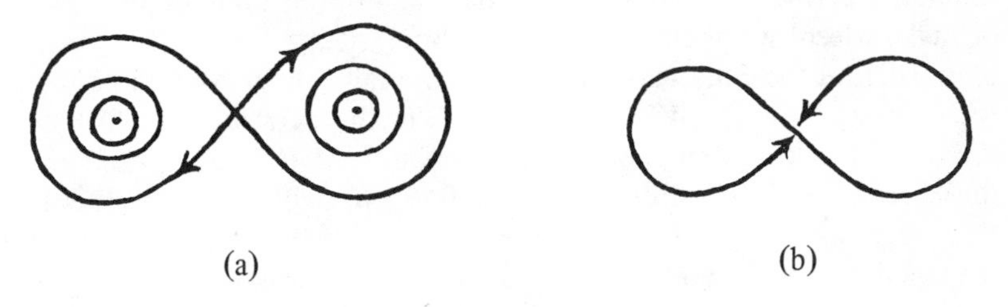

FIGURE 4.

The scope of the stable manifold theorem as an analytic result is much broader than the results we have stated above. The theorem has been generalized to various infinite dimensional contexts. Some of these generalizations have important applications to the theory of dynamical systems. By considering the differential equations induced by a vector field on various function spaces, sets similar to the stable and unstable manifolds of an equilibrium point are produced for other invariant sets of a flow (e.g., a periodic orbit) which satisfy the proper hypotheses. We shall say more about this later.

One of the most important features of the stable and unstable manifolds of a hyperbolic equilibrium or fixed point p is that they intersect *transversally* at p. This means that their tangent spaces together span the tangent space to $\mathbf{R}^n$ at p. Transverse intersections of submanifolds are important because they persist under perturbations of the submanifolds. Related to this is the observation that hyperbolic equilibria and fixed points persist under perturbation of a dynamical system. Arguments based upon transversality will later tell us a great deal about the "structural stability" of the phenomena we describe. Structural stability is a concept which will be defined later. It concerns the persistence of the qualitative dynamics of a vector field or map under perturbation of the vector field of map. To illustrate the ideas of this section, we close it with an example.

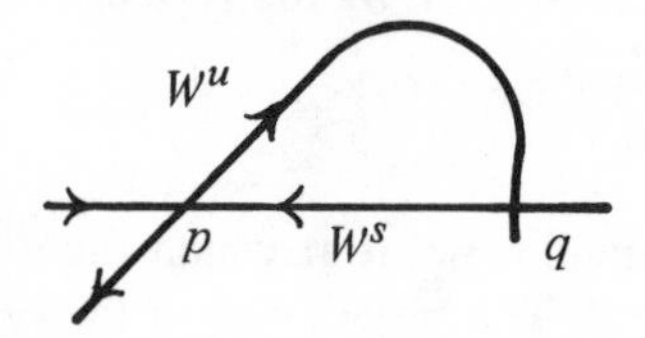

FIGURE 5.

Consider an invertible map f of the plane which has a saddle (i.e., a hyperbolic fixed point with one dimensional stable and unstable manifolds W^s and W^u). Assume further that W^s and W^u intersect transversally at a point q different from p as in Figure 5. We assert that this implies that the diffeomorphism f has an infinite number of periodic orbits! Indeed, we shall find a "horseshoe" sitting inside the dynamics of f. To find the horseshoe, start with a small box B containing p and with sides parallel to W^s and W^u. Look at the iterates $f^n(B)$. If $n > 0$, the images of B are stretched in the direction of W^u and contracted in the direction of W^s. For some $n > 0$, $f^n(B)$ will contain q because $f^n(q) \to p$ as $n \to -\infty$. Similarly, for $m < 0$ the boxes $f^m(B)$ are stretched along W^s and contracted along W^u. For some $m < 0$, $f^m(B)$ contains q. The picture is now given by Figure 6. Notice that the map f^{n-m} on the set $f^m(B)$ looks much like horseshoe in that $f^n(B)$ cuts across $f^m(B)$ in the same way that the horseshoe map cut across the square on which it was defined. The arguments used to describe the horseshoe map can be employed to prove that f^{n-m} (and hence f) has an infinite number of periodic orbits.

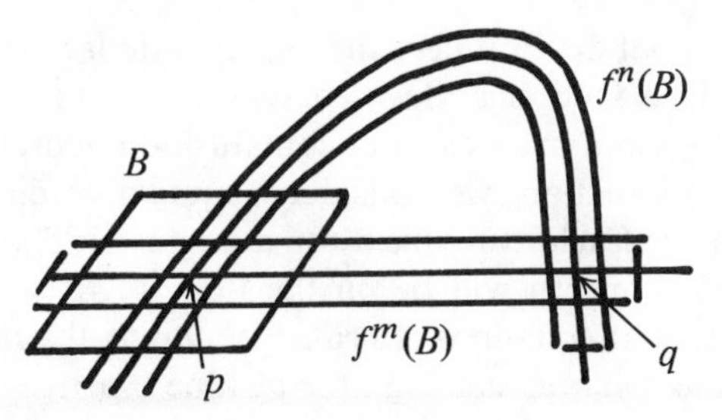

FIGURE 6.

An equilibrium point for a vector field or a fixed point for a diffeomorphism whose stable and unstable manifolds intersect at another point is called *homoclinic*. The example considered above generalizes to the following statement.

PROPOSITION. *If p is a hyperbolic fixed point for a diffeomorphism f such that the stable and unstable manifolds of p intersect transversally at a point different from p, then f has an infinite number of periodic orbits.*

Can one determine the existence of a homoclinic point of a diffeomorphism numerically? It is possible sometimes. If one knows the location of a fixed point p, then the approximate directions of its stable and unstable manifolds can be computed. One can then build a small box B as in the example and follow some of its trajectories for a finite number of iterates. If one finds a picture like that illustrated in Figure 6, then one can be confident that the fixed point is homoclinic. By using our "symbolic" arguments, limited amounts of computation can establish the existence of an infinite number of periodic orbits.

There is a large gap between finding an infinite number of periodic orbits and having a firm grasp of the asymptotic behavior of a dynamical system. This is a gap which is not easily closed. The theory of dynamical systems gives one a language in which to begin describing the full asymptotic behavior of trajectories. In suitable circumstances, one can say a great deal. We now head in this direction.

V. Invariant sets. In dealing with electrical circuits or mechanical systems, one often speaks of the transient response of the system. Starting in some state, the circuit or system will move to an equilibrium, periodic oscillation, or some more complicated limiting behavior. Often times, the limiting behavior is the most important feature of the system. Approach to near this limiting behavior occurs quickly and is relatively unimportant to the functioning of the system. For example, this is evident in a radio in which the warm-up period of the radio is a minor feature of its functioning.

The description of the dynamics of a model requires that we have the means for working with the sets which describe the limiting behavior of trajectories. We shall develop here the appropriate language for describing limit sets. This section is devoted largely to introducing this language of "topological dynamics" and contains no important theorems.

Throughout this section, we consider a system of differential equations defined on a Euclidean space (or a vector field on a manifold). The focus of our concern will be on the flow $\Phi: \mathbf{R}^n \times \mathbf{R} \to \mathbf{R}^n$ of the system of differential equations. Recall that this is the map defined by $\Phi(x, t) = y$ if the solution starting at x at time 0 arrives at y at time t. Most of the sets we shall describe are invariant. A set S is *invariant* if it is a union of solution curves. This means that if $x \in S$, $t \in \mathbf{R}$, and $\Phi(x, t)$ is defined, then $\Phi(x, t) \in S$. In other words, a solution which intersects S lies entirely within it. The simplest example of an invariant set is a single trajectory of Φ. Other examples are the stable and unstable manifolds defined in the preceding section.

Consider a point x and the trajectory $\{\Phi(x, t) | t \in \mathbf{R}\}$ through x. We are interested in the limiting behavior of the trajectory as $t \to \infty$. The *ω-limit set* of the trajectory through x is the set of all points y with the following property: there is an increasing sequence $\{t_i\} \subset \mathbf{R}$ such that $t_i \to \infty$ and $\Phi(x, t_i) \to y$. The ω-limit set of a trajectory captures all of the points that the trajectory approaches as time t tends to ∞. Equilibria and periodic orbits are their own ω-limit sets and are also the ω-limit sets of trajectories in their stable manifolds.

To show the connection of the two ideas introduced thus far in this section, we state an easy lemma:

LEMMA. *The ω-limit set of a trajectory is an invariant set of Φ.*

PROOF. Let y be in the ω-limit set of the trajectory through x. To prove that the ω-limit set of the trajectory is invariant, we need to prove that if z is on the trajectory through y, then z is in the ω-limit set of the trajectory. If z is on the trajectory through y, there is an s such that $z = \Phi(y, s)$. Since y is in the ω-limit set of x, there is an increasing sequence $\{t_i\}$ such that $t_i \to \infty$ and $\Phi(x, t_i) \to y$. Now the sequence $\{t_i + s\}$ is increasing and unbounded and $\Phi(x, t_i + s) = \Phi(\Phi(x, t_i), s) \to \Phi(y, s) = z$. Therefore, z is in the ω-limit set of the trajectory through x. Hence the ω-limit set of this trajectory is invariant. This proves the lemma.

By reversing the direction of time, the concept of the α-limit set of a trajectory is obtained. The *α-limit set* of the trajectory through x is the set of points y such that there is a decreasing sequence $\{t_i\} \subset R$ satisfying $t_i \to -\infty$ and $\Phi(x, t_i) \to y$. The α-limit set of a trajectory represents the set of points from which a trajectory emerged in its primordial past. Like ω-limit sets, α-limit sets are invariant.

Periodic orbits literally go around in circles, but there are orbits which almost do so without quite hitting the target on each cycle. These recurrent trajectories are important because they have no transient response. Formally, a trajectory is *recurrent* if it is contained in its own α- and ω-limit sets. A recurrent orbit is "almost periodic" in the sense that it continues to return as close as one would like to its initial point. (*Warning*. The term "almost periodic" is customarily used in mathematics with a more restrictive meaning than "recurrent.") Recurrence is too strong an idea for many of our purposes, however.

The idea which will capture for us the right properties of "returning" embodies the intuition that it suffices for some nearby points to return. A *wandering point* x is one for which there is a neighborhood U of x with the property $\Phi(U, t) \cap U = \varnothing$ whenever $|t| \geqslant 1$. Intuitively, after a short initial time, no point near x returns to a point near x. The wandering points are those which do not even have returning neighbors. The points which are not wandering are called *nonwandering*. A nonwandering point x has the property that if U is a neighborhood of x, there is a $t \geqslant 1$ such that $\Phi(U, t) \cap U \neq \varnothing$. The collection of nonwandering points of Φ is called the *nonwandering set* of Φ and is denoted by Ω. The nonwandering set Ω plays a central role in dynamical systems. It captures a very weak notion of recurrence or return. The wandering trajectories of a flow all wander from one piece of Ω (containing its α-limit set) to another piece of Ω (containing its ω-limit set).

The way in which we approach the task of describing a flow is to first describe the nonwandering set together with its dynamics, and then to describe the way in which trajectories flow from one piece of the nonwandering set to another. This is a vague indication of the direction in which we are heading, but much more needs to be described before a clearer picture emerges. In particular, we need to understand a few basic properties of Ω and what the "pieces" of Ω should be. We proceed by listing two basic properties of the nonwandering set and then discuss its "pieces".

PROPOSITION. *The set of wandering points of a flow is open; the set of nonwandering points of a flow is closed. The α- and ω-limit sets of a trajectory are contained in the nonwandering set.*

The proof of the first two statements is an immediate consequence of definitions. The second is a good exercise for the puzzled reader.

In describing a "piece" of Ω we would like to be able to feel that the dynamics of the flow is intertwined within each piece in an inextricable way. A concept which embodies this idea is that of topological transitivity. A closed, invariant set S for the flow Φ is *topologically transitive* if there is a trajectory γ inside S with the property that every point of S is the limit of a sequence of points selected from γ. In other words, S is the closure of γ and γ is *dense* in S. A topologically transitive set for a flow is tied together by a dense trajectory. If a set is nonwandering and topologically transitive, then there is no good way of further decomposing it. Thus our first objective in describing the solutions of a system of differential equations will be to decompose the nonwandering set of its flow into disjoint topologically transitive subsets. These subsets will function as the "pieces" of Ω referred to above.

In many applications there are certain subsets which have particular importance because of their stability properties. These are the attractors. A topologically transitive subset S of Ω is an *attractor* for the flow if S has small neighborhoods U such that the flow carries U into itself with increasing time and every point near S asymptotically tends to S as $t \to \infty$. This expresses the fact that small changes in initial conditions from those in S do not send a trajectory far away from S in the state space.

For well-behaved systems of differential equations one expects that most trajectories will asymptotically tend to an attractor. The "steady state" behavior of the system is then described by an attractor with its dynamics. In the case of a complicated attractor, one hopes that some kind of statistical description of the dynamics of the attractor is possible. We shall see later that this can be done in the appropriate circumstances.

We note that everything in this section makes sense for discrete time systems without change. The only difference is that t is restricted to lie in the integers.

To end this section, we mention how the ideas put forward can be applied to one of the examples considered before, the horseshoe. The nonwandering set of the horseshoe consists of two pieces, the Cantor set which lies inside the square S and an attracting fixed point. The ω-limit set of most points is the attracting fixed point. Determining the α- and ω-limit sets of trajectories in the Cantor set is difficult. There are trajectories which are dense in the Cantor set and have the entire Cantor set as their α- and ω-limit sets. There are also periodic orbits in the Cantor set which are their own α- and ω-limit sets. Still other trajectories have α- and ω-limit sets which are larger than their own orbits but smaller than the entire Cantor set. Analysis of what possibilities do occur is done most easily by using the symbolic description of the Cantor set.

VI. Hyperbolic structures. Describing qualitatively the solutions to all systems of differential equations is a hopeless task. Arbitrarily complicated behavior can occur in highly degenerate situations. For example, one can construct systems of differential equations whose set of equilibria is any predetermined closed set. A choice must be made between a general theory which says little about a wide range of equations and a theory which says much more about a restricted class of equations. We have adopted a compromise between these two positions by describing a theory which states a good deal about a large class of equations (though not as large as we might like). About 1960, Peixoto proved a theorem which served as a model for the development of dynamical systems over the next ten years. Roughly, Peixoto proved that "almost all" systems of differential equations in two dimensions could be described in the rather simple terms of equilibria, periodic orbits, and their stable and unstable manifolds. Complicated limit sets which occur in higher dimensions do not occur in most two dimensional systems of differential equations. This spurred efforts to find a reasonable theory which applied to "almost all" systems. For the most part, these efforts were unsuccessful. We return to a more extensive discussion of this topic later. Here we develop hypotheses to be satisfied by "good" systems for which an extensive theory has been developed.

The approach which has been fruitful is to make assumptions about the behavior of the solutions of a system of differential equations which can be used to infer additional structure about the solutions. The prototype of this sort of assumption occurs in the stable manifold theorem. By making assumptions about linearized equations at an equilibrium, one can deduce qualitative information about the set of trajectories for which the equilibrium is the α- or ω-limit set. This section is devoted primarily to extensions of the sort of assumptions made in the stable manifold theorem to more complicated invariant sets.

These assumptions are often much more difficult to verify for specific examples than those made in the stable manifold theorem. One needs to know a lot about the solutions of a system of equations to verify these properties. Hardly any examples of specific equations with complicated limit sets have been shown to satisfy the assumptions we make. This is an area which warrants significant further exploration. A gap exists between the abstract theory dealt with in this survey and the solution of specific differential equations. It is easy to find particular solutions to a system of differential equations numerically, but often difficult to build a coherent picture of the "phase portrait" which shows the behavior of typical solutions. The language of dynamical systems should prove to be a suitable framework for this endeavor.

We want to state conditions which generalize the assumptions made at an equilibrium point in the stable manifold theorem. The statement of these conditions involves a short detour to introduce a new concept. A tangent vector to $\mathbf{R}^n$ (or a manifold) corresponds to the concept of a "free vector" often used in calculus. At each point $x \in \mathbf{R}^n$ we attach a copy of $\mathbf{R}^n$ which represents the collection of all possible free vectors or tangent vectors to curves at x. If we take the union of all of these copies of $\mathbf{R}^n$, they form a space $\mathbf{R}^n \times \mathbf{R}^n = T(\mathbf{R}^n)$ called the *tangent space* (or tangent bundle) of $\mathbf{R}^n$. An element of $T(\mathbf{R}^n)$ consists of a pair (x, v) where x is a point of $\mathbf{R}^n$ and v is a tangent vector based at x. The n-dimensional space of tangent vectors at x is denoted $T_x\mathbf{R}^n$.

Now let X be a vector field and $\Phi: \mathbf{R}^n \times \mathbf{R} \to \mathbf{R}^n$ be its flow. Fix t and consider the map $\Phi(\cdot, t) = \Phi_t: \mathbf{R}^n \to \mathbf{R}^n$. The derivative of Φ_t at a point x (denoted $(D\Phi_t)_x$) is a linear map from $\mathbf{R}^n$ to itself, but it should be thought of as a map from the tangent space at x to the tangent space at $\Phi_t(x)$. Formally, $(D\Phi_t)_x: T_x\mathbf{R}^n \to T_{\Phi_t(x)}\mathbf{R}^n$. The map $(D\Phi_t)_x$ measures the following: consider a curve γ through x. Follow γ with the map Φ_t to obtain a new curve $\Phi_t \circ \gamma$ through the point $\Phi_t(x)$. Then $D\Phi_t$ maps the tangent vector to γ at x to the tangent vector to $\Phi_t \circ \gamma$ at $\Phi_t(x)$. The derivative $D\Phi_t$ measures the infinitesimal stretching and twisting of the flow Φ around its trajectories.

Without reference to eigenvalues, we can use $D\Phi_t$ to state the hypotheses of the stable manifold theorem at an equilibrium x. The assumption that the linearization of the vector field has negative eigenvalues on a subspace V is equivalent to the following condition: if $v \in V$ is a tangent vector at x, then $\|(D\Phi_t)_x v\| \to 0$ as $t \to \infty$. Indeed, in this special case, if the length of $(D\Phi_t)v$ tends to 0, then it does so exponentially fast. There are constants $c > 0$, $\lambda < 0$ such that $\|(D\Phi_t)_x v\| < ce^{\lambda t}\|v\|$ for all $t > 0$. The reason for this is that if A is the Jacobian matrix of the vector field at x, then $(D\Phi_t)x = e^{tA}$.

This statement of the hypotheses for the stable manifold theorem is definitely more complicated and harder to use in practice. Its virtue is that it can be extended to situations other than those involving an equilibrium. For example, the sufficient condition for asymptotic stability of a periodic orbit can be stated as the requirement that there are constants $c > 0$, $\lambda < 0$ such that if v is any tangent vector at a point of the closed orbit which is not tangent to the orbit, then $\|D\Phi_t(v)\| \to 0$ as $t \to \infty$. This definition avoids the introduction of a Poincaré map for the periodic orbit. By insisting that estimates of this sort should be uniformly valid on pieces of the nonwandering set we shall obtain analogs of stable and unstable manifolds for complicated limit sets. These stable and unstable manifolds provide a wealth of information about the dynamics of a flow.

Let Λ be an invariant set for the flow Φ containing no equilibria. We define a *hyperbolic structure* on Λ. (An equilibrium x has a hyperbolic structure if the derivative of the vector field set x has no pure imaginary eigenvalues.) The data of the hyperbolic structure consist of three linear subspaces E_x^0, E_x^s, E_x^u of each tangent space $T_x\mathbf{R}^n$, $x \in \Lambda$. These subspaces are to satisfy the following properties:

(1) E_x^0, E_x^s, E_x^u depend continuously on x. (This means that bases can

be chosen in small regions of Λ so that each basis vector depends continuously on x.)

(2) $\dim E_x^0 = 1$ and E_x^0 is the tangent space to the trajectory through x.

(3) E_x^0, E_x^s, and E_x^u together span $T_x\mathbf{R}^n$ and $\dim E_x^s + \dim E_x^u = n - 1$.

(4) $(D\Phi_t)_x(E_x^s) = E^s_{\Phi_t(x)}, (D\Phi_t)_x(E_x^u) = E^u_{\Phi_t(x)}$.

(5) There are constants $c, \lambda > 0$ such that

(a) if $v \in E_x^s$, then $\|(D\Phi_t)_x v\| < ce^{-\lambda t}\|v\|$ for all $t > 0$;

(b) if $v \in E_x^u$, then $\|(D\Phi_t)_x v\| < ce^{\lambda t}\|v\|$ for all $t < 0$.

If the subspaces E_x^0, E_x^s, and E_x^u have these properties, they define a hyperbolic structure for Λ. For discrete time systems, the definition is similar. The only differences are that there is no E_x^0 and $\dim E^s + \dim E^u = n$.

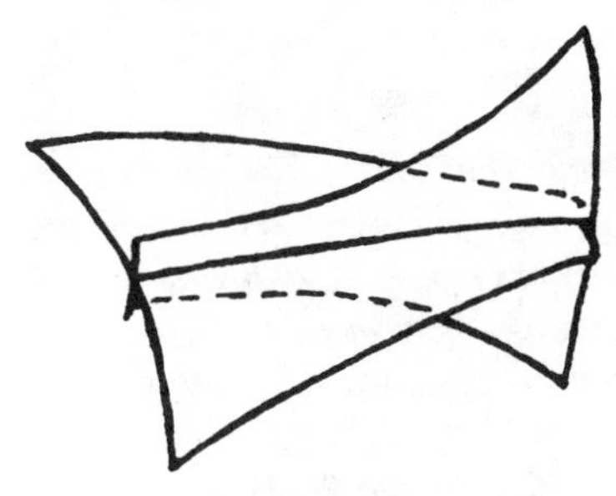

FIGURE 7.

What does a hyperbolic structure entail? Apart from the direction of the flow itself, it says that the transverse directions split into those in which the flow expands away from trajectories in Λ and those in which it contracts toward trajectories in Λ. A "local" picure of the flow is like that in Figure 7. The fourth condition says that these directions are consistent with the flow. The fifth condition requires that the expansion and contraction occur at a uniform rate after a certain amount of time. Note that the Cantor set of the horseshoe has a hyperbolic structure.

We will want to assume the existence of a hyperbolic structure on the nonwandering set of a flow throughout most of the monograph. The hypothesis we shall often assume was formulated by Smale as Axiom A. A flow Φ satisfies *Axiom A* if the following two conditions are met:

(a) The nonwandering set Ω of Φ has a hyperbolic structure.

(b) The equilibria of Φ are isolated in Ω. The nonwandering set Ω is the closure of the periodic orbits and equilibria of Φ.

The statement that the equilibria of Φ are isolated in Ω means that each equilibrium has a neighborhood containing no other points of Ω.

The qualitative analysis of flows satisfying Axiom A has been one of the striking achievements of the recent theory of ordinary differential equations. The reason why this analysis seems to have been so successful is that Axiom A embodies those properties of a flow which guarantee the stability and persistence of nonwandering points. This "structural stability" contrasts with the complicated ergodic behavior within pieces of the nonwandering sets of Axiom A flows. Is there some kind of philosophical lesson to be learned from this persistence of randomness?

VII. Dynamics of Axiom A flows. In this section we shall begin to describe the qualitative behavior of solutions of systems of differential equations satisfying Smale's Axiom A. *We assume throughout the section that all flows satisfy Axiom A.*

The first major result concerning Axiom A flows is a generalization of the stable manifold theorem stated previously. This generalized stable manifold theory is contained in the invariant manifold theory of Hirsch, Pugh, and Shub. We first give a statement of the theorem and then discuss its meaning.

THEOREM [**13**]. *Assume that Φ is a flow which satisfies Axiom A and has a bounded (compact) nonwandering set. Then there are collections of submanifolds $W^s(x)$, $W^u(x)$, $x \in \Omega$, called stable and unstable manifolds, respectively, with the following properties.*

(1) *$E^s(x)$ is the tangent space to $W^s(x)$ at x, $E^u(x)$ is the tangent space to $W^u(x)$ at x;*

(2) *$\Phi_t(W^s(x)) = W^s(\Phi_t(x))$ and $\Phi_t(W^u(x)) = W^u(\Phi_t(x))$;*

(3) *$W^s(x) = \{y \mid \|\Phi(y, t) - \Phi(x, t)\| \to 0$ as $t \to \infty\}$; $W^u(x) = \{y \mid \|\Phi(y, t) - \Phi(x, t)\| \to 0$ as $t \to -\infty\}$; in both cases, convergence is at an exponential rate;*

(4) *if y is a point such that the distance from $\Phi(y, t)$ to Ω tends to 0 as $t \to \infty$ $[-\infty]$, then $y \in W^s(x)$ $[W^u(x)]$ for some point $x \in \Omega$.*

This theorem presents us with the following picture. The set of points on trajectories which do not go to ∞ with increasing time are partitioned into stable manifolds. Each stable manifold is a smooth surface consisting of points with the same forward asymptotic behavior. Points on the same stable manifold eventually approach each other at an exponential rate as one follows their trajectories. Different points of Ω may lie on the same stable manifold if they have the same forward asymptotic behavior. The stable and unstable manifolds are permuted by the flow Φ.

The geometrical structure produced by the decomposition of $\mathbf{R}^n$ into stable and unstable manifolds leads to a rich insight into the dynamics of a flow by identifying sets of points with the same asymptotic behavior. Still more theory is necessary, however, before a complete picture emerges. We need more information about the dimensions of the stable and unstable manifolds and the ways in which they fit together. One of the fundamental theorems which leads to such an understanding is the following theorem of Smale:

SPECTRAL DECOMPOSITION THEOREM [7]. *Suppose that* Φ *is a flow satisfying Axiom A with a compact nonwandering set* Ω. *Then* Ω *can be written as a finite disjoint union* $\Omega = \Omega_1 \cup \cdots \cup \Omega_m$ *with each* Ω_i *topologically transitive for* Φ.

Recall that a set is topologically transitive if it is a closed invariant set which has a dense orbit. The sets $\Omega_1, \ldots, \Omega_m$ are called *basic sets* of Φ. They are the "pieces" of the nonwandering set with which much of the additional theory deals. The invariance properties of stable and unstable manifolds imply that their dimensions are constant along each trajectory. Continuity and the topological transitivity of the basic sets imply that the dimensions of stable and unstable manifolds are constant on each basic set. This contributes more to our picture of the dynamics of an Axiom A flow. All of the recurrence of the flow occurs within the basic sets, and each basic set is indecomposable. Trajectories outside the nonwandering set flow from one basic set to another (or to ∞). Each basic set may be complicated within itself, but trajectories which approach a basic set do so at an exponential rate. The transverse behavior of trajectories to the basic set resembles the flow near a hyperbolic equilibrium.

In our later discussions of stability, it will be necessary to make one more fundamental assumption about the flows we deal with beyond Axiom A. This hypothesis involves the way in which the stable and unstable manifolds of different basic sets fit together.

Strong Transversality Property. If $y \in W^s(x_1) \cap W^u(x_2)$, then the tangent spaces to $W^s(x_1)$, $W^u(x_2)$, and the trajectory through y together span the tangent space $T_y(\mathbf{R}^n)$ of $\mathbf{R}^n$ at y.

One of the consequences of the Strong Transversality Property is called the No Cycle Property. If Ω_i is a basic set, we denote

$$W^s(\Omega_i) = \bigcup_{x \in \Omega_i} W^s(x) \quad \text{and} \quad W^u(\Omega_i) = \bigcup_{x \in \Omega_i} W^u(x).$$

We introduce a relation on the basic sets of Axiom A flow by saying that $\Omega_i \leqslant \Omega_j$ if there is a trajectory that goes from Ω_i to Ω_j. In other words, $W^u(\Omega_i) \cap W^s(\Omega_j) \neq \varnothing$. Using this relation we state the No Cycle Property as the following theorem:

THEOREM [7]. *If* Φ *is a flow satisfying Axiom A and the Strong Transversality Property, then* $\Omega_{i_0} \leqslant \Omega_{i_1} \leqslant \cdots \leqslant \Omega_{i_k} \leqslant \Omega_{i_0}$ *implies that* $\Omega_{i_j} = \Omega_{i_0}$ *for* $j = 1, \ldots, k$. *Moreover* $\Omega_i = W^u(\Omega_i) \cap W^s(\Omega_i)$.

Thus the No Cycle Property states that the relation $\leqslant$ is a partial ordering of the basic sets; it has no "cycles". This partial ordering allows us to establish the idea of "downhill" along a flow, with plateaus at the basic sets. At the "top" of the flow are basic set(s) from which nearly all wandering orbits flow. These basic sets are called sources or repellors. Some of the orbits get trapped at an intermediate basic set, a plateau which can be thought of as a "generalized saddle". However, most of the trajectories flow downhill to a basic set into which nearly all trajectories are flowing. These basic sets are called sinks or attractors. In a chain of basic sets $\Omega_{i_0} \leqslant \Omega_{i_1} \leqslant \cdots \leqslant \Omega_{i_k}$, the dimensions of the stable manifolds $W^s(\Omega_i)$ are a nondecreasing function.

This yields a picture of the overall dynamics of a flow satisfying Axiom A and the No Cycle Property which is similar to a gradient flow. The intuitive picture is that one sits on a surface subject to the influence of gravity pulling one downward. One then descends on the surface in the steepest possible way at each instant until one reaches a horizontal point. There one sits, being unable to decide which way is the steepest descent from the saddle (if one is not at the bottom of a well). (See Figure 8.) For a flow satisfying Axiom A and the No Cycle Property, the basic sets play the role of horizontal points in this scenario. There is a "downhill" until one hits a basic set at which point one gradually comes to rest (if the basic set is an equilibrium) or searches madly in vain around the basic set for a way to descend further.

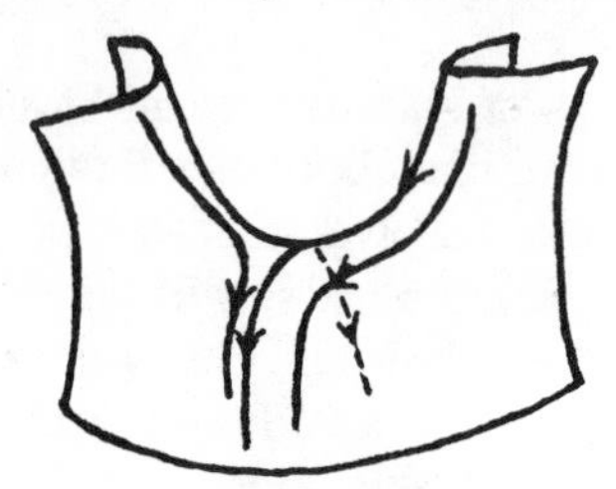

FIGURE 8.

Our discussion has reached a saddle point of its own, with two quite different directions of pursuit open to us. On the other hand, we want to inspect the basic sets in more detail. Inside the basic set is where the action is for a well-behaved flow. Axiom A imposes conditions on the topology and dynamics of the basic sets. This is something to be explored in more detail. On the other hand, we want to take a step away from the analysis of individual flows to consider how the qualitative behavior of Axiom A + Strong Transversality flows varies under perturbation. This is the domain of structural stability. If we are not to be stuck at this saddle, we must choose to deal with one topic before the other. So let us now take a closer look at the basic sets themselves and only later consider the property of structural stability.

VIII. Basic sets. In this section and the next, we tackle the problem of describing the behavior of a flow on a basic set. The theory we present is due primarily to Bowen and Sinai. To begin with, we shall work with discrete time systems because the description of their basic sets is simpler than the corresponding description for continuous time flows. The results are adapted well to the determination of the kinds of periodic and recurrent trajectories which occur for a system. They also display clearly the random features of the flow on a basic set and aid us in the analysis of statistical properties of the flow. The prototype for the sort of description of basic sets which we give is the horseshoe example. Early analysis of this sort can already be found in the work of Morse and Hedlund on geodesic flows of surfaces of negative curvature.

Consider a discrete time system of difference equations given by a smooth invertible map f: $\mathbf{R}^n \to \mathbf{R}^n$. We assume that f satisfies Axiom A and has a compact nonwandering set. We focus our attention on one basic set $\Lambda \subset \Omega$. Assume that, for each point $x \in \Lambda$, $s = \dim E^s(x)$ and $n - s = u = \dim E^u(x)$. We want to give a symbolic description of the dynamics of f on Λ in the same way that we were able to describe the horseshoe. The first step in this process will be to cut Λ into pieces so that the images of these pieces under f fit back onto themselves in a particularly nice way. When this is done, a symbol will be assigned to each piece. We will be able to label points by sequences of symbols related to the dynamic behavior of f on Λ.

For the most part, we can work strictly inside Λ and forget the space around Λ. As a compact subset of Euclidean space, Λ inherits a distance function d which makes it a compact metric space. Inside Λ, we can still speak of stable and unstable manifolds. The stable manifold is defined by

$$\Lambda \cap W^s(x) = \{y \in \Lambda | d(f^n(x), f^n(y)) \to 0 \text{ as } n \to \infty\}.$$

The unstable manifold is defined similarly with $n \to -\infty$. It will be useful to deal with *local* stable and unstable manifolds also. These are defined by

$$W_\varepsilon^s(x) = \{y | d(f^n(x), f^n(y)) < \varepsilon \text{ for all } n \geqslant 0\}$$

and

$$W_\varepsilon^u(x) = \{y | d(f^n(x), f^n(y)) < \varepsilon \text{ for all } n \leqslant 0\}.$$

The hyperbolic structure of Λ implies that $W_\varepsilon^s(x) \subset W^s(x)$ and $W_\varepsilon^u(x) \subset W^u(x)$ for ε sufficiently small.

An additional consequence of the stable manifold theorem is the existence of "canonical coordinates" for Λ. These are provided by the following theorem.

THEOREM. *Let Λ be a hyperbolic basic set for a diffeomorphism. Then there are constants ε, $\delta > 0$ with the following property: if $x,y \in \Lambda$ and $d(x, y) < \delta$, then $W_\varepsilon^s(x) \cap W_\varepsilon^u(y)$ contains exactly one point z, and z is in Λ.*

This theorem establishes a "product" structure on basic sets. Small regions inside a basic set can be decomposed into expanding and contracting directions. These can be used to construct *canonical coordinates* in the following way. Choose a point $x \in \Lambda$ and $\varepsilon, \delta > 0$ appropriately. If $y \in W^s_\delta(x) \cap \Lambda$ and $z \in W^u_\delta(x) \cap \Lambda$, then $W^u_\varepsilon(y) \cap W^s_\varepsilon(z)$ contains a single point which is in Λ. The set of points obtained in this way is a neighborhood of x in the metric space Λ. Thus the sets $W^s_\delta(x) \cap \Lambda$ and $W^u_\delta(x) \cap \Lambda$ can be regarded as the axes in a "canonical" coordinate system for a neighborhood of x in Λ (see Figure 9). The horseshoe example provides a good prototype for this construction. In the horseshoe, the basic set is given as the intersection of a Cantor set of horizontal segments and a Cantor set of vertical segments. These segments contain the local stable and unstable manifolds of points in the basic set.

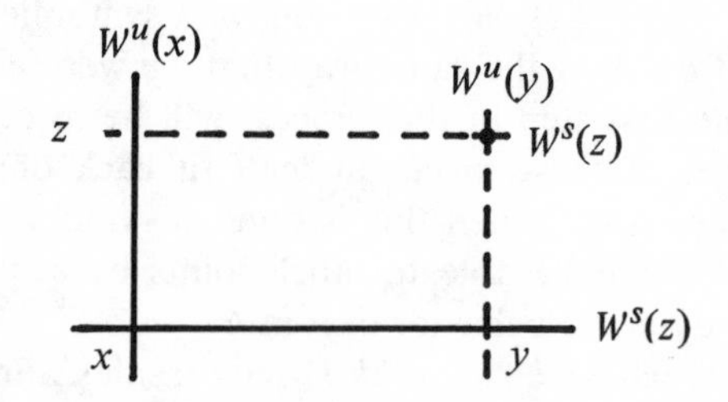

FIGURE 9.

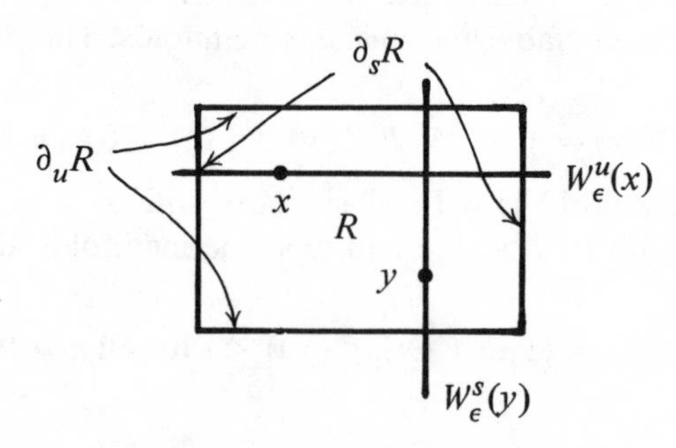

FIGURE 10.

Closely connected with the concept of a canonical coordinate system is the idea of a "rectangle" in a basic set. A rectangle is to be roughly the product of a local stable manifold and a local unstable manifold in a basic set. Specifically, choose $\varepsilon > 0$, $\delta > 0$ so that if $x,y \in \Lambda$ satisfy $d(x, y) < \delta$, then $W^u_\varepsilon(x) \cap W^u_\varepsilon(y)$ contains exactly one point. Then a *rectangle* is a closed set $R \subset \Lambda$ with the property that $x,y \in R$ imply that $W^u_\varepsilon(x) \cap W^u_\varepsilon(y) \in R$. The rectangle R is *proper* if it is the closure of its interior (as a subset of Λ). (See Figure 10.)

The boundary of a rectangle R comes in two kinds of pieces (which may intersect). One piece, denoted $\partial_s R$, consists of points x with the property that for every $\delta > 0$ there are points of $W^u_\delta(x) \notin R$. The other piece, denoted by $\partial_u R$, consists of points x with the property that for every $\delta > 0$, there are points of $W^s_\delta(x) \notin R$. Canonical coordinates and the definition of a rectangle imply that if $x \notin \partial_u R \cup \partial_s R$, then x is an interior point of R. We call $\partial_s R$ the *stable boundary* of R and $\partial_u R$ the *unstable boundary* of R.

Rectangles are essential in carrying out the symbolic description of basic sets given in this section. Their definition is somewhat involved, but that appears to be unavoidable because the basic sets often have complicated topology. The next step in carrying out the symbolic description of basic sets is to divide the basic sets into a collection of rectangles with particularly nice properties. This division of the basic set is called a Markov partition.

A *Markov partition* for a hyperbolic basic set Λ of a map f is a finite collection of proper rectangles $R_1, \ldots, R_k$ with the following properties:

(1) $\Lambda = R_1 \cup \cdots \cup R_k$.
(2) If $i \neq j$, the interiors of R_i and R_j are disjoint.
(3) $f(\bigcup_{i=1}^k \partial_s R_i) \subset \bigcup_{i=1}^k \partial_s R_i$.
(4) $\bigcup_{i=1}^k \partial_u R_i \subset f(\bigcup_{i=1}^k \partial_u R_i)$.

The last two properties in this definition give Markov partitions their special character. They force the image of one rectangle to cut across another rectangle in a particular way, as indicated in Figure 11.

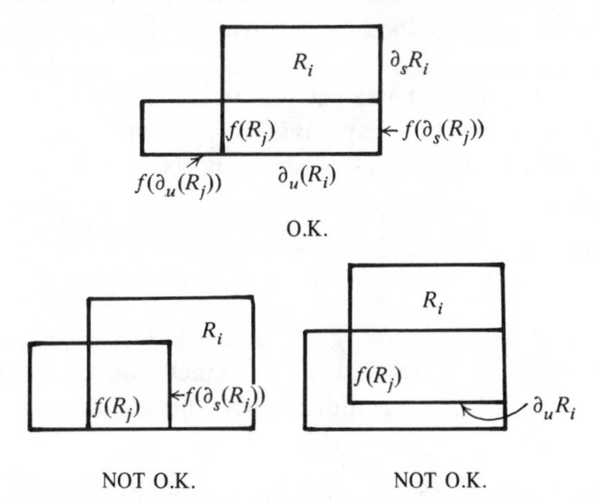

FIGURE 11.

When the image of a rectangle intersects another, it must go all the way across the unstable directions. In the stable directions, the cross-section of the image must be contained in the rectangle which it is intersecting. In the horseshoe, a Markov partition exists with two rectangles, one being the portion of the horseshoe in the left half of the square and the other being the portion of the horseshoe in the right half of the square.

As one iterates the map f, the images of the rectangles become longer in the unstable directions and shorter in the stable directions. This happens at an exponential rate. Similarly, the images of the rectangles under the iterates of f^{-1} become longer in the stable directions and shorter in the unstable directions. This is the kind of "hyperbolic" behavior we shall soon exploit.

Before proceeding further, however, we should state the fundamental theorem of this section.

THEOREM [8]. *If f is an invertible map with a hyperbolic basic set Λ and if $\varepsilon > 0$, then Λ has a Markov partition consisting of rectangles of diameter at most ε.*

This theorem, due to Sinai in a special case and Bowen generally, leads to the powerful and intuitive technique of symbolic dynamics. This is the topic of the next section.

IX. Symbolic dynamics. The object of symbolic dynamics is to use the properties of a Markov partition in order to keep track of which orbit will occur in a basic set. This is done by introducing an abstract space together with a "shift" map which mimics the dynamics of the map on the basic set. Before proceeding with a formal description of this space and its dynamics, we examine how the symbol space is obtained from a Markov partition.

Let Λ be a hyperbolic basic set for an invertible map f. Let $R = (R_1, \ldots, R_k)$ be a Markov partition for Λ. Now construct a $k \times k$ matrix of 0's and 1's, called the transition matrix T of R by setting

$$T = (T_{ij}), \qquad T_{ij} = \begin{cases} 1 & \text{if int } R_i \cap f^{-1}(\text{int } R_j) \neq \varnothing, \\ 0 & \text{if int } R_i \cap f^{-1}(R_j) = \varnothing. \end{cases}$$

Next consider a sequence $\{a_i\}_{i=-\infty}^{\infty}$ such that $a_i \in \{1, \ldots, k\}$ and $T_{a_i a_{i+1}} = 1$ for all i. We assert that there is exactly one point $x \in \Lambda$ such that $f^i(x) \in R_{a_i}$ for all i. Let us indicate why this should be true.

Consider the sets

$$R_{a_0}, \quad R_{a_0} \cap f^{-1}(R_{a_1}), \qquad R_{a_0} \cap f^{-1}(R_{a_1}) \cap f^{-2}(R_{a_2}),$$

etc. The properties of a Markov partition imply that if these sets are nonempty, then they extend all the way across R_{a_0} in the stable direction. Their unstable boundaries cannot lie in the interior R_{a_0}. We assert that these sets are indeed nonempty. We have assumed that $R_{a_i} \cap f^{-1}(R_{a_{i+1}})$ is nonempty, so $f^{-1}(R_{a_i}) \cap f^{-i+1}(R_{a_{i+1}})$ is also nonempty. In particular, $R_{a_i} \cap f^{-1}(R_{a_1})$ extends across R_{a_0} in the stable directions. The set $f^{-2}(R_{a_2})$ extends across $f^{-1}(R_{a_1})$ in the stable directions, and hence $R_{a_0} \cap f^{-1}(R_{a_1}) \cap f^{-2}(R_{a_2})$ is a nonempty set extending across R_{a_0} in the stable directions. Similarly, the sets $\bigcap_{i=0}^{n} f^{-i}(R_{a_i})$ are nonempty. The diameter of these sets in the unstable directions is exponentially decreasing to 0. It follows that $\bigcap_{i=0}^{\infty} f^{-i}(R_{a_i})$ is a single local stable manifold. Reversing time, $\bigcap_{i=-\infty}^{0} f^{-i}(R_{a_i})$ is a single local unstable manifold extending across R_{a_0}. Canonical coordinates then imply that $\bigcap_{i=-\infty}^{\infty} f^{-i}(R_{a_i})$ is a single point $x \in R_{a_0}$. Look at the orbit of x. Since $x \in f^{-i}(R_{a_i})$, we conclude that $f^i(x) \in R_{a_i}$. Thus x is the point we seek. If there were another point y with $f^i(y) \in R_{a_i}$, then

$$y \in \bigcap_{i=-\infty}^{\infty} f^{-i}(R_{a_i})$$

also. We know that this last set contains a single point, so there is only one point with the desired properties.

The point of this discussion is that symbolic sequences $\{a_i\}_{i=-\infty}^{\infty}$ with $a_i \in \{1, \ldots, k\}$ and $T_{a_i a_{i+1}} = 1$ for all i give us a good description of the trajectories of f inside Λ. This description can be formalized by giving a name to the set of sequences of the sort considered here. If $T = (T_{ij})$ is a $k \times k$ matrix of 0's and 1's, denote by Σ the set of sequences $\{a_i\}_{i=-\infty}^{\infty}$ with $a_i \in \{1, \ldots, k\}$ and $t_{a_i a_{i+1}} = 1$ for all i. The map σ: $\Sigma \to \Sigma$ is defined by $\sigma(\{a_i\}) = \{b_i\}$ where $b_i = a_{i+1}$. The set Σ together with the map σ is called a *subshift of finite type*. The matrix T is called the *transition matrix* of the subshift.

The shift map σ is closely related to the dynamics of f on Λ. If $\{a_i\} \in \Sigma$ is a sequence associated to x, then $\sigma(\{a_i\}) \in \Sigma$ is a sequence associated to $f(x) \in \Lambda$. This is easily seen. If

$$x = \bigcap_{i=-\infty}^{\infty} f^{-i}(R_{a_i}) = \bigcap_{i=-\infty}^{\infty} f^{-(i+1)} R_{a_{i+1}},$$

then

$$f(x) = \bigcap_{i=-\infty}^{\infty} f^{-i}(R_{a_{i+1}}) = \bigcap_{i=-\infty}^{\infty} f^{-i}(R_{b_i})$$

where $b_i = a_{i+1}$. Thus σ acts on the "symbols" in Σ in the same way that f acts on point in Λ. The symbols in Σ give us a lot of information about the asymptotic behavior of trajectories in Λ. They tell us in which sequences of rectangles of the Markov partition for Λ we will be able to find trajectories of f. Moreover, there is at most one trajectory in any sequence of rectangles.

We have not yet specified how many sequences can be assigned to a specific point of Λ. In the horseshoe, the correspondence between symbols and points of Λ is one-to-one. This continues to be true for 0-dimensional basic sets, but it will not always be true. Indeed, it will never be true for "strange" attractors (that is, for attractors larger than a single periodic orbit). However, it is true generally that the correspondence is finite-to-one. There are only a finite number of sequences assigned to any point. In addition, "most" points are assigned to only one sequence. The term "most" can have different meanings in this context. We give it a topological meaning here.

The set Σ can be given a complete metric space structure with the following definition of a distance function $d: \Sigma \times \Sigma \to \mathbf{R}$:

$$d(\{a_i\}, \{b_i\}) = \sum_{i=-\infty}^{\infty} \frac{\delta_i}{2^{|i|}} \quad \text{with } \delta_i = \begin{cases} 0 & \text{if } a_i = b_i, \\ 1 & \text{if } a_i \neq b_i. \end{cases}$$

The reader can easily verify that this does indeed define a distance function. Two sequences are close to one another if they agree in a long "central block" of terms. Using this metric space structure, we now summarize the discussion of this section the following theorem:

THEOREM [8]. *Let Λ be a hyperbolic basic set for an invertible map f. Let $\mathfrak{R}$ be a Markov partition with transition matrix T. If Σ is the subshift of finite type with transition matrix T, then there is a continuous map φ: $\Sigma \to \Lambda$ such that*

(1) $\varphi\sigma = f\varphi$,

(2) *φ is finite-to-one, and there is an upper bound on the number of sequences in $\varphi^{-1}(x)$, $x \in \Lambda$,*

(3) *the set of sequences $\{a_i\}$ in Σ for which $\varphi^{-1}(\varphi(\{a_i\}))$ consists of just one sequence is a countable intersection of open, dense sets.*

The first condition of the theorem is stated graphically by saying that the following diagram commutes.

$$\begin{array}{ccc} \Sigma & \xrightarrow{\sigma} & \Sigma \\ \varphi\downarrow & & \downarrow\varphi \\ \Lambda & \xrightarrow{f} & \Lambda \end{array}$$

This means nothing more than $f\varphi = \varphi\sigma$. Condition (1) of the theorem implies that $\varphi\sigma^n = f^n\varphi$. As a result, φ maps σ-trajectories of Σ to f-trajectories in Λ. In this sense, σ acting on Σ is an abstract description of the dynamics of f on Λ.

The redundancy in describing points by their symbol sequences is analogous to the redundancy in describing numbers by infinite decimal expansions. The number 1 has two decimal expansions: $1.000 \cdots$ and $0.999 \cdots$. Similarly points which lie on the boundaries of more than one rectangle may have more than one symbol. If so, the entire trajectories of these points will have more than one symbol. These are the only points for which this sort of redundancy can occur.

Just as in the horseshoe, the symbolic description of a basic set Λ can tell us a great deal about its internal dynamics. For example, if a subshift corresponding to Λ has an infinite number of periodic orbits, then f has an infinite number of periodic orbits in Λ. If the subshift has a dense trajectory, then f has a trajectory which is dense in Λ. Symbolic dynamics has proved itself a most useful technique for studying a variety of properties of basic sets. As an example of its power, we state the following theorem which can be proved using symbolic dynamics.

THEOREM. *Let Λ be a hyperbolic basic set for an invertible map f. Denote the number of fixed points of f^i in Λ by N_i. Then the power series* $\exp(\Sigma_{i=1}^{\infty} (N_i/i)S^i)$ *is the quotient of two polynomials in the variable s.*

The expression in this theorem is called the zeta function of f (restricted to Λ). It summarizes all of the information about the number of periodic orbits of f in Λ in one formula. The impact of the theorem is that the sequence $\{N_i\}_{i=1}^{\infty}$ actually contains only a finite amount of information about the dynamics of f in Λ. After a certain time, the N_i can be calculated from the preceding N_i. For a 0-dimensional basic set, the zeta function is the reciprocal of a polynomial. Indeed, if Σ is a subshift of finite type with transition matrix T describing the 0-dimensional basic set Λ, then N_i is the trace of ith power of T.

The most striking applications of symbolic dynamics have been to the "statistical mechanics" of Axiom A dynamical systems. This is the topic of the next section. However, before proceeding we want to describe the analogue of subshifts of finite type for continuous time flows.

There is a general construction, called *suspension*, which allows one to construct a continuous time flow from a discrete time flow so that the dynamic properties of the two are closely related. The drawback of this construction is that the state spaces upon which the flows are defined are different. The state space constructed for the continuous time flow often has a complicated topology.

Here is the construction. Let $f\colon M \to M$ be an invertible map. On $M \times \mathbf{R}$, consider the flow Ψ which moves points uniformly in the $\mathbf{R}$ direction with velocity 1. If (x, s) are coordinates for $M \times \mathbf{R}$, then $\Psi(x, s; t) = (x, s + t)$. To change Ψ into a flow with dynamical properties related to f, we form a new space from $M \times \mathbf{R}$. This new space N is obtained by identifying the points $(x, s + 1)$ and $(f(x), s)$ in $M \times \mathbf{R}$. The result is a new space which looks much like a twisted cylinder. For example, if M is the unit interval I and $f\colon I \to I$ is defined by $f(x) = 1 - x$, then the new space N is a Möbius band. Now Ψ induces a flow Φ

on N. The flow Φ does the same thing Ψ does in terms of coordinates of N which are the projections of the coordinates we have used on M.

How are Φ and f related? Notice that if we consider the set $M \times \{0\} \subset M \times \mathbf{R}$, then after 1 unit of time, Ψ maps $M \times \{0\}$ to $M \times \{1\}$. But $M \times \{0\}$ and $M \times \{1\}$ give the same subset of N. Thus the flow Φ on N has a cross-section which looks just like M. What is the Poincaré map of this cross-section? If we follow the trajectory through $(x, 0)$ for unit of time, then we arrive at $(x, 1)$. But $(x, 1)$ and $(f(x), 0)$ are the same point in N. Thus the return map is just f. So we have constructed from $f: M \to M$ a continuous time flow on a space of one higher dimension which has f as a Poincaré return map. Another way of thinking of this construction is to form the space $M \times I$ and then use the map f to "glue" the two ends $M \times \{0\}$ and $M \times \{1\}$ together. The flow Φ then follows the direction along I (see Figure 12).

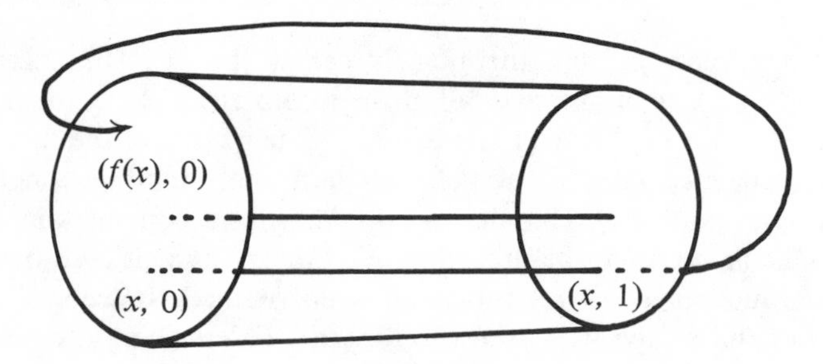

FIGURE 12.

Our immediate goal is the description of the continuous time analogue of a subshift of finite type. A first guess for what the analogue should be is the suspension of a subshift of finite type. The difficulty with this is that the suspension has a "global" cross-section which is mapped into itself by the flow after some *specific* time which is the same for all points of the cross-section. There are hyperbolic basic sets for continuous time flows for which such a cross-section could not possibly exist. Thus we need something more general than the suspension of a subshift of finite type for our analogue.

The extension which we need is obtained by allowing the time of return to vary in the suspension construction. A *special flow* $\Lambda(T, \psi)$ is defined as follows: let Σ be a subshift of finite type with transition matrix T, and let $\psi: \Sigma \to \mathbf{R}$ be a continuous, positive function. Then the special flow is defined on the space $\Sigma \times \mathbf{R}$ modulo the identifications of the points $(x, \psi(x))$ and $(\sigma(x), 0)$. The special flow Γ is the one induced on this space by flowing in the direction of $\mathbf{R}$ with unit velocity. With this definition, the analogue for continuous flows of a Markov partition is provided in the following theorem.

THEOREM. *Let Λ be a hyperbolic basic set for a flow Φ. Then there is a subshift of finite type with transition matrix T and a continuous positive function ψ: $\Sigma \to \mathbf{R}$, and a surjective map ρ: $\Lambda(T, \psi) \to \Lambda$ such that $\rho\Gamma(x, t) = \Phi(\rho(x), t)$. Here Γ is the special flow on $\Lambda(T, \psi)$ constructed from Σ and ψ.*

We remark that Bowen proves a theorem stronger than the one we have stated. He proves that the map ρ and the function ψ have additional properties.

The study of special flows and subshifts of finite type is one of the principal tools for understanding the "orbit structure" of dynamical systems. Special flows and subshifts of finite type display a great deal of the dynamical information about a flow in a particularly accessible fashion. This was already seen in the horseshoe example. In passing from a basic set to a special flow, much of the geometry of the flow has been removed. For many purposes, this is not a serious loss. The loss is felt most severely when dealing with topological questions related to a flow. This is not our principal concern, so we move on to the next topic.

X. Statistical mechanics. Trajectories lying inside a basic set can look very complicated. They may have some features which appear regular and some features which appear random. When confronted with specific problems involving the future or asymptotic behavior of trajectories, it may be unreasonable to give a precise quantitative solution. The best that can be hoped for in some instances is statistical information about the probability of various sorts of behavior occurring. A familiar physical situation which presents the need for such treatment is the statistical mechanical description of a gas. The difficulty encountered here is one of balancing the features of regularity in the trajectories with those that vary in an irregular way. We are interested in studying the dynamics of a hyperbolic basic set which lies somewhere in a large phase space. The problem is to deal with the restriction to trajectories that lie inside the basic set, while at the same time taking account of the complicated internal dynamics of the basic set.

It is already a difficult problem to deal with the internal dynamics of basic sets. We shall deal with this aspect of the problem first. It is best treated by substantial use of the Markov partitions introduced in the previous sections. From a certain point of view, the flow within a hyperbolic basic set is as random as it could be. Consider the horseshoe example. If we divide the square into right and left halves, then specifying in which half one starts gives no information at all about which half of the square one will be in after one unit of time. There is no more correlation between these two pieces of information than there is between the outcome of two consecutive tosses of a coin. A narrower specification of initial data gives some information about the behavior of a trajectory for a certain amount of time, but eventually all of the information one had about initial conditions is useless for predicting where one is going in the horseshoe at specific times. The difficulty is a serious one because the precision of initial data required to predict the future increases exponentially with the length of the desired prediction. Contrast this behavior with the asymptotic approach to an equilibrium is which one's future behavior is channeled independently of moderate changes in initial data.

The situation we have described for the horseshoe is typical of that for a full shift; i.e., a subshift of finite type whose transition matrix consists of all 1's. In dealing with subshifts of finite type, one has a slightly more complicated situation much like a Markov chain in probability theory. If Σ is the subshift of finite type with $k \times k$ transition matrix T, the k symbols $\{1, \ldots, k\}$ can be thought of as a finite set of states. The matrix T tells us which states are allowed to follow which other states. Determining the states accessible from a given state at a given time can be done without a knowledge of past states. This is the characteristic property of a Markov chain in probability theory. There are now some restrictions on the possible sequences of states of trajectories, but a great deal of the randomness remains.

We want to apply probability theory here to derive statistical information about the flow within a basic set. Doing so will allow us to go beyond these intuitive statements about randomness to quantitative statements which can be analyzed for specific systems (at least in principle). The framework for probability theory requires that one specify a set of possible outcomes to an experiment and then assign numbers to subsets of these possible outcomes.

The numbers determine the probability that the real outcome will lie in a given subset. Measure theory is the mathematical formalism within which this framework is developed. Before proceeding, we pause to review a few concepts of measure theory for subsequent use.

A *measure space* consists of a set M, a collection $\mathfrak{B}$ of subsets of M, and a function $\mu\colon \mathfrak{B} \to \mathbf{R}$. The collection $\mathfrak{B}$ is required to be closed under the operations of forming countable unions and intersections and of taking complements. Furthermore it is assumed $M \in \mathfrak{B}$. The function μ is required to be a nonnegative function which is additive in the sense that if $\{B_i\} \subset \mathfrak{B}$ is a countable collection of disjoint sets in B, then $\Sigma_i \mu(B_i) = \mu(\bigcup_i B_i)$. In applications to probability theory, it is usual that $\mu(M) = 1$. One then says that μ is a probability measure and interprets $\mu(B)$ as the probability that an experimental result will lie in B. If M is a metric space, one customarily requires that open and closed sets lie in the collection $\mathfrak{B}$.

In the situation we are considering, the set M of the measure space will be a basic set, a subshift of finite type, or a special flow. The collection $\mathfrak{B}$ will seldom be specified but is assumed to contain the open and closed sets of M. There are various measures μ on these underlying spaces which will be considered. One measure to work with on a basic set in Euclidean space is the ordinary Lebesgue measure which assigns to a set its volume. Often times, this is not suited to our purposes for a couple of reasons. The most serious of these is that basic sets often have Lebesgue measure zero, so that Lebesgue measure cannot capture any of the internal structure of the basic set. The second drawback is that Lebesgue measure need not be nicely related to the dynamics of a flow. Thus, it is important to consider other measures on basic sets as well as measures on subshifts of finite type and special flows.

We illustrate these ideas by describing a class of measures on subshifts of finite type. Let Σ be the subshift of finite type with $k \times k$ transition matrix $T = (T_{ij})$. Each open set in Σ can be written as a countable disjoint union of sets of the form

$$U = \{\{a_i\} | a_r = b_r, \ldots, a_s = b_s\}.$$

Here $r < s$ are integers, $b_i \in \{1, \ldots, k\}$ and $A_{b_i b_{i+1}} = 1$ for $i = r, \ldots, s - 1$. Such sets U are called *cylinder* sets. Choose nonnegative numbers m_{ij}, $1 \leqslant i \leqslant k$, $1 \leqslant j \leqslant k$, such that $\Sigma_{j=i}^{k} m_{ij} = 1$ for each i and $T_{ij} = 0$ implies $m_{ij} = 0$. We assign measure $\mu(U_{ij}) = m_{ij}$ to the set

$$U_{ij} = \{\{a_i\} | a_0 = i, a_1 = j\}.$$

If U is the cylinder set $\{\{a_i\} | a_r = b_r, \ldots, a_s = b_s\}$, then U is nonempty precisely when $T_{b_i b_{i+1}} = 1$ for $i = r, \ldots, s - 1$. For such nonempty U, we set $\mu(U) = m_{b_r b_{r+1}} \cdot m_{b_{r+1} b_{r+2}} \cdots m_{b_{s-1} b_s}$. This completely determines a measure μ on the smallest collection $\mathfrak{B}$ of subsets containing all open sets in Σ. The measure μ is *invariant* under σ in the sense that $\mu(B) = \mu(\sigma^{-1}(B))$ for all sets $B \in \mathfrak{B}$. It is determined by being the invariant measure for which $\mu(U_{ij}) = m_{ij}$ and the probabilities of being in U_{ij} and $\sigma^t(U_{rs})$ are independent of one another for all $t \neq 0$. This last statement means that if $U_{ij} \cap \sigma^t(U_{rs}) \neq \varnothing$, then its measure is $\mu(U_{ij}) \cdot \mu(\sigma^t(U_{rs}))$.

The choice of a matrix m_{ij} and the corresponding measure μ on Σ makes Σ into a Markov chain in the sense of probability theory. The numbers m_{ij} represent the transition probabilities of passing from state i to state j in one unit of time. This is not the place to delve deeply into the theory of these Markov chains, but we do make a few remarks. One usually studies subshifts of finite type which are topologically transitive. This is equivalent to the condition that T^t is a matrix of all positive entries for some $t > 0$. For a transitive subshift, there is a real valued function, called entropy and denoted $h(\mu)$, defined on the set of measures. The entropy measures the ergodicity and mixing properties of the shift map on Σ with respect to the measure μ. A measure with positive entropy is one for which sets of positive measure spread themselves rather uniformly across the shift space with increasing time. Under mild hypotheses on the measures involved, the entropy of transitive subshifts is positive.

In statistical mechanics, one is often interested in measures for a system which have maximal entropy. In dealing with the dynamics of Axiom A dynamical systems, there are various measures which are singled out because of their entropy properties. We shall describe the two of these which have been studied in the most depth. Much of the techniques used in studying these measures is carried over directly from statistical mechanics via Markov partitions and subshifts of finite type.

The first measure we look at is the one with maximal entropy. For a subshift of finite type, Parry proved that there is a unique probability measure with maximal entropy. This measure is one of the class described above. The matrix (m_{ij}) determining the measure can be described directly in terms of the transition matrix of the subshift. Bowen proved that this measure has a very interesting property: It can be described directly in terms of the periodic orbits of the shift. Define a measure μ on open sets U by the formula

$$\mu(U) = \lim_{t\to\infty} \frac{(\text{number of fixed points of } \sigma^t \text{ in } U)}{(\text{number of fixed points of } \sigma^t \text{ in } \Sigma)}.$$

This limit exists for all open sets U, and the function μ extends to a probability measure on a collection of subsets of Σ. The measure defined in this way is the unique probability measure with maximal entropy on Σ. A similar statement is true for hyperbolic basic sets.

THEOREM [**1**]. *Let Λ be a hyperbolic basic set for an invertible map f. Define the measure μ on open sets of Λ by*

$$\mu(U) = \lim_{t\to\infty} \frac{(\text{number of fixed points of } f^t \text{ in } U)}{(\text{number of fixed points of } f^t \text{ in } \Lambda)}.$$

Then μ extends to a probability measure on Λ which is the unique f-invariant probability measure on Λ with maximal entropy. Furthermore, μ is positive on open sets.

We remark that there is a definition of topological entropy which does not involve measures. The topological entropy of a hyperbolic basic set is always the entropy of the measure μ defined in this theorem. There is an analogous theorem to the one above for continuous time flows.

The second measure on a basic set which we examine is more subtle in its definition, and it has even more striking properties. This second measure is the one which allows us to genuinely "do" statistical mechanics on a basic set. The measure we construct will relate the dynamics of an attractor to the metric properties of the space in which it is embedded.

This measure is first obtained by generalizing the theorem stated above. If Λ is a hyperbolic basic set, μ a measure on Λ, and $\varphi\colon \Lambda \to \mathbf{R}$, then we can form the expression $h(\mu) + \int \varphi \, d\mu$ (denoted by $P_\varphi(\mu)$) on Λ. Here h is the entropy of μ and $\int \varphi \, d\mu$ is the integral of φ with respect to the measure μ. The integral is defined in much the same way as the Lebesgue integral in Euclidean space. The functions φ which we consider will be assumed to satisfy a condition intermediate between continuity and differentiability. The function $\varphi\colon \Lambda \to \mathbf{R}$ is said to be (uniformly) Hölder continuous if there are constants c, $\alpha > 0$ such that $|\varphi(x) - \varphi(y)| < c(d(x, y))^\alpha$ for all $x,y \in \Lambda$. A differentiable function in Euclidean space is Hölder continuous with $\alpha = 1$. We will apply the following theorem to a Hölder continuous function.

THEOREM. *Let Λ be a hyperbolic basic set for a discrete or continuous flow. If $\varphi\colon \Lambda \to \mathbf{R}$ is a Hölder continuous function on Λ, then there is a unique invariant probability measure $\mu(\varphi)$ on Λ which maximizes the function $P\varphi(\mu) = h(\mu) + \int \varphi \, d\mu$. The measure $\mu(\varphi)$ is positive on open sets.*

The measure $\mu(\varphi)$ is also ergodic. This means that there are no invariant, measurable sets $B \subset \Lambda$ for which $0 < \mu(B) < 1$. The previous theorem is obtained by setting φ to be the function which is identically zero.

We want to apply this theorem to a specific function φ defined on an *attractor* Λ. Recall that an attractor is a basic set which has small neighborhoods that are mapped into themselves by the flow. The function which is used in the application measures the expansion of the flow in the unstable directions.

If f is the time t map of a flow Φ with an attractor Λ, define $\lambda_t(x)$ to be the Jacobian determinant of the linear map $Df\colon E_x^u \to E_{f(x)}^u$. Here E_y^u denotes the tangent space to the unstable manifold at a point $y \in \Lambda$. Then we have the result.

THEOREM [9]. *Let Λ be a hyperbolic attractor for a flow Φ. Let U be a neighborhood of Λ such that $\Phi(U, t) \subset U$ for all $t > 0$ and such that $\Lambda = \bigcap_{t \geqslant 0} \Phi(U, t)$. Let φ be the function constructed above, and let μ be the measure which maximizes $P\varphi(\mu)$. Then, for almost all points $x \in U$ with respect to Lebesgue measure,*

$$\lim_{T\to\infty} \frac{1}{T} \int_0^T g(\Phi(x, t)) \, dt = \int g \, d\mu$$

holds for all continuous functions g. (*For discrete time flows, the integral on the left is replaced by a sum*.)

This theorem is the basic result which gives a statistical mechanics for attractors. It implies that the asymptotic behavior of almost all trajectories approaching the attractor is statistically the same. Most trajectories wind around the attractor in such a way that the portion of their time spent near a given subset of the attractor is determined by the measure μ. While it is impossible to predict where a trajectory will be at some time far in the future, one can assign probabilities to its being near different subsets of the attractor. This is the statistical sense in which it is possible to describe the asymptotic behavior of trajectories. Note the strong ergodicity of the attractor in that almost all (in the sense of Lebesgue measure) trajectories asymptotically approach a trajectory which is dense in the attractor.

There is one additional theorem of a statistical nature which is important for Axiom A flows.

THEOREM [9]. *Let Φ be a flow which satisfies Axiom A. Then almost all trajectories* (*in the sense of Lebesgue measure*) *tend to an attractor. Moreover, unless an attractor Λ is an open set of the state space, Λ has Lebesgue measure zero.*

In other words, those points which are not in the stable manifolds of attractors ("basins of attractions") form a set of Lebesgue measure zero. Thus, from a statistical point of view, one may assume that there is probability one of any specific trajectory tending to an attractor.

These two theorems present us with a picture which illustrates the asymptotic statistical properties of Axiom A flows. Asymptotically, almost all trajectories end at attractors. Moreover, they do so in an ergodic way which is described statistically by certain measures on the attractors. It is on the attractors that the ergodic, apparently random, behavior occurs. In a sense one cannot detect the presence of "noise" inside one of these attractors. The flow inside an attractor already appears so chaotic that the addition of small amounts of genuine randomness to the system of differential equations makes no apparent difference in the observed trajectories. In principle, one should be able to compute numerically the location of the attractors and their invariant measures. In practice, doing this with much precision is probably a time consuming and difficult task which is seldom justified. Yet the information contained in the last two theorems is useful as a guide to one's intuition in evaluating numerical studies of difference or differential equations with strange attractors.

We close this section with just a brief remark about the proofs of these theorems. An essential aspect of their proofs is the symbolic dynamics introduced in the previous section. Subshifts of finite type and special flows are objects which are abstract on the one hand but explicitly described in a useful way on the other hand. It is often much easier to use Markov partitions and the corresponding subshift or special flow to analyze a basic set than it is to work with the basic set directly. This is illustrated by the problem of finding dense or periodic orbits in a basic set. The map from a subshift or special flow to a basic set obtained by assigning a point to its symbolic representations preserves statistical properties defined in terms of reasonable measures. The techniques of statistical mechanics are themselves analytically sophisticated. They are much more easily grafted onto the quite explicit subshifts and special flows, but it is much more difficult to introduce these techniques into geometrically complicated basic sets. So one makes extensive use of Markov partitions in proving the theorems of this section.

This concludes our main discussion of the statistical mechanics of Axiom A flows. Before proceeding, we reiterate that the internal dynamics of a basic set is as strongly ergodic as it could be. Basic sets have finite partitions so that the passage of a trajectory through the sets in the partition looks statistically like a random variable in a Markov chain.

XI. Structural stability. In this section we begin a completely new topic. The theory of Axiom A dynamical systems developed thus far is independent of what follows, but it was developed with the theory of structural stability in mind. In this section, we shall give a general introduction to the concept of structural stability as motivation for the theory presented here.

When one gives a model for a biological or physical system, one usually selects values for certain parameters describing a class of models. For example, in describing a linear relationship between two variables, one has the choice of the slope of the relationship at one's disposal. Typically, the choice of parameters is based upon "curve fitting" experimental results. One is seldom certain of the specific "correct" values, so one hopes that the predictions which one makes do not depend upon the exact values of the parameters chosen. If this is the case, one says that the model is robust.

Within the context of ordinary differential equations, the idea of robustness was formalized by the Russian mathematicians Andronov and Pontryagin in the middle 1930's. They defined the concept of a "rough" dynamical system. Later the term "structurally stable" came into common usage. The early development of the theory of structural stability dealt with two dimensional dynamical systems. Two dimensional structurally stable systems could be described easily in explicit terms involving singular points, periodic orbits and their stable and unstable manifolds. This work culminated in the work of Peixoto in the late 1950's.

Following Peixoto's work, there were efforts made in the Soviet Union and in the United States to extend the theory beyond two dimensions. It was quickly discovered that the simpler theory of two dimensions becomes much more complicated in higher dimensions. Smale's horseshoe was one of the first examples of the kind of complexity which can occur in three dimensional structurally stable systems. At the same time Anosov was studying systems for which the entire state space was nonwandering and possessed a hyperbolic structure. He proved that these systems are structurally stable. Since then a fairly complete picture of the implications of structural stability for systems of differential equations has emerged. These developments have depended upon an interplay between geometric insight into the properties of flows and techniques of functional analysis which have been necessary to give efficient proofs of many results. There are a number of technical questions which do remain in completing the picture.

A precise definition of structural stability requires that one specify the universe of discourse. Whether or not the predictions of a model are robust depends upon the class of models allowed and the concept of perturbation within this class. In other words, we must specify a set of allowable models together with the meaning of two models being close to one another. The mathematical object which includes just this information is a topological space.

In dealing with structural stability, we will want to specify a topological space of vector fields (or invertible maps). The perturbations of a vector field X will then consist of a neighborhood of X in this topological space. A property of X will be *robust* if the property holds for all the fields in some neighborhood of X. The extent of the robustness depends upon how large a neighborhood consists of vector fields with the specified property. The definition of structural stability requires that we specify (1) a set of vector fields, (2) a topology on the set of vector fields, and (3) the properties which we want to remain the same for perturbations of a vector field. We shall take up each of these three topics in turn.

There are two kinds of issues involved in the choice of a space of vector fields. The first involves the smoothness of the vector fields allowed. This is primarily a technical matter and is discussed below when we consider topologies on the space of vector fields. The second issue involves choosing a space of vector fields so that the defining equations obey certain constraints. This can influence which parameters are considered robust.

A good illustration of how the choice of a space of vector fields is made is provided by models for a single predator population and a single prey population. The Lotka-Volterra model is the classic one. This is a system of two differential equations of the particular form

$$\frac{dP}{dt} = -aP + bPH, \qquad \frac{dH}{dt} = cH - bPH.$$

There are three parameters to be chosen, so one is dealing with a three dimensional space of vector fields. For any positive choices of these parameters, the system of equations will have a family of periodic orbits which fills up the positive quadrant of $\mathbf{R}^2$ except for a single singular point. Thus the existence of periodic solutions is robust for the Lotka-Volterra models. If, however, one expands the class of models by allowing the coefficients of PH and $-PH$ in the two equations to differ, then one finds that the periodic solutions exist only when the two coefficients are equal. Thus, in the extended class, the prediction of periodic solutions is no longer robust. One can go still further and expand the class of allowable systems to include *all* differential equations in the plane. In this biological setting one might require of the systems that $dP/dt = 0$ when $P = 0$ and $dH/dt = 0$ when $H = 0$, reflecting the fact that neither predator nor prey generates spontaneously. With either of these two choices we need to consider the matter of topologies. As long as the set of models depends on only a finite number of parameters, this is no difficulty. However, when we consider all differential equations, the matter becomes more delicate as we see below.

One might ask whether there is any advantage to considering robustness within the class of restricted models. The answer is a definite yes. There are many situations which arise in which the natural set of vector fields giving models for a particular situation is restricted in such a way that the concept of robustness changes. The most striking example of this is the theory of classical Newtonian mechanics. The equations of motion for a conservative mechanical system can always be put in Hamiltonian form

$$\frac{dq_i}{dt} = \frac{\partial H}{\partial p_i}, \qquad \frac{dp_i}{dt} = -\frac{\partial H}{\partial q_i}$$

Here $(q_1, \ldots, q_n, p_1, \ldots, p_n)$ are the "generalized" position and momentum variables of the system (having n degrees of freedom) and $H: \mathbf{R}^{2n} \to \mathbf{R}$ is the Hamiltonian function of the system. There are many dynamical properties of Hamiltonian vector fields on $\mathbf{R}^{2n}$ which are not robust within the class of all vector fields on $\mathbf{R}^{2n}$. This gives the study of Hamiltonian vector fields quite a different flavor from the theory of structural stability considered here [**33**].

A general philosophy regarding when one should deal with restricted classes of vector fields is that the class of vector fields used in a problem should be determined by the natural constraints which operate. Thus, a natural property of models for closed ecological systems is that if a species is extinct, then it remains extinct. This imposes restrictions on the equations which are allowed. Another example from population dynamics is the "Leslie model" for an age-structured population. If one divides a population into age classes with the same age span, and considers a discrete model with this span as the time unit, then the resulting equations have a particular form. Each individual has only two possibilities each unit of time: he can advance one age class or die. This places severe restrictions on the form of the equations which determine such a model.

Sometimes the restrictions on the class of model do change those properties which are robust and sometimes they do not. It is a matter which must be dealt with in each particular situation. Indeed, this is a drawback in trying to base arguments about the way real systems work upon the structural stability or robustness of models. If the real system does not behave in a structurally stable way, it may be that there are constraints which prevent it from doing so. This is what happens in mechanics. There is often no way of determining *a priori* whether a constraint is operating which limits the qualitative behavior of models which should be used. The theory which we discuss in this paper deals for the most part with the class of all differential equations and places no restrictions (other than differentiability) on the equations which are allowed.

The second issue to be dealt with in defining structural stability deals with the topology to be imposed on a set of vector fields. If one deals with a set of vector fields which can be described by a finite set of parameters, then there is usually little problem in choosing the topology. The finite dimensional space of parameters has a natural topology as a Euclidean space: Two vector fields are nearby if the corresponding parameter values for the two vector fields are close to one another. In the set of all vector fields, an infinite number of parameters is required to determine a particular vector field. This produces a situation in which there are different choices for a topology which may yield different results.

Let us illustrate such a situation with an example. Consider the property which is the number of zeros of a function $f\colon \mathbf{R} \to \mathbf{R}$. We want to know when "nearby" functions have the same number of zeros. If we study this problem in, say, the space of polynomials of degree $\leqslant n$, then there is no difficulty in specifying when two polynomials are close: they are close when their corresponding coefficients are close. However, if we consider this problem in the space of all analytic or infinitely differentiable functions, there are many possible choices for the concept of

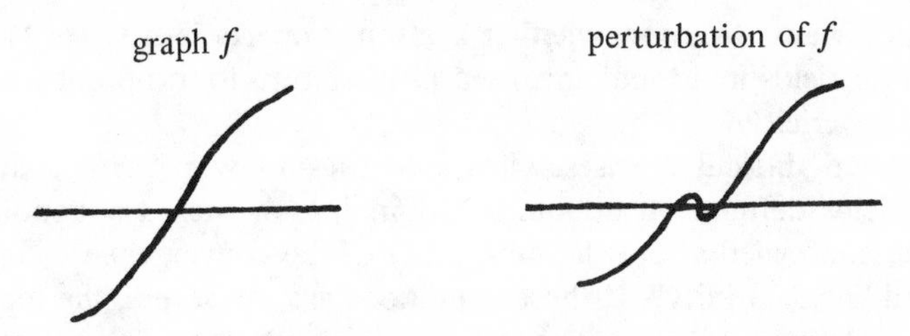

FIGURE 13.

closeness for two functions. One definition of closeness is that two functions f and g are close if $|f(x) - g(x)|$ is small for all x. With this choice, any function f which has a zero is not robust with respect to the property being considered. Indeed, "small" changes in f near a zero can produce a function which has additional zeros. (See Figure 13.) On the other hand, we might define two functions to be close if both $|f(x) - g(x)|$ and $|f'(x) - g'(x)|$ are small for all x. This new definition gives f much smaller neighborhoods in the function space. Now, if f has the property that the equations $f(x) = 0$ and $f'(x) = 0$ have no simultaneous solutions, then f is robust with respect to the property considered. Nearby functions will have the same number of zeros as f. Introducing "wiggles" as we did with a perturbation with respect to the old definition is no longer allowed because large changes in the derivative are necessary to produce a wiggle. One can go on in this way. If the property being considered is the number of points of inflection of a function, then there will not be robust functions with respect to this property unless the topology defining "closeness" involves the third and fourth derivatives of the functions. Hopefully, we have made our point: the concept of robustness does depend in substantive ways on how one chooses a topology for a function space.

There is another problem involved in the choice of a topology for a space of vector fields. The problem concerns the "completeness" properties of the topological space. Consider the space $\chi^\infty(M)$ of infinitely differentiable vector fields, defined on some bounded domain Ω in $\mathbf{R}^n$. On this space one can define a distance function, called the C^1 metric, by

$$d(X, Y) = \sup_{x \in \Omega} (\|X(x) - Y(x)\|, \|DX(x) - DY(x)\|).$$

Here DX and DY are $n \times n$ matrix-valued functions on Ω and we use some norm on the space of $n \times n$ matrices. It is often awkward to use the topology defined by this metric on $\chi^\infty(M)$. The difficulty is that there are Cauchy sequences with respect to this metric in $\chi^\infty(M)$ which do not converge to vector fields, in $\chi^\infty(M)$. The limits of such sequences need only be once differentiable. This can cause difficulty, particularly when one wants to know whether a given property is true for "almost all" vector fields in a topological sense. We return to this point when we discuss genericity.

A related difficulty arises when one tries to work with spaces of vector fields defined on unbounded domains. In such a situation, one must be concerned about the uniformity of the convergence conditions imposed "near infinity". If these conditions are too severe, the topology which is defined will be awkward to work with for technical reasons involving point-set topology. On the other hand, if severe conditions are not imposed, then essentially nothing will be robust. The neighborhoods of each vector field will all be too large. There is no easy way out of these dilemmas. The theory of dynamical systems has dealt almost exclusively with vector fields defined on compact manifolds to avoid this last problem. The problem of differentiability properties is usually dealt with by assuming that one deals with the space of vector fields with the weakest differentiability properties for which the theory works. Virtually all of the theory described in this paper prior to §XIV requires that vector fields be at most twice differentiable. In the space $\chi^r(M)$ of r times differentiable vector fields on a closed bounded domain M for $1 \leqslant r \leqslant \infty$, one usually uses the topology given by the C^r metric:

$$d(X, Y) = \sup_{x \in M, 0 \leqslant i \leqslant r} (\|D^i X(x) - D^i Y(x)\|).$$

In χ^s, if $r < s$, then neighborhoods with respect to the C^r topology are C^s neighborhoods but not vice versa. Thus, it is a stronger statement to say that a property holds in a C^r neighborhood than it is to say that the property holds in C^s neighborhood. Often results in dynamical systems prove that a property holds in a C^1 neighborhood but requires that the vector fields involved be C^2. This is all a complicated and somewhat confused situation. The qualitative differences between C^r and C^s vector fields are not well understood. In practical terms, these issues are usually irrelevant. One ordinarily works with analytic (or at least C^∞) vector fields and has no qualms about making strong smoothness assumptions. What is important are the ideas (1) that the topology chosen for a space of vector fields does matter in discussions of robustness and (2) there is a subtle interaction between the space of vector fields selected, the topology used, and those properties which are "generic".

After this lengthy discussion of abstract topology, we can return now to the matter of structural stability. The idea behind the definition of structural stability is that the qualitative dynamics of a structurally stable vector field should be robust. There are a number of pitfalls of topologies and differentiability to be avoided in the definition, and we point these out as the discussion proceeds. In order to define structural stability, we need to specify now those properties which are to be robust. Formally, this is done by specifying an equivalence relation between vector fields such that equivalent vector fields have the same properties. Unlike the examples we considered above, one cannot give a lot of the properties to be evaluated. Rather, it is possible to compare two vector fields and determine whether or not they are the same. Robustness of a vector field then corresponds to its equivalence class containing a neighborhood of the vector field.

The strongest equivalence relation embodying the idea that equivalent vector fields have the same qualitative dynamics is the following: two vector fields X, Y defined on M are differentiably equivalent if there is a smooth change of coordinates M whose derivative maps X to Y. In other words, $(Dh)Xh^{-1}(x) = Y((x))$. Unfortunately, there are seldom (if ever) vector fields which are robust with respect to these properties. There are at least two obstructions to vector fields being stable with respect to these properties. The first involves singular points. If X has a singular point at p, then the derivative of X at p in a new coordinate system is replaced by a matrix which is similar to $DX(p)$. If the map h describes the change of coordinates, the new derivative is $(Dh)(DX)(Dh)^{-1}$. Linear algebra tells us that the eigenvalues of a square matrix do not change when it is replaced by a similar matrix. Consequently, if one perturbs X so that the eigenvalues of DX at a singular point change, then the perturbed vector field will not be differentiably equivalent to X.

The second obstruction to robustness for this equivalence relation involves periodic orbits. The existence and uniqueness theorem for ordinary differential equations implies that a coordinate change of the sort considered above will map trajectories of the vector field X to trajectories of the vector field Y in a way which preserves the parametrizations of the vector fields. Thus if one arrives at the point y by following the X trajectory starting at x for time t, then one arrives at the point $h(y)$ by following the Y trajectory starting at $h(x)$ for time t. When $x = y$ and the two trajectories are periodic, they must have the same period. Consequently, perturbations of X which change the lengths of the periodic trajectories will not be equivalent to X.

These two obstructions force upon as a weaker concept of equivalence which represents the similarity of the dynamics of vector fields. First of all, one has to drop the condition that the coordinate change between the vector fields be differentiable. If $h\colon M \to M$ is only a continuous coordinate change, then it does not have a derivative and hence does not induce a map of vector fields. But it still makes sense to speak of the action of h on trajectories. To deal with the second difficulty, we cannot require that the parametrizations of trajectories be preserved. Thus, we make the following definitions. Two vector fields

X, Y defined on M are *topologically equivalent* if there is a continuous change of coordinates $h: M \to M$ such that h maps trajectories of X to trajectories of Y, preserving the orientation coming from time.

A vector field X is *structurally stable* if it has a neighborhood consisting of vector fields which are topologically equivalent to X. Customarily, the neighborhood is required to be a neighborhood in the C^1 topology. This is the basic definition around which a theory is built. A structurally stable vector field is one whose "phase portrait" does not change from a continuous point of view under smooth perturbations. Note the mixture of topologies and differentiability, nondifferentiability involved in the definition. This is an aspect of the subject which makes the analytic study of structural stability quite hard.

Before developing the theory, we add a bit more history. There was hope for a few years that "almost all" vector fields would be structurally stable. This is indeed the case for two dimensional systems. After Smale discovered a counterexample, a vector field which had a neighborhood of structurally unstable vector fields, he proposed a weaker property and conjectured that almost all vector fields might be robust for this weaker property. The property he proposed, called Ω-stability, was based upon topological equivalence restricted to the nonwandering sets of the corresponding vector fields. This conjecture lasted but a few brief years before counterexamples were discovered. A number of other equivalence relations were proposed for the search for a property or equivalence relation embodying most of the qualitative dynamics of a vector field on the one hand which would be "generic" on the other. Abraham has called this the Yin-Yang problem in dynamical systems. There has not been a satisfactory solution, and it is doubtful whether one exists. It seems that we must be content with describing large classes of vector fields which are reasonably well behaved on the one hand, and describing relatively weak general properties which are generic on the other. We return to discuss genericity at some length later.

One of the central problems of dynamical systems has been to characterize structurally stable dynamical systems in a useful way. It is a problem which has not been completely solved, but there is a rather satisfying partial solution at this time. On the one hand, there is a set of conditions, Axiom A and Strong Transversality, which imply that a vector field is structurally stable. This is the topic of the next section. At the same time, there are strengthened concepts of structural stability due to Franks which imply Axiom A and the Strong Transversality Hypothesis. To round out this picture, Franks proves that Axiom A and the Strong Transversality Hypothesis do imply the strengthened concepts of structural stability. It is unresolved as to whether there are structurally stable vector fields which are not structurally stable with respect to the strengthened concepts.

XII. Structural stability theorems. In this section we consider theorems whose conclusion is that certain vector fields are structurally stable. To begin the discussion we look at vector fields on two dimensional manifolds. There the situation is simpler and structurally stable vector fields are readily characterized. After reviewing this theory, we see how the various conditions necessary for structural stability in two dimensional systems can be generalized to higher dimensions.

Let X be a vector field defined in a region M of the plane. Under what circumstances is X structurally stable? We want to determine now conditions which are certainly necessary for the structural stability of X. If Y is a perturbation of X, then a homeomorphism h: $M \to M$ is called a *topological equivalence* or *conjugacy* from X to Y if h maps trajectories of X out to trajectories of Y. If X and Y are topologically equivalent, then the conjugacy h maps equilibria of X to equilibria of Y. It also maps the stable manifolds of the equilibrium points of X to the stable manifolds of the equilibria of Y. This implies that the stable manifolds of X and Y at corresponding equilibria must have the same dimension. If X has an equilibrium point whose derivative has a pure imaginary eigenvalue, then there are perturbations of X whose equilibria have different numbers of eigenvalues with negative real parts. These perturbations will have corresponding stable manifolds of different dimensions, implying that X is not structurally stable. Thus the equilibria of a structurally stable vector field are hyperbolic. A similar argument shows that the periodic orbits of a structurally stable vector field are also hyperbolic. We can summarize these properties in a lemma:

LEMMA. *Let p be an equilibrium point and γ a periodic orbit of a vector field X. If p is hyperbolic, then each perturbation of X has a unique equilibrium point near p. These equilibria are hyperbolic and their stable and unstable manifolds have the same dimension as those of p. If γ is hyperbolic and has period τ, then there is an $\varepsilon > 0$ such that each perturbation of X has a unique periodic orbit near γ with period in the interval $(\tau - \varepsilon, \tau + \varepsilon)$. These periodic orbits are hyperbolic and their stable and unstable manifolds are the same dimension as those of γ.*

Even for two dimensional systems, hyperbolicity of equilibrium points and periodic orbits is not enough to imply structural stability. There are two other conditions which are needed. A *saddle point* is an equilibrium which has one dimensional stable and unstable manifolds. The trajectories in the stable and unstable manifolds of a saddle (other than the saddle) are called *separatrices*. It may happen that two separatrices for one saddle or for two different saddles coincide as a trajectory which is both the stable manifold of a saddle and the unstable manifold of a saddle. Such a trajectory is called a *saddle connection*. Saddle connections are preserved under topological equivalence. If X is a vector field with a saddle connection, then there are perturbations of X which do not have a saddle connection. Thus structurally stable vector fields do not have saddle connections.

The final condition for structural stability of a two dimensional vector field is necessary only when the state space of the vector field is not contained in the plane. It is common, for example, that the state space for a vector field involve two angular variables. Each angular variable comes from a point on the circle, so such a state space is a *torus*, the topological product of two circles. The torus can be thought of as a square with opposite sides identified. In the plane, one has the *Poincaré-Bendixon theorem* which implies that the nonwandering set of a vector field in the plane with hyperbolic equilibria consists of equilibria, periodic orbits, and saddle connections. On the torus this is no longer true. There are vector fields with large nonwandering sets and no equilibria or periodic orbits.

An example of such a vector field is the *irrational flow* of the torus. If (θ, φ) are angular variables, then the equations

$$\dot{\theta} = a, \qquad \dot{\varphi} = b,$$

define a flow of the torus. If a/b is a rational number, then the trajectories are all periodic orbits which wrap around the θ direction a times while they wrap around the φ direction b times in 2π units of time (see Figure 14). On the other hand, if a/b is irrational, then the trajectories are all dense in the torus. They wind around without ever returning to their original initial condition. In this case, each point of the torus is nonwandering.

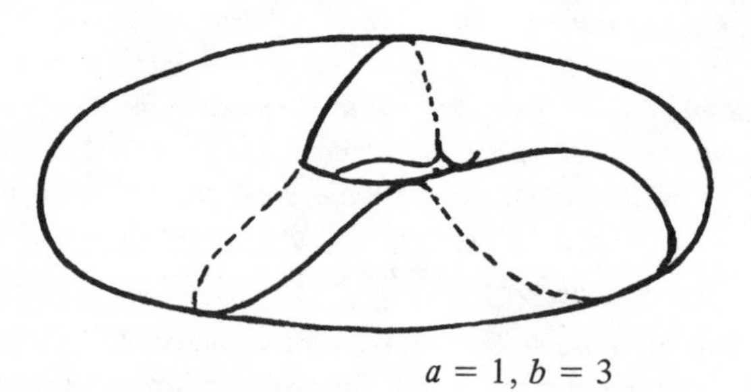

$a = 1, b = 3$

FIGURE 14.

Behavior like the irrational flow on the torus is not robust for two dimensional vector fields. Indeed, one can prove that for structurally stable two dimensional vector fields, the nonwandering set consists of equilibria and periodic orbits. These observations are combined in the next theorem.

THEOREM (PEIXOTO). *Let X be structurally stable two dimensional vector field, then* (1) *the equilibria and periodic orbits of X are hyperbolic,* (2) *X has no saddle connections, and* (3) *the nonwandering set of X consists of the equilibria and periodic orbits.*

Conversely, if X is a vector field defined on a two dimensional compact manifold M which satisfies (1)–(3), *then X is structurally stable. If M is orientable, the structurally stable vector fields form an open and dense set of $\chi(M)$.*

If one imposes suitable conditions at the boundary of a domain, a similar theorem can be stated for vector fields defined on bounded domains in Euclidean space. Rather than pursuing these boundary conditions, we turn to the more central issue of extending this theorem to higher dimensions insofar as possible. Condition (1) of the theorem is satisfied for a vector field satisfying Axiom A. Indeed, Axiom A was developed as a generalization of (1) with structural stability in mind. Generalizing condition (2) to higher dimensions requires that we make a brief digression.

One of the fundamental concepts of differential topology is transversality. Two linear subspaces of Euclidean space intersect *transversally* if together they span the Euclidean space. Two submanifolds N and P of a third manifold M intersect transversally if $x \in N \cap P$ implies that the tangent spaces to N and P at x together span the tangent space to M. Note that one says N and P intersect transversally at a point x for which $x \notin N \cap P$. If N and P are submanifolds which do not intersect transversally, then N can be perturbed so that the intersection of N and P is transversal. One of the reasons transversality is so important is the following theorem:

THEOREM. *Let N and P be submanifolds, n- and p-dimensional, respectively, of the m-dimensional manifold M. If N and P intersect transversally, then $N \cap P$ is an $(n + p) - m$ dimensional submanifold of M.*

Another important result concerning transversality is Sard's theorem. It can be stated in many forms. Roughly, its conclusion is that almost all submanifolds intersect transversally.

Both conditions (1) and (2) of Peixoto's theorem can be stated in terms of transversality. For example, condition (2) can be stated as the requirement that the stable and unstable manifolds of saddle points intersect transversally. For two-dimensional vector fields satisfying Axiom A, saddle connections are the only intersection of stable and unstable manifolds which might not be transversal. So condition (2) is also equivalent to the requirement that the intersections of the stable and unstable manifolds of each trajectory are transversal. (Here, we want to include the whole stable or unstable manifold of a periodic orbit rather than just consider the stable or unstable manifold of one of its points.)

It is in this form that condition (2) generalizes to vector fields of higher dimension. If X is a vector field which satisfies Axiom A, then one says that X satisfies the *Strong Transversality Property* if every intersection of the stable manifold of a trajectory with the unstable manifold of a trajectory is transversal. The basic theorem, due to Robbins for discrete time flow and Robinson for continuous time flow, is the following:

THEOREM. *If X is a vector field defined on a compact manifold which satisfies Axiom A and the Strong Transversality Property, then X is structurally stable* (*with respect to C^1 perturbations*).

This theorem represents one of the most striking achievements of the theory of dynamical systems within the past few years. The proof is quite complicated, but it may be helpful to give a few remarks about how the theorem was proved.

The proof begins with an analysis of the robustness of the equilibria and periodic orbits of a vector field satisfying Axiom A. The equilibria and periodic orbits of an Axiom A vector field X vary smoothly with perturbations. There is a neighborhood U of X in the space of vector fields so that if $Y \in U$, then Y has an equilibrium point corresponding to each equilibrium point of X and a periodic orbit corresponding to each periodic orbit of X. The perturbation Y could have more in its nonwandering set than does X. If it does, one says that X has suffered an Ω-explosion. The Strong Transversality Property rules this out. It implies that one cannot return to a basic set that one has left by following trajectories forward to their ω-limit sets, then following these trajectories, and so on. The basic sets can be ordered in such a way that if the α- and ω-limit sets of a trajectory are Ω_1 and Ω_2, then Ω_1 precedes Ω_2 in the ordering. With Axiom A and such an ordering for X, Smale proved that perturbations of X have nonwandering sets which are topologically equivalent to the nonwandering set of X.

For Axiom A vector fields whose nonwandering set is the entire state space, this result was proved geometrically by Anosov around 1960. Moser then (1970) gave a much more elegant proof of this result for discrete time flows by using functional analysis. The subsequent proofs of structural stability have been based upon the functional analytic techniques introduced by Moser together with the geometric decomposition of the nonwandering set into basic sets introduced by Smale. The idea behind Moser's proof is that an invertible map induces a map on the infinite-dimensional vector space of continuous (or bounded) vector fields on the state space. Under the conditions studied here, this induced map is hyperbolic. The hyperbolicity implies that another map is invertible; its inverse is used in constructing conjugacies. In the general case, one does not end up here with an invertible map but only one having a right inverse. This right inverse is sufficient for constructing the conjugacies, however. Moser's paper is an excellent introduction to these more complicated theorems.

The theorem we have been discussing gives sufficient conditions for a vector field to be structurally stable. One has quite a good picture of the dynamics of these vector fields. Their nonwandering sets split into isolated basic sets. The dynamics of these basic sets was considered earlier. We saw then that within each basic set the flow is strongly ergodic and can be described quite well in statistical terms. With the Strong Transversality Property, one can order the basic sets so that the flow moves among them in a way consistent with the order. Once one leaves a neighborhood of a basic set, one can never return to that basic set or any of the ones preceding it in the ordering. In that sense, all of the recurrence for the vector field is taking place within the basic sets themselves.

We have not yet dealt with the analogue of the first part of Peixoto's theorem for vector fields of higher dimension. This is better left to the next section.

XIII. Necessary conditions for stability and genericity. In the preceding section we stated conditions which imply that a vector field is structurally stable. In this section we want to consider a converse question: what does structural stability imply about a vector field? The theory surrounding this question is less developed than any other dealt with in these lectures. Good techniques for treating structurally stable vector fields directly are still lacking.

All of the properties considered in this section will be ones which are compatible with the relation of topological equivalence. The nature of this compatibility is vague and will not be specified in a technical sense. An example of the sort of property we consider is whether the equilibria of a vector field are hyperbolic. If such a property holds for a dense set of vector fields, then it holds for the structurally stable vector fields because the structurally stable vector fields form an open set in the space of vector fields. Thus, one way of finding conditions implied by structural stability is to look for properties enjoyed by "almost all" vector fields.

Here we encounter difficulties once again with topologies. Suppose one has an infinite, but countable, list of properties satisfied by "almost all" vector fields. It is reasonable to hope that almost all vector fields will then have all of this list of properties. When studying Lebesgue measure on the line this happens: a countable intersection of sets of measure one in $[0, 1]$ is a set of measure one. There is no natural measure on the space of vector fields, so we seek a topological interpretation of this sort of phenomenon. The topological interpretation we seek is based upon the Baire Category Theorem which states that a countable intersection of open dense subsets of a *complete* metric space is a dense subset of the metric space. Thus a countable intersection of open dense sets can be thought of as comprising "almost all" of the metric space in a topological sense. A space of vector fields will be a complete metric space only if one chooses the right topology–the C^r topology for the space of C^r vector fields ($1 \leqslant r < \infty$). In this space, one says a property is *generic* if it is satisfied by vector fields in a countable intersection of open dense sets.

One of the goals of dynamical systems has been to describe as many "generic" properties of vector fields as possible. Unfortunately, the list of such properties is quite a short one. There are a few properties which are readily proved to be generic by using techniques of transversality. These are embodied in the Kupka-Smale theorem:

THEOREM. *The following properties are generic in the space of C^r vector fields* ($1 \leqslant r < \infty$):

(1) *the equilibrium points of the vector field are hyperbolic and have pairwise disjoint neighborhoods*;

(2) *the periodic orbits of the vector field are hyperbolic*;

(3) *the stable and unstable manifolds of the equilibria and periodic orbits always intersect transversally*.

There is one more notable example of a generic property. It is based upon the closing lemma of Pugh which has been proved only for the C^1-topology. The closing lemma states that a vector field with a recurrent trajectory can be given a C^1 perturbation which creates a periodic orbit near the former recurrent trajectory. Settling the closing lemma for C^r perturbations with $r > 1$ is one of the difficult outstanding questions in dynamical systems. The problem here is not one of the differentiability of the perturbation but one of topology. The perturbations used in proving the closing lemma are highly differentiable. They are close to the original vector field in the C^1 topology but not close to the original vector field in the C^r topology when $r > 1$. A corollary of the closing lemma is the following theorem:

THEOREM. *For a generic set in the space of C^1 vector fields, the nonwandering set of a vector field is the closure of its periodic orbits and equilibria*.

As far as I know, this is an essentially complete list of generic dynamical properties of vector fields.

It is certainly not true that these generic conditions are sufficient to imply that a vector field be structurally stable. Various examples have been given of open sets of vector fields which contain no structurally stable ones. For the most part, these examples are also based upon transversality. One usually has a family of submanifolds with the property that structural stability implies that another submanifold is transverse to each submanifold in the family. It may happen that all perturbations leave the single submanifold with a nontransverse intersection with one of the submanifolds in the family.

The question remains as to what conditions on a vector field are implied by structural stability. One suspects that structural stability implies that a vector field satisfies Axiom A, but there is no proof of this conjecture. As a result, there is not a great deal more that one can say about structurally stable vector fields beyond the generic properties listed above. Instead, John Franks has made some progress with this question by strengthening the concept of structural stability. The strengthened concept of structural stability does imply both Axiom A and the Strong Transversality Property. The proof of the structural stability theorem is easily amended to prove that Axiom A and the Strong Transversality Property imply the strengthened concept of structural stability. Thus, with the strengthened structural stability one has a rather complete theory.

The concept of structural stability arises from considering perturbations of vector fields (or maps in the discrete case). These perturbations correspond to slight changes in the equations of motion. The strengthened concept of structural stability has a more stochastic flavor. Instead of allowing the equations to change once, one allows the equations to change slightly at each instant of time. This is more easily described in the discrete time case. An invertible map f is *time dependent structurally stable* if for every set $g_1, \ldots, g_k$ of perturbations of f, $g_k \circ \cdots \circ g_1$ is topologically equivalent to $f \circ \cdots \circ f$. With this definition, Franks proves the following theorem.

THEOREM. *An invertible map f is time-dependent structurally stable if and only if f satisfies Axiom* A *and the Strong Transversality Property.*

This concludes our review of the theory of structurally stable vector fields. There are portions of the theory which have been totally neglected. For instance, there is a great deal known about the topology of some basic sets. Williams has given a detailed description of "expanding attractors". Another portion of the theory which has been ignored concerns the relationship between the topology of a state space and the dynamics of the vector fields which exist on the state space. There are also relations between the topological structure of a map and its dynamics. The results here are rather fragmentary.

What is missing from the theory from the point of view of applications is the analysis of specific classes of differential equations arising in practice in terms of the dynamical system theory which has been described in this review. Most of the examples of dynamical systems which have been analyzed in detail have been constructed in topological and geometric ways rather than by solving equations. Solving equations approximately is quite easy with modern computers. However, it is not always easy to reconstruct the full dynamical behavior of a vector field from the numerical results. The theory of dynamical systems should prove to be a useful tool in the interpretation of numerical results. It also gives one a perspective on the richness of dynamical behavior which does occur in systems of differential equations.

XIV. Bifurcation theory. Most models have a set of coefficients or parameters which can be varied to make the model "fit" a set of data as well as possible. In the simplest instance, one chooses the slope and intercept of a straight line when trying to establish a linear correlation between two quantities. Here we are interested in the change of dynamic behavior as parameters are varied. The study of the qualitative changes in the dynamics of a system of differential or difference equations is called *bifurcation theory*. The origins of the subject are to be found in the work of Poincaré on celestial mechanics. Recent qualitative work on the subject has been carried out by Sotomayor, Takens, Newhouse, Palais, Arnold and others within the context of dynamical systems. From a somewhat different point of view, Thom's theory of elementary catastrophes falls within the scope of bifurcation theory.

Throughout most of our discussion we have emphasized the study of generic or nondegenerate systems of differential equations. The concept of genericity must be broadened if we hope to study generic systems which depend upon parameters. Consider the following simple example to illustrate the kinds of difficulties which may occur. Let $X = \lambda x$ with $\lambda \in \mathbf{R}$ a parameter we are free to vary. For $\lambda \neq 0$, this differential equation has a nondegenerate equilibrium point at 0. It is stable when $\lambda < 0$ and unstable when $\lambda > 0$. A bifurcation occurs when $\lambda = 0$ and the system is degenerate. Thus, we must study at least some degenerate systems of differential equations to understand how qualitative behavior of generic systems changes. We can continue to play the game of genericity by being flexible enough to expand our horizons. We demand that the "families" of equations that we study are "generic" in whatever sense is appropriate for the problem under consideration. (Soon, we discover that the example considered above is highly degenerate from our point of view.)

There is a price to be paid for broadening our viewpoint from the study of a single system of equations to a family of systems depending upon parameters. The primary penalty is that much of the coherent elegance of the theory described in previous chapters deteriorates into a bewildering array of complicated and poorly understood phenomena. There are a few landmarks which guide us to a partial understanding of what we observe, but the details of the landspace are often obscure. This becomes increasingly true when there are several parameters in a system. With three or four parameters available, phenomena occur which seem hopelessly complicated from the point of view of qualitative dynamical systems. There is a second difficulty involved with bifurcation theory. The nondegeneracy conditions which occur for bifurcations of periodic orbits and equilibria involve higher derivatives of the system of differential equations. The larger the number of parameters, the larger the number of derivatives one needs. In practice, this is an unpleasant fact of life because the evaluation of the relevant derivatives is often a formidable task.

Why, then, do bifurcation theory at all? The answer is simple. In spite of its complexity, bifurcation theory provides a very useful tool for the analysis of systems of differential equations. Certainly, one of the easiest ways to detect periodic orbits comes from a study of the "Hopf bifurcation". As we explain below, there is a limited dictionary of "generic" bifurcations for systems depending upon a single parameter. By starting with a value of the parameter for which the dynamical behavior of the system is easily understood and then varying the parameter, one can sometimes analyze the behavior of systems with complex behavior. In practice, this is a particularly useful aspect of bifurcation theory.

Let us begin to compile the "dictionary" of generic bifurcations of 1-parameter families of vector fields. This dictionary has been compiled by Sotomayor [**26**] for families of two dimensional vector fields. The same phenomena plus additional ones occur for families of vector fields of higher dimensions. The generic bifurcations can be split into four categories as they affect (1) an equilibrium point, (2) a periodic orbit, (3) the connection between equilibria and/or periodic orbits, and (4) more complicated basic sets. The "Hopf bifurcation" falls in between the first and second categories. We consider each of the four different categories in turn.

Suppose that $X(\tau)$ is a vector field depending upon a parameter τ. What does it mean for $X(\tau)$ to have an equilibrium point which undergoes a bifurcation at $\tau = 0$? Previously, we have seen that an equilibrium point p of a vector field Y is nondegenerate if $DY(p)$ is a matrix which has no pure imaginary eigenvalues. This condition yields two kinds of useful information about (C^1) perturbations of Y. In the first place, the condition that zero is not an eigenvalue of $DY(p)$ implies that a perturbation of Y has a unique equilibrium point near p. This is implied by the implicit function theorem of advanced calculus. Intuitively, the reason that this is true is that the graph of Y as a vector-valued function has no directions of tangency with the graph of the function which is identically zero (cf. Figure 15). The second kind of information we obtain from the fact that $DY(p)$ has no pure imaginary eigenvalues is that contained in Hartmann's theorem. The flow near p is topologically conjugate to a linear flow with no periodic orbits. This situation also persists under perturbation with the perturbed flow topologically conjugate to that of Y near p.

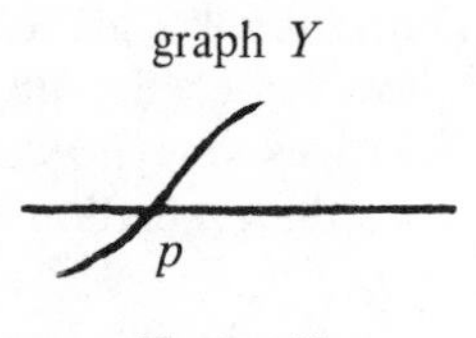

FIGURE 15.

In seeking "generic" bifurcations of the 1-parameter family $X(\tau)$, we want to violate the nondegeneracy conditions in as mild a way as possible. This can happen either (a) by one eigenvalue of $X(0)$ being 0, or (b) by a pair of eigenvalues at $X(0)$ being pure imaginary but not 0.

Let us examine these two cases, starting with (a). We want to consider what happens when one eigenvalue of X becomes 0 in the simplest possible way. This will occur when there is a vector v at the singular point p for which the directional derivative of X along v, $DX(v)$, vanishes, but for which the second directional derivative of X along v, $D^2X(v, v)$, is not orthogonal to v. If coordinates are chosen so that v lies along the first coordinate axis, then these conditions state that $\partial X(p)/\partial x_1 = 0$ and $\partial^2 X(p)/\partial x_1^2 \neq 0$. The second condition guarantees that the eigenvalue of X has become 0 in the most nondegenerate way possible. If X is allowed to depend upon a parameter τ and $\partial X/\partial\tau \neq 0$ in the above coordinates, then the local geometry of singular points of X_τ near p is determined. Figure 16 illustrates the graph of the singular points X_τ projected along the axis of v. There is a smooth curve γ of singular points which is tangent to the plane $\tau = 0$ along the direction v. The tangency is nondegenerate in the sense that the curvature of γ at p is nonzero. This implies that, near p, the curve γ projects onto a curve which looks like a parabola in the (v,τ)-plane.

Geometrically, this is a representation of a process by which two singular points coalesce and disappear (or are born, depending upon the

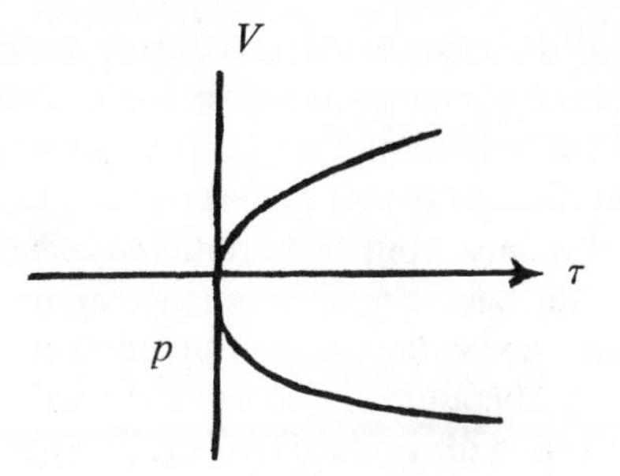

FIGURE 16.

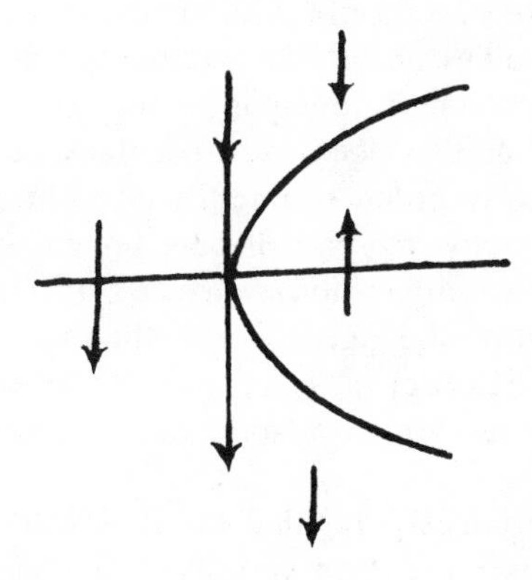

FIGURE 17.

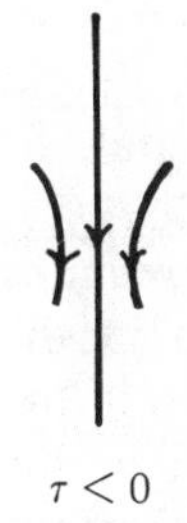

$\tau < 0$

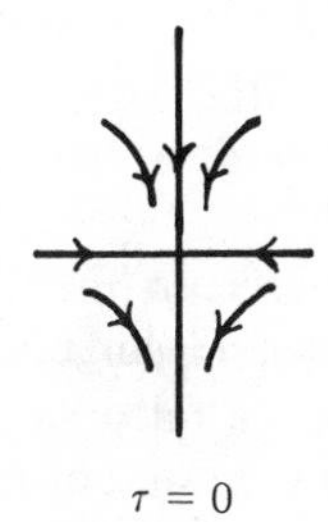

$\tau = 0$

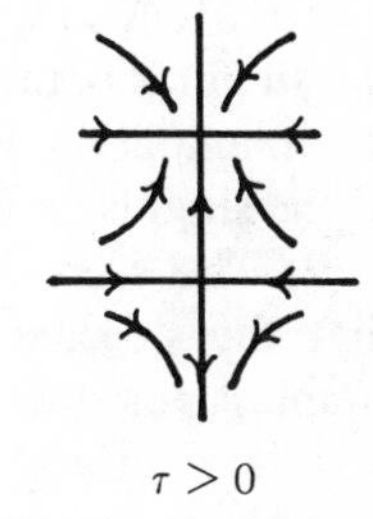

$\tau > 0$

FIGURE 18. The "unfolding" of a saddle node.

orientation of τ) as the parameter is varied. If one considers the flow in the direction along the v axis, it looks like Figure 17. From this diagram, it is seen that the stability of the two singular points will be different. In directions complementary to v, the flow will look like that near a nondegenerate hyperbolic point. This produces a picture $d\tau = 0$ which Sotomayor calls a saddle node (see Figure 18). This bifurcation is the simplest way in which singular points for a flow can appear or disappear. Note that the dimensions of the stable manifolds of the two nondegenerate singular points which exist for a fixed value of τ must differ by one. Thus a source and a sink for a two dimensional flow cannot coalesce and disappear in a generic one-parameter family.

As a final remark concerning the saddle-node bifurcation, we note that its occurrence can have dramatic impact upon the dynamics of a flow. Sotomayor has shown that the occurrence of a saddle node can change an Axiom A dynamical system with an infinite number of periodic orbits into one with no periodic orbits at all. This is behavior which is possible for continuous flows, but not discrete flows.

The second way in which a singular point p can be degenerate occurs when there is one pair of nonzero, pure imaginary eigenvalues of $DX(p)$ and all other eigenvalues lie off the imaginary axis. This leads to the *Hopf* bifurcation. For generic one-parameter families, we want to impose a restriction that the eigenvalue becomes nondegenerate in the simplest possible way. In order to state these conditions, one first proves that if $DX(p)$ has exactly one pair of pure imaginary eigenvalues, then there is an invariant two dimensional surface S through p whose tangent space at p is spanned by the eigenspace of the pure imaginary eigenvalues. The bifurcation behavior of X at p can be studied by examining this two dimensional plane. We now state the nondegeneracy conditions imposed on X at p.

(1) The first condition requires that the eigenvalues on the imaginary axis cross the imaginary axis transversally as the parameter is varied. If X depends upon the parameter τ, and if $\lambda(\tau)$, $\bar{\lambda}(\tau)$ are the eigenvalues which cross the imaginary axis when $\tau = \tau_0$, then we require that $[(d/d\tau)\mathrm{Re}(\lambda(\tau))]|_{\tau=\tau_0}$ be nonzero. This insures that the degeneracy of the singular point p is restricted to the parameter value τ_0. As λ, $\bar{\lambda}$ cross the imaginary axis, the dimensions of the stable and unstable manifolds of the singular point change by 2.

(2) The second condition imposed in the Hopf bifurcation requires that the degeneracy of $X(\tau_0)$ at the singular point p occurs in the simplest possible way. To state it, we introduce polar coordinates in the invariant surface S introduced above, so that $DX(p)$ is a rotation in this coordinate system. With respect to these coordinates, the requirement imposed upon X is that the third derivative of its radial component in the radial direction be nonzero. If this derivative is negative, then X is a "vague attractor" in the surface S. Nearby orbits in S spiral toward p, but they do so at much less than an exponential rate. If the derivative is positive, nearby orbits in S spiral away from p. In practice, the computation of this derivative is tedious. The formulas in Cartesian coordinates are quite complicated.

What are the dynamics of the family of vector fields X associated with the Hopf bifurcation? Depending upon the signs of the third radial derivative and $[(d/d\tau)/\text{Re}\,\lambda]|_{\tau=\tau_0}$, there will be a one-parameter family of periodic orbits for $\tau > \tau_0$ or $\tau < \tau_0$. The stability of these orbits depends upon whether p is a vague attractor in S for $X(\tau_0)$. The family of periodic orbits together with p forms a smooth two dimensional surface in the space which includes the parameter. The picture of a Hopf bifurcation for two dimensional flows is shown in Figure 19. Orbits spiral toward or away from the bowl, staying in the same vertical plane.

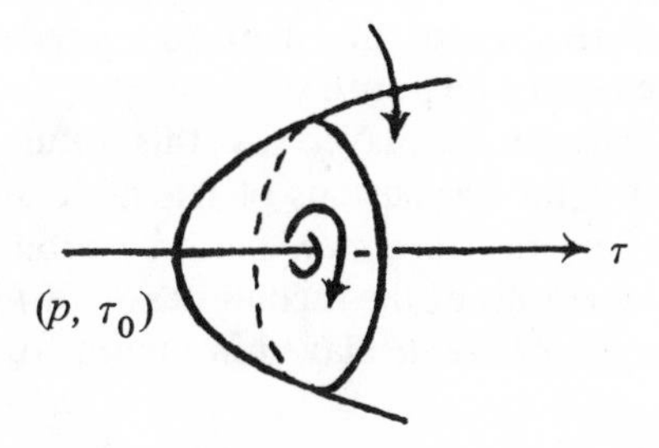

FIGURE 19.

A great deal of attention has been devoted to the Hopf bifurcation in applied mathematics. It represents a mechanism by which a stable equilibrium can give rise to a stable oscillation with changes in parameters. A number of hydromechanical phenomena are explained by applying the Hopf bifurcation theorem to the Navier-Stokes equations.

The second category of bifurcations which occurs in generic one-parameter families involves the bifurcation of periodic orbits. The bifurcation phenomena which occur for periodic orbits are similar for continuous and discrete flows. The nondegeneracy condition for a periodic orbit is that none of the eigenvalues of the Poincaré map has absolute value one. Thus the generic bifurcations will occur in one of three situations: (a) an eigenvalue of $+1$, (b) an eigenvalue of -1, and (c) a pair of nonreal eigenvalues of absolute value 1. Let us consider these three situations in turn. Throughout the discussion, it is easier to speak in terms of a cross-section and its Poincaré map. This reduces the discussion to the case of a discrete flow, so we shall speak in terms of iterating a map.

The case of an eigenvalue of $+1$ at a periodic orbit is similar to that of a singular point with an eigenvalue of 0. With the appropriate nondegeneracy hypotheses, this bifurcation describes the phenomena of two periodic orbits coalescing and disappearing or of the birth of two periodic orbits. As a discrete flow, the periodic orbits look like the saddle-node singular points described above. The appropriate nondegeneracy conditions are (1) that the derivative of the eigenvalue with respect to the parameter be nonzero, and (2) that the second derivative of the (Poincaré) map in the direction of the eigenvector with eigenvalue one be nonzero.

The second type of bifurcation of a periodic orbit occurs when there is an eigenvalue of -1. This case does not have a counterpart for the bifurcation of singular points. As usual, we want to specify the appropriate nondegeneracy conditions which guarantee that the bifurcation will happen in the simplest possible way. This requires that (1) the derivative of the eigenvalue with respect to the parameter be nonzero, and (2) the third derivative of the map composed with itself in the direction of the eigenvector be nonzero.

The geometric behavior displayed by this bifurcation is that the periodic orbit changes the dimensions of its stable and unstable manifolds by 1 while giving rise to a new periodic orbit of approximately twice the period. The graphs of the second iterate of f in the direction of the bifurcating eigenvector are displayed in Figure 20.

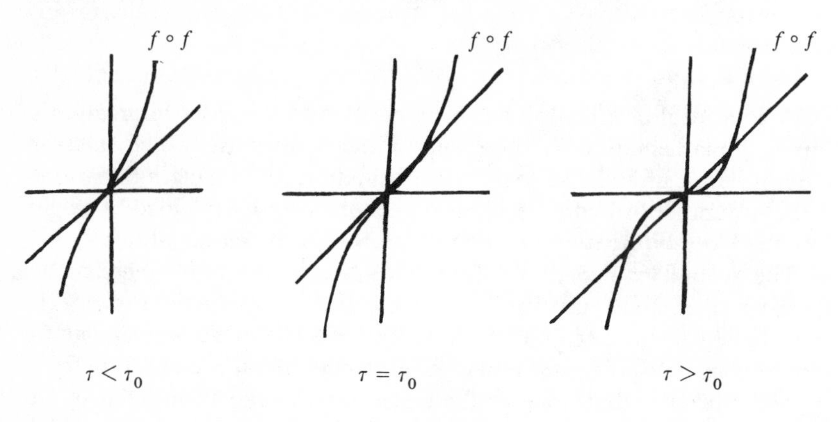

FIGURE 20.

The final type of bifurcation for a periodic orbit occurs when a pair of complex eigenvalues crosses the unit circle. This is again called a *Hopf bifurcation* (for a periodic orbit.) Once again, there is an invariant circle of orbits generated around the periodic orbit as it changes its stability. There is a real difference between the dynamics encountered here and those present in a Hopf bifurcation of a singular point.

We are studying here a discrete flow (which may be generated by a Poincaré map for a continuous flow). Consequently, the invariant circle which arises will not be a single periodic orbit but will contain many orbits. To gain a full picture of the dynamics, one must study the dynamics of families of diffeomorphisms of the circle to itself.

Poincaré was the first person to carry out such a study. If $f\colon S^1 \to S^1$

is a C^2 diffeomorphism, Poincaré defined a rotation number $\rho(f)$ which characterizes some of the dynamics of f. If $\rho(f)$ is irrational, then f has no periodic orbits and is topologically conjugate to the map which rotates the circle by the angle $2\pi\rho(f)$. On the other hand, if $\rho(f)$ is rational, say m/n (m, n relatively prime integers) then there are periodic orbits of prime period n. All periodic orbits of f have prime period n, but the rotation number does not determine how many such orbits are present. The rotation number depends continuously on the map f.

Thus, after a Hopf bifurcation has occurred and there is an invariant circle, many (an infinite number) bifurcations can occur near the Hopf bifurcation. These bifurcations are the changes in the dynamics of the map on the invariant circle. The description of this set of bifurcations is subtle. There are features of this situation which are poorly understood. On the one hand, in a topological sense, almost all of the flows in a family may have rational rotation numbers on their invariant circles. On the other hand, the set of flows with irrational rotation numbers has positive measure. Sorting out what is to be expected generically in this case depends upon one's viewpoint.

There is still a third category of bifurcation phenomena occurring in one-parameter families of flows. This deals with bifurcations which are much more "global" in character than the bifurcations of singular points and periodic orbits described above. These are bifurcations which involve nontransverse intersections between the stable and unstable manifolds of singular points and/or periodic orbits. Some of the geometry connected with these bifurcations can be exceedingly complicated, as illustrated especially in the work of Newhouse and Palis. It is perhaps safe to say that there is no comprehensive picture of the phenomena arising in some nondegenerate situations. One must be able to deal simultaneously with the appearance and disappearance of infinite numbers of periodic orbits. One can only dream still of reasonable ways of understanding these situations.

We can close our brief synopsis of bifurcation theory with a few remarks about bifurcations of higher dimensional families. The most successful undertaking has been Arnold's description of how singular points bifurcate, without considering the dynamics around these singular points [**10**]. Takens has shown that there can be very complex dynamical behavior associated with mildly degenerate singular points [**28**]. Even a primitive dictionary of phenomena seems to be out of the question except for families of one or two parameters, or families of vector fields in the plane.

XV. Epilogue. Our rapid survey of dynamical systems theory is ended. As promised in the beginning much has been left out. A good way to gain some perspective on the extent of the theory of dynamical systems is to look through the symposium volumes listed in the references. Each of these presents a rather broad view of the state of the theory at the time these meetings were held. Rather than dwell upon these matters in this epilogue, I shall discuss briefly once again the role of the theory in applications. This is a highly subjective discussion, presenting my own views and prejudices more than incontrovertible fact.

Most of the development of dynamical systems theory has started from geometric descriptions of flows and diffeomorphisms. The conclusions one has obtained are also geometric statements about limit sets and statements about the asymptotic behavior of trajectories. There are at least two major pieces of information missing from this analysis which would make the theory considerably more useful.

In the first place, one would like to have some means for proceeding from a system of equations–written as actual equations–to a geometric description of its flow. This is certainly not an easy task. Stated in the most demanding way, this task includes Hilbert's 16th problem as a small special case. (Hilbert's problem calls for an upper bound $m(n)$ to the number of limit cycles a system of equations $\dot{x} = P(x,y)$, $\dot{y} = Q(x,y)$ in $\mathbf{R}^2$, can have when P and Q are polynomials of degree n.) As difficult as this problem may be, it is also clearly of great practical importance. For example, it includes the problem of describing chemical reactions which lead to oscillatory or chaotic behavior (assuming an algorithm which allows one to write the equations from a specification of the reactions). A similar task holds for electrical circuits where the differential equations describing the currents and voltages in the circuit are relatively easy to obtain from a description of the circuit.

What kinds of partial solutions to this problem may be possible? At the risk of being wrong, I would suggest that the correct approach involves the analysis of a substantial number of model equations in some detail. It is difficult to find "naturally occurring" systems of equations with complicated behavior which fit exactly into the theory of Axiom A dynamical systems. As with the Lorenz system, the examination and analysis of individual examples should provide clues as to the practical scope of the current theory and the directions in which the theory can be extended to be more useful.

The examination of examples in this way is a task which would be virtually impossible without the powerful numerical capability of modern computers. It seems to me that the analysis of dynamical systems is one area in which the computer can be extremely useful as a "translator" of equations into pictures. However, the computer is not an answer by itself. It is still necessary to interpret the results of numerical computations, and here the theory of dynamical systems can prove to be very useful. There would be much less need for the theory if the results of numerical computations were always readily understandable.

The second direction in which I think there is a real need for better information involves turning qualitative information into more quantitative information. Let me illustrate this with an example. In §X we stated a theorem of Ruelle (and Bowen and Ruelle) which gave the existence of an invariant measure on an Axiom A attractor which describes the asymptotic behavior of time averages along a particular trajectory. In practice, one cannot compute asymptotic quantities, one can only compute long time averages. How long is long enough? Suppose one wants to compute this measure on an attractor to a certain degree of accuracy by computing time averages. When can one stop? These seem to me to be important practical questions. Unfortunately, I have no suggestions as to how they should be approached.

Finally, let me mention one additional problem. The equations which arise in classical mechanics (Hamiltonian systems) have special properties which make their typical behavior quite special when viewed in the context of examining all systems of differential equations. The dynamics of Hamiltonian systems of ordinary differential equations is a much studied subject which we have avoided here. In other fields, one encounters special classes of differential equations. For example, the equations describing electrical circuits and chemical reactions each form restricted classes of differential equations. Within the context of one of these classes, is the typical behavior of solutions different from that exhibited by a "generic" system of differential equations? If it is different, in what ways is it different? It may be that the typical behavior within the restricted class is no different. This was observed to be the case by Smale for a class of differential equations describing the interactions of several competing species in a model ecosystem.

With these comments I hopefully leave the reader with a feeling of inadequacy about his acquaintance with the theory of dynamical systems. In the back of my mind there is a feeling that a great many interesting and subtle phenomena have been ignored because of their dynamic complexity. If these notes are inspiration for others to investigate some of these phenomena, they will have served their purpose well.

References

I. Surveys and Symposia

1. S.-S. Chern and S. Smale, eds., Proc. Sympos. Pure Math., vol. 14, Amer. Math. Soc., Providence, R. I., 1970.
2. D. Chillingworth, ed., Proc. Sympos. Differential Equations and Dynamical Systems (Univ. Warwick, Coventry, 1968/1969), Lecture Notes in Math., vol. 206, Springer-Verlag, Berlin, 1971.
3. A. Manning, ed., *Dynamical Systems–Warwick*, 1974, Lecture Notes in Math., vol. 468, Springer-Verlag, Berlin, 1975.

4. J. Marsden and M. McCracken, *The Hopf bifurcation and its applications*, Applied Math. Sci., vol. 19, Springer, New York, 1976.

5. Z. Nitecki, *Differentiable dynamics*, MIT Press, Cambridge, 1971.

6. M. Peixoto, ed., *Dynamical systems*, (Proc. Sympos. Univ. of Bahia, Salvador, 1971), Academic Press, New York, 1973.

7. S. Smale, *Differentiable dynamical systems*, Bull. Amer. Math. Soc. **73** (1967), 747–817.

7a. M. Shub, *Stabilité globale des systèmes dynamiques*, Asterisque **56** (1978).

II. **Papers and Books**

8. R. Bowen, *Equilibrium states and the ergodic theory of Anosov diffeomorphisms*, Lecture Notes in Math., vol. 470, Springer-Verlag, Berlin, 1975.

9. R. Bowen and D. Ruelle, *Ergodic theory of Axiom A flows*, Invent. Math. **29** (1975), 181–202.

10. V. I. Arnold, *On matrices depending upon a parameter*, Russian Math. Surveys **26** (1971), no. 2, 29–43.

11. J. Guckenheimer, *A strange, strange attractor*, in The Hopf Bifurcation, J. Marsden and M. McCracken, eds., Applied Mathematical Sciences, vol. 19, Springer-Verlag, New York, 1976.

12. J. Hadamard, *Les surfaces à courbures opposées et leur lignes géodèsiques*, J. Math. Pures Appl. **4** (1898), 27–73.

13. M. Hirsch, C. Pugh and M. Shub, *Invariant manifolds*, Lecture Notes in Math., vol. 583, Springer-Verlag, New York, 1977.

14. M. W. Hirsch and S. Smale, *Differential equations, dynamical systems, and linear algebra*, Academic Press, New York, 1974.

15. E. Lorenz, *Deterministic nonperiodic flow*, J. Atmospheric Sci. **20** (1963), 130–141.

16. R. May, *Stability and complexity in model ecosystems*, Princeton Univ. Press, Princeton, N. J., 1973.

17. M. Morse, *A one to one representation of geodesics on a surface of negative curvature*, Amer. J. Math. **43** (1921), 33–51.

18. M. Peixoto, *Structural stability on two dimensional manifolds*, Topology **1** (1962), 101–120.

19. H. Poincaré, *Les méthodes nouvelles de la mécanique céleste*. vols. I, II, III, Paris, 1892, 1893, 1899; reprint, Dover, New York, 1957.

20. J. Robbin, *A structural stability theorem*, Ann. of Math. (2) **94** (1971), 447–493.

21. R. C. Robinson, *Structural stability of vector fields*, Ann. of Math. (2) **99** (1974), 154–175.

22. D. Ruelle, *A measure associated with Axiom A attractors*, Amer. J. Math. **98** (1976), 619–654.

23. Ya. G. Sinai, *Gibbs measures in ergodic theory*, Russian Math. Surveys **166** (1972), no. 4, 21–64.

24. S. Smale, *Diffeomorphisms with many periodic points*, Differential and Combinatorial Topology, Princeton Univ. Press, Princeton, N. J., 1965, pp. 63–80.

25. ______, *On the differential equations of species in competition*, J. Math. Biol. **3** (1976), 5–7.

26. J. Sotomayor, *Bifurcations of vector fields on two dimensional manifolds*, Inst. Hautes Études Sci. Publ. Math. No. 43 (1974), 5–46.

27. M. Spivak, *Calculus on manifolds*, Benjamin, New York, 1965.

28. F. Takens, *Singularities of vector fields*, Inst. Hautes Études Sci. Publ. Math. no. 43 (1974), 47–100.

29. R. Thom, *Structural stability and morphogenesis*, (translated by D. Fowler), Addison-Wesley, Reading, Mass., 1974.

30. R. F. Williams, *Expanding attractors*, Publ. I.H.E.S. No. 43 (1974), 169–203.

31. A. Winfree, *Patterns of phase compromise in biological clocks*, J. Math. Biol. **1** (1974), 73–95.

32. J. Guckenheimer and R. Williams, *Structural stability of Lorenz attractors*, Publ. I.H.E.S. No. 50 (1979).

33. J. Moser, *Stable and random motions in dynamical systems*, Princeton Univ. Press, Princeton, N. J., 1973.

34. R. Williams, *The structure of Lorenz attractors*, preprint.

35. J. Moser, *On a theorem of Anosov*, J. Differential Equations **5** (1969), 411–424.

DEPARTMENT OF MATHEMATICS, UNIVERSITY OF CALIFORNIA, SANTA CRUZ, CALIFORNIA 95060